CONSTRUCTION
systems

by
Doug Polette
Department of Technology Education
Montana State University
Bozeman, Montana
and
Jack M. Landers
Department of Manufacturing and Construction
Central Missouri State University
Warrensburg, Missouri

South Holland, Illinois
THE GOODHEART-WILLCOX COMPANY, INC.
Publishers

Library of Congress Catalog Card Number 94-4464
International Standard Book Number 1-56637-041-8

1 2 3 4 5 6 7 8 9 10 95 98 97 96 95 94

Library of Congress Cataloging-in-Publication Data

Polette, Doug
Construction Systems/by Doug Polette and Jack M. Landers.

p. cm.
Includes index.
ISBN 1-56637-041-8
1. Building. 2. Construction industry. [1. Building. 2. Construction industry.] I. Landers, Jack M. II. Title.

TH149.P65 1995
690--dc20 94-4464
CIP
AC

INTRODUCTION

CONSTRUCTION SYSTEMS provides you with an understanding of how construction impacts your life, socially and professionally. These impacts are explored using the systems method.

CONSTRUCTION SYSTEMS explores in detail:

- Construction's origins, its present condition, and the future of construction.
- The relationship between construction and the three other technology clusters–manufacturing, communication, and transportation.
- The seven elements of construction–people, information, time, materials, capital, tools, and energy.
- The various types of construction enterprises.
- The applications of specific construction systems and utilities.

As you work with CONSTRUCTION SYSTEMS, you will learn that construction is a managed system that draws upon many resources. You will explore these resources and see how they affect, and are affected by, construction activities. You will gain a firm foundation of knowledge of the construction world, and you will be introduced to many employment possibilities. Communication and leadership are stressed.

CONSTRUCTION SYSTEMS is heavily illustrated with photographs, drawings, and diagrams. The material has been carefully selected to make construction easy to understand and enrich the text.

Each chapter begins with a list of objectives to inform you of what will be covered. Key words are in bold to help make you aware of them. They are also listed at the end of each chapter. Each chapter concludes with review questions and a set of activities to improve your understanding of the material covered.

CONSTRUCTION SYSTEMS is a basic text for use by anyone interested in the many career opportunities available in the cluster of construction trades and related jobs. Construction is a provocative field of study. As you study, you will begin to see the consequences, both positive and negative, that construction has on your daily life. With a solid background in construction systems, you will be able to make sound decisions regarding your housing needs and your professional needs.

Doug Polette
Jack Landers

CONTENTS

section one:

Technology and the Construction System

section two:

Relationship of Construction to Other Technology Systems

section four:

Types of Construction Enterprises

section five:

Application of Construction Systems

section 1

Technology and Construction Systems

AMERICAN PLYWOOD ASSOC.

Construction is a broad subject. It keeps many people busy at many different occupations. A study of its processes and the people who work to make it an industry is interesting and well worth the effort.

The knowledge and skills of many persons go into construction projects. Among them are architects, consultants, engineers, surveyors, carpenters, concrete workers, typists, estimators, and bricklayers. These are only a sample of the careers connected with the construction industry.

Homes, bridges, dams, highways, factories, and airports start from ideas. They are made real by the labors of people who earn their living through building. This labor continues to serve the needs of people long after the project is done.

Construction is an activity that is very old. In fact, from the beginning of civilization, people have built shelters for themselves. Today we construct homes, places of worship, commercial buildings, transportation systems, and industrial complexes. We also build electrical generating plants and hydroelectric dams. These are necessary to generate the electrical energy that lights and heats these structures.

Some buildings are erected to honor great persons or to express noble ideas. These are called symbols or monuments. Some monuments, such as the great pyramids of Egypt, have challenged modern builders to create wonders like the Gateway Arch in St. Louis, Missouri. Complex interchanges, railroads, and airports are other examples of efforts to answer transportation needs through building. Every day we depend on such structures to serve our needs.

Technology allows society to follow new paths.

1
CHAPTER

INTRODUCTION TO CONSTRUCTION

The information provided in this chapter will enable you to:
- *Identify a few of the vast number of careers that are part of construction.*
- *Understand that construction activities are part of all societies.*
- *Define construction and list major construction projects.*
- *List career opportunities available in the construction field.*
- *Draw and explain the four major parts of the Universal Systems Model.*
- *Identify the major types of construction.*

IMPORTANCE OF CONSTRUCTION

As you begin reading this textbook, stop and take a look around you. Notice the many things that you depend on to make your life easier and better. Many of these devices and structures, Fig. 1-1, have come about only because skilled engineers, architects, and builders worked many hours to produce them. All of these elements are available to us because construction workers have built buildings, highways, railroads, communications towers, and many other structures. These structures allow us to be more comfortable and live a better life.

It is not difficult to trace back in time and discover all of the structures built through the ages. We also know that within the next 25 years more projects will be constructed than have been built in all the previous years of human existence, Fig. 1-2. The construction industry has been around since the beginning of human history and will continue to exist and grow in the future. We can also be quite confident that construction enterprises will continue to change in order to keep up with technology and the demand of people for a better life. However, most major

Fig. 1-1. Well-constructed buildings and bridges are important. The talents of many workers went into these projects.

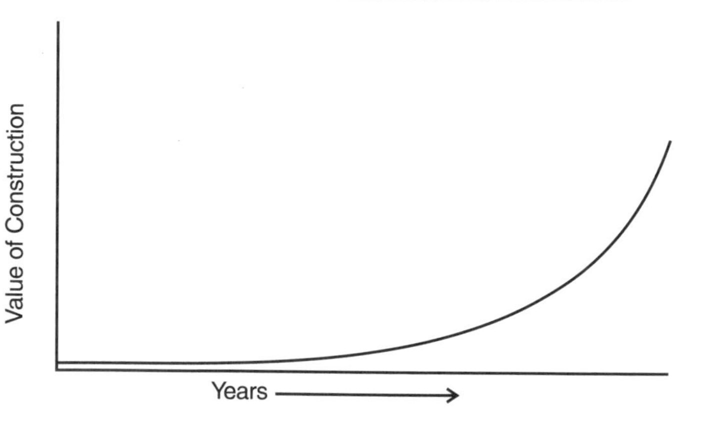

Fig. 1-2. Over the past few decades, tremendous leaps have been made in equipment and manufacturing technology. Construction projects that were once very difficult are conquered regularly.

concepts of construction will remain the same, even though the methods and materials may change considerably.

WHAT IS CONSTRUCTION

Not all items and processes fall under the classification of construction. Other than construction, there are also communication, transportation, and manufacturing technology clusters. **Construction** is the technology that deals with the design and building of structures. This includes more than just residential structures. It also includes commercial buildings, roads, dams, bridges, transmission lines, space stations, and canals.

Construction represents one of the single largest employment areas, Fig. 1-3. Jobs available in the construction field range from simple hand labor jobs, such as digging or hammering a nail, Fig. 1-4, to operating the most sophisticated computer controlled robots. The robots might be manufacturing structures in a factory, or they could be building structures in space. The career opportunities are nearly endless, with new areas of employment

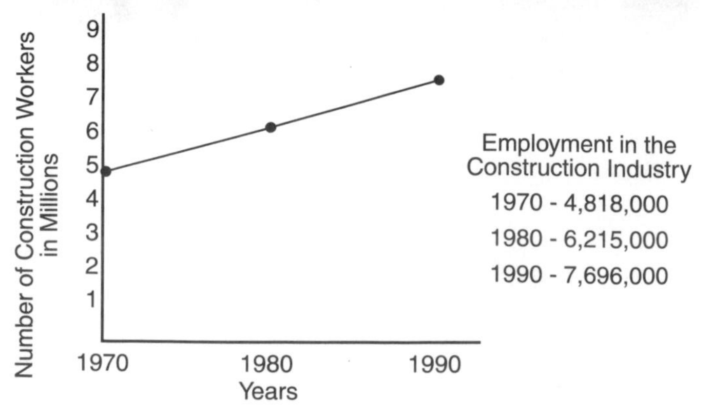

Fig. 1-3. From short one-lane roads to elaborate buildings like the one shown, all construction projects employ great numbers of people to design and build.
(The National Data Book, U.S. Department of Commerce)

Fig. 1-4. Even simple tasks, such as digging a trench, are vital to the success of a job.

being developed every day. The construction industry is one of the most dynamic and rewarding areas to enter. Few other areas can provide the satisfaction that construction workers feel when they complete a day's work, look back, and say, "I built that."

WHY STUDY CONSTRUCTION

All individuals need to have a basic understanding of how items are constructed. Whether it be a building, a bridge, a dam, a pipeline, or a road, you must interact with these items on a daily basis. In the future, you, as a voter, will have to decide on the value and importance of many public projects, Fig. 1-5. The place you choose to call home, whether it is a rented apartment or a house that you own, will depend to some extent on the knowledge you have about the construction industry.

Even though some structures are very complicated, Fig. 1-6, all structures can be broken down into individual parts and groupings. These groupings will become clear to you as you study the construction industry. A large skyscraper, a house, a dam, a road, or a transmission tower all have a foundation system, as shown in Fig. 1-7. These structures also have other systems that are similar.

If you break down any constructed project into its component parts, you can then understand it better. With this understanding, you will be able to make intelligent decisions. A broad understanding of construction will provide you with this knowledge. This knowledge will allow you to make decisions about future construction. You will also be better able to assess the impacts construction projects may have on humans and the environment.

CONSTRUCTION AS PART OF ALL SOCIETIES

Just as all societies have a need for a communications system, a transportation system, and a manufacturing system, societies also have a need for a construction system. This was as true for ancient societies as it is today. Wherever you live on this world, or whatever society you visit, you will see a system of construction. If you were to travel to the deep areas of the jungle, you would find structures built by humans. If you step back in time, the early structures were little more than caves that people had modified to make them more comfortable.

People are able to use the materials at hand to construct items that shelter the family from the elements, Fig. 1-8. In our society, we often think of buildings built out of

Fig. 1-5. Every construction project has some effect on the surrounding environment. Some effects are positive, some are negative. (Montana Highway Dept.)

Fig. 1-6. This building is being built in stages. Large projects are broken into a series of smaller assignments. (Morrison Knudsen Corp.)

Fig. 1-7. Shown are a variety of foundations. A stable foundation is vital to any structure. (Montana Power, Morrison Knudsen Corp.)

wood, concrete, masonry, or steel. In other societies, it is just as common to see structures built from grass, clay, stones, snow, or animal skins. Regardless of the building material, all societies have some system of construction to make life better.

Even many animals have developed construction systems. Think of the ants and bees. They build elaborate

Fig. 1-8. The wood frame house is a common shelter for people today.

Fig. 1-9. Industrious beavers engineer and construct their own impressive structures.

structures. Think of birds who build nests to shelter their young. Beavers are perhaps best known for their ability to construct dams, Fig. 1-9.

CONSTRUCTION AND ITS RELATIONSHIP TO POPULATION

Demographics is the study of population and its characteristics such as size, growth, and the density of people. We know from demographic studies that the world population will continue its rapid growth into the foreseeable future, Fig. 1-10. Therefore, in the years to come we will need more homes, roads, pipelines, and dams. In addition, construction projects of the past will have to be rebuilt, Fig. 1-11, and remodeled from time to time in order to keep them up to standards that are acceptable to society. All of these are construction activities.

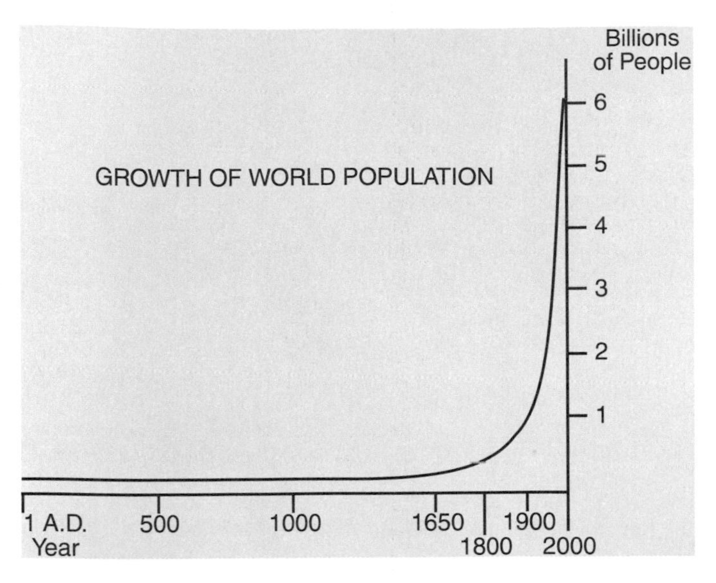

Fig. 1-10. World population is growing at a tremendous rate. (United Nations Estimates)

You have probably noticed, and been delayed by, roads and bridges being rebuilt or replaced by new modern structures. Sometimes, this construction is due to new safety standards. Often, new structures have to be built to replace old ones that are not adequate for the current population. This remodeling and replacing of existing structures provides additional jobs for individuals knowledgeable in the construction system. As the population grows, the demand for new homes, factories, and utility systems provides work for a great number of construction workers.

The construction field provides the opportunity to obtain work in a large variety of careers. Construction careers range from simple but important jobs, that require a great deal of manual labor to jobs requiring extensive experience in a large number of technical areas. All of these careers are open to individuals who are interested in the construction field.

CONSTRUCTION SYSTEMS

To fully understand any complicated technology or construction activity, you must first understand the major components that make up the technology or structure. One way to divide larger components into smaller understandable parts is through the use of the **systems approach.** This approach is similar to the use of "systems" when we talk about our digestive system, our home heating system, our political system, our transportation system, or our solar system.

Any complicated technology can be better understood with the use of some system that allows you to look at, and understand, the relationships between the various components of the entire technology. Once you see the interrelationships between the various parts, you can better understand the total workings of the completed whole. The

Fig. 1-11. Construction projects do not last forever. If a bridge deteriorates to a point where it is no longer safe, it must be rebuilt. (Montana Highway Dept.)

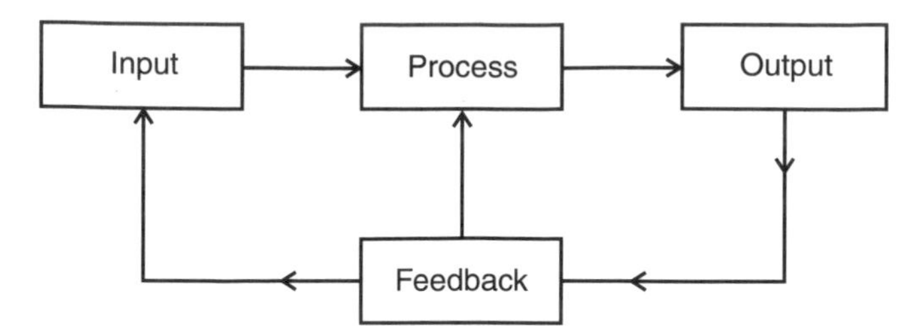

Fig. 1-12. The Universal Systems Model is a useful tool in solving construction problems.

system most commonly used to study a technology today is the Universal Systems Model as shown in Fig. 1-12.

UNIVERSAL SYSTEMS MODEL

The **Universal Systems Model** consists of four main parts. These four main parts work together to meet our needs, wants, and goals. This systems model will help you better understand how you arrive at a quality solution to any need. As you continue through this text, you will become very familiar with the Universal Systems Model and how it relates to the construction field. The four basic parts of the system consist of inputs, processes, outputs, and feedback.

INPUTS

The first part of the Universal Systems Model involves inputs to the system. The first step in the process is to establish a goal (societal or individual need) that you wish to accomplish. Assume you are interested in building a doghouse for the family pet. You will need information (plans), materials, tools, labor, capital, time, and energy to construct the doghouse. These components are called **inputs,** Fig. 1-13.

Regardless what you build, you will always require the same seven inputs in order to complete a construction project:

- **Information:** The flow of data needed to construct a project.
- **Materials:** The parts and physical resources used in a structure.
- **Tools:** The devices and machines used to aid in the construction process.
- **Labor:** The human resource needed for construction.
- **Capital:** The financial backing of a project.
- **Time:** The interval in which construction takes place.
- **Energy:** The force that drives a construction project.

Fig. 1-13. These are some of the inputs needed to build a doghouse. Rolled up near the center of the materials are the instructions.

In fact, you will need these seven inputs for the advancement of any technology. These input areas are the same whether you are building a doghouse or a 100-story skyscraper. The only thing that changes is the amount of each input that you will need. Naturally, the skyscraper will require a great deal more material, labor, and money than the doghouse.

PROCESSES

The next major part of the Universal Systems Model is processes, as shown in Fig. 1-14. **Processes** consist of purposeful action. In other words, this is the part of the systems model that assembles the various inputs into a desired product. In the case of the doghouse, this is where you begin to assemble the various pieces: lumber, nails, shingles, and paint, into the doghouse itself.

Management plays an extremely important part in converting resources into purposeful action. Even with all of the inputs ready to be processed into a project, careful and proper management is needed to make sure the project is completed. Management is a critical ingredient to any successful enterprise.

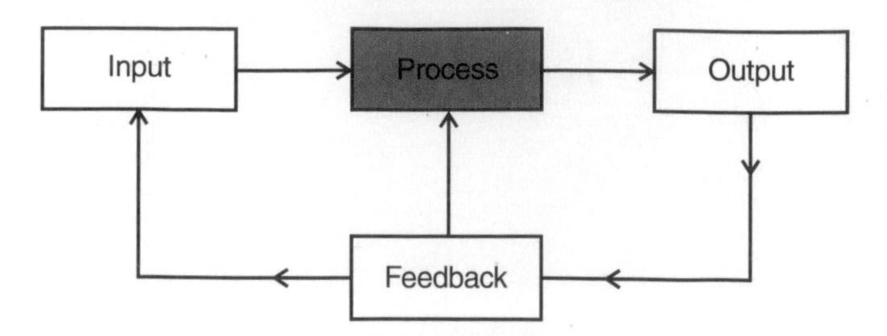

Fig. 1-14. The process area is where the action takes place.

OUTPUT

As shown in Fig. 1-15, the next major component to the Universal Systems Model is the output. The **output** phase is the completed project. In our example, the output would be the completed doghouse. In a larger context, the output could be a completed skyscraper, road, dam, or bridge.

FEEDBACK

The fourth part of the Universal Systems Model is the feedback. **Feedback** provides a monitoring system that helps us to determine if our original goal has been achieved and to what degree.

In the case of the doghouse, feedback would reflect how the completed house meets the needs of the family pet. Is the house too large or too small? Is it warm enough? Of course, all of these items should have been considered before you began to build the structure. However, some items may have been overlooked or may not have worked out as planned. Some modification may be required to correct these problems.

Feedback is also used while a project is under construction. Changes are made to make the project better meet the original goals. Notice in Fig. 1-16, the feedback loop has an arrow pointing to the processes block, which represents this situation.

TYPES OF CONSTRUCTION

Construction can be classified in a variety of ways. The most common method in use today breaks all construction into two major classifications. One class of projects contains buildings used primarily for the purpose of housing people. This class is called building construction. The other class contains structures that are primarily used for purposes other than housing individuals. This class is called industrial and civil construction. Each of these two major classifications can be further divided into sub-categories, Fig. 1-17.

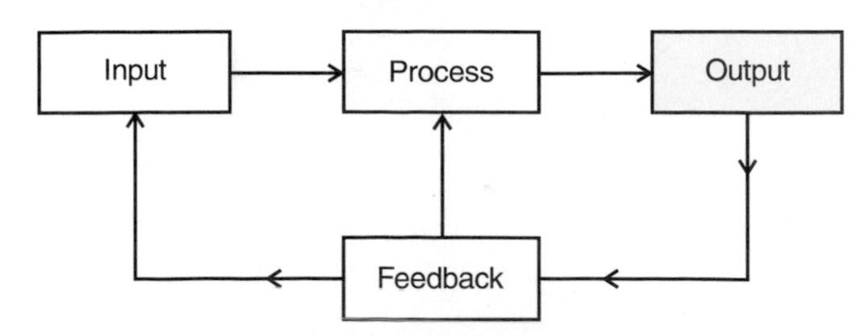

Fig. 1-15. The output is the result of the construction action.

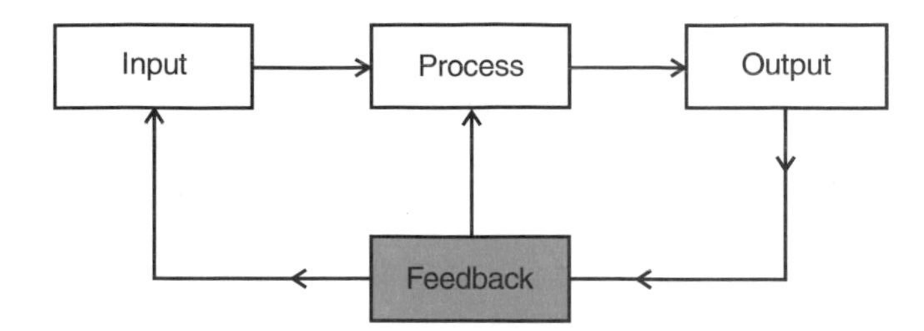

Fig. 1-16. Feedback allows you to correct problems that occur

Under the heading, building construction, are light construction and heavy building construction. **Light construction** consists of homes and structures such as small stores and offices. Large buildings that are used for commercial purposes, such as hospitals, churches, and schools, are called **heavy building construction.**

Structures that are used for purposes other than providing shelter to individuals can be classified either as industrial construction or civil construction. **Industrial construction** consists of structures such as electrical power plants and various materials processing plants. Materials processing plants include steel mills, saw mills, and chemical processing plants. **Civil construction** consists of highways, bridges, dams, canals, and utility structures, such as pipelines and electrical transmission lines. Generally, civil construction activities require a great deal more earthmoving equipment than the other construction categories.

These construction classifications evolved because the major equipment used in one category is considerably different from the equipment used in others. For example, the civil construction field requires the use of crawler dozers and scrapers. The light construction field has little use for this type of equipment. Light construction relies a great deal more on hand labor using saws, hammers, and finishing materials. Heavy building construction and industrial

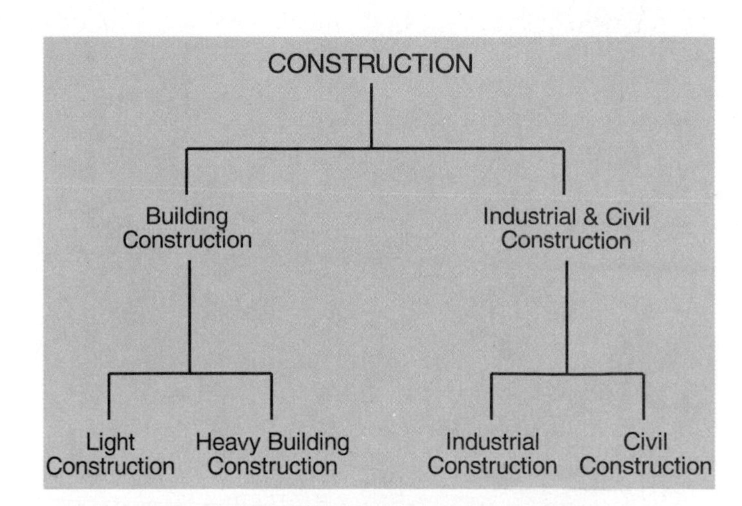

Fig. 1-17. Construction branches into different areas.

construction may require similar materials and much of the same equipment. Yet, the industrial construction industry generally does not require careful attention to human comfort, such as carpeting and built-in cabinets. Each of these construction classifications has its own separate characteristics, but all achieve their goals in a similar way.

SUMMARY

Construction activities and projects form part of all societies–past, present, and future. Construction plays an important part in our society today. We depend on constructed items, such as our homes, workplaces, roads, and pipelines to provide us with our desired standard of living, now and for the future. As our society grows, the need for construction grows.

The construction field can be broken down into light building construction, heavy building construction, industrial construction, and civil construction. All types of construction rely on basic tools, techniques, and concepts that are similar.

The construction field provides many opportunities for rewarding careers. Some of these careers require very simple skills and knowledge. Other careers require a great deal of expertise and training.

KEY WORDS AND TERMS

All of the following words and terms have been used in this chapter. Do you know their meaning?

Capital
Civil construction
Construction
Demographics
Energy
Feedback
Heavy building construction
Industrial construction
Information
Input
Labor
Light construction
Material
Output
Process
Systems approach
Time
Tool
Universal Systems Model

TEST YOUR KNOWLEDGE

Do not write in this book. Please write your answers on a separate sheet of paper.

1. List the four technology clusters.
2. To understand a complicated technology, it helps to use a _____ approach.
3. Make a diagram of the Universal Systems Model.
4. List and give a brief description of the seven inputs required for all technological activities.
5. Identify four major types of construction and give two examples of each.

APPLYING YOUR KNOWLEDGE

1. Identify several construction projects being constructed in your community. Classify each as to its type of construction.
2. Take photos of several construction projects. Make a five minute presentation to your class about the projects.
3. Locate two systems used in your home. Describe how the systems work. Diagram the inputs, processes, outputs, and feedbacks.
4. With other students in your class, make a bulletin board showing the various types of construction. Try and include at least one photograph for each category.

2

CHAPTER

HISTORY OF CONSTRUCTION

The information provided in this chapter will enable you to:
- *Discuss the early history of construction and how it developed.*
- *List three ancient human construction efforts that still exist today.*
- *Explain how individual trades developed.*
- *List five large construction efforts completed within the last 200 years.*

EARLY HISTORY OF CONSTRUCTION

Any study of early civilization includes the use of tools in the development of civilization. Technology was born when early humans first started using tools to make their lives easier, Fig. 2-1. At first, people used stones that were shaped by the forces of nature in such a way that the tool would serve a particular use. This use of tools can be considered the first use of technology by humans. **Technology** is the science of using tools and techniques in their most efficient manner.

Humans soon found that they could shape stone, Fig. 2-2. They used this new knowledge to produce better and more useful tools. This production was the beginning of **manufacturing.** Next, it was discovered that plants could be grown and cultivated. This was the beginning of **agriculture,** Fig. 2-3.

Up to this point, materials used to make the tools were limited to those materials that were provided by nature. Soon, however, people would discover ways to refine natural materials and minerals in such a way that even better tools could be built.

Fig. 2-1. Sharp stones aided early humans in the tasks they needed to survive. (Ellen McLatchy)

DEVELOPMENT OF EARLY SHELTERS

Until the development of agriculture, humans roamed from place to place in search of animals and plants for food and clothing. Because they were always on the move, there was little time or need to develop a complex construction system. Once people learned to cultivate the land and to domesticate animals, the opportunity for erecting shelters was enlarged.

Shelters were designed to protect people from the sun and storms. The first construction efforts, Fig. 2-4, were probably simple shelters made from the skins of animals. The people who roamed over mountains had the additional possibility of finding caves in which to protect themselves from the weather.

Fig. 2-2. Using rocks to break or form other objects is one of the earliest examples of humans using tools. (Ellen McLatchy)

Fig. 2-3. With the ablility to grow and cultivate plants, people no longer needed to roam in search of food. (Ellen McLatchy)

DEVELOPMENT OF TOWNS AND VILLAGES

Often, families banded together for safety. These groups began cultivating fields. There was less need to roam in search of food and a greater need to work together and support each other. These aspects led to the development of villages and meetings between families regarding the common good of the entire village, Fig. 2-5. This development required the construction of larger structures that were used as town halls and, often, as places of worship.

Since permanent villages were being established, a need arose for a road system to travel between villages. A need for bridges, for easier travel, and dams, to store water for the fields, then followed. The construction industry was well established even in these very early times, thousands of years ago. New discoveries helped in the development of new methods to do jobs more efficiently. This process continues today. In fact, with each new method, Fig. 2-6, more information is provided that leads us to even more new technologies.

Fig. 2-4. Tents made of animal skins are still used as shelter in some cultures.
(Connecticut Department of Economic Development)

DEVELOPMENT OF TRADES

As civilization progressed, it was not necessary for people to spend all of their time gathering food. Some people were able to spend their time doing tasks that were more interesting to them. Individuals became very skillful at particular tasks. These people began to **specialize,** they did those things they enjoyed most or that they were the

Fig. 2-5. Families banded together for safety. (Ellen McLatchy)

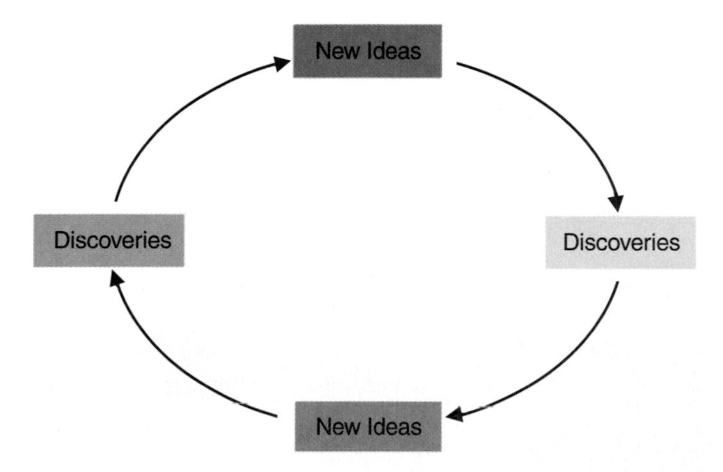

Fig. 2-6. New discoveries feed on previous discoveries and ideas.

best at. As any individual gains practice at a particular task, they get better and more efficient at doing that task.

Some individuals became very good at hunting, while others developed their skills at building shelters. These specialized workers then traded or sold their work to others. Those individuals got what they wanted from this trading or selling. Even today, in our modern society, this economic system of specialization works very well.

RESOURCES OF EARLY CIVILIZATIONS

As you learned in Chapter 1, there are seven common inputs used to construct structures that meet our wants and needs. If you think back to the time periods when these great construction projects were being built, you realize that the builders of those times had the same seven inputs for their projects as we have today.

Today, however, we have more advanced tools and equipment, more advanced materials, and more powerful sources of energy. We also have greater construction knowledge, a different labor force, and an advanced financial system. The only input that has remained completely constant is time. We still have the same number of hours in the day and the same number of days in the year as the ancient builders.

MAJOR CONSTRUCTION EFFORTS OF EARLY CIVILIZATIONS

Some of the great structures of the past still stand as monuments to those ages, Fig. 2-7. The extent and size of some of these ancient structures staggers our imagination, even today, when we have huge earthmoving equipment.

The **pyramids** of Egypt, Fig. 2-8, built approximate 4500 years ago, were originally 482 feet high. The pyramids are constructed of thousands of limestone blocks weighing from 2000 to 4000 pounds each. In the time that the great pyramids were constructed, the Egyptians had no steel tools or any large earthmoving equipment. They had to rely on their problem-solving skills to cut the giant blocks of

Fig. 2-7. Some structures created thousands of years age still stand in some form today. Stonehenge was erected around 4000 years ago.

Fig. 2-8. The pyramids in Egypt are some of the oldest human constructions still in existence.

limestone, transport the heavy blocks, and put them in their proper place. The work force needed to complete the pyramids was in the many thousands. They toiled for a number of decades to complete each one of the great pyramids.

There are many other notable construction accomplishments of the past. The **cliff dwellings** of Arizona and New Mexico, Fig. 2-9, built about 1000 years ago are an example. Some of these buildings had as many as four floors. The dwellings were built into the cliffs for protection from the weather and from enemies.

The Romans are remembered as the great road and bridge builders, Fig. 2-10, of ancient times. They designed stone bridges, and they designed aqueducts to transport their water. Often, they used stone arches to support heavy loads. The Romans knew that stone had great compressive strength. Therefore, the arches they built could stand a heavy load when the load pressed the stone blocks together. This principle is what makes the arches so strong.

The early Chinese built the **Great Wall of China,** Fig. 2-11. The great wall was begun approximately 2500 years ago. The wall is so massive that it can be easily seen by astronauts circling the earth.

Fig. 2-9. Cliff dwellings are an example of humans creating shelters out of the materials they have available.

Fig. 2-10. The sketch is of a Roman aqueduct. Similar aqueducts are nearly 2000 years old and still stand today.

Fig. 2-11. The Great Wall of China ranges about 4000 miles through northern China. (Morrison Knudsen Corp.)

These are but a few of the great construction projects of the past. You can be assured that the construction industry will continue to construct great structures in the future.

MANAGEMENT OF EARLY CONSTRUCTION ACTIVITIES

The management system used in constructing these great projects was considerably more harsh than management systems today. The workers, who were for the most part slaves, had little choice about their working conditions. Many lives were lost in the construction of the projects during those times.

Today, workers have more input into their working conditions. In addition, many of the more dangerous tasks are performed by machines.

RECENT CONSTRUCTION EFFORTS

Within recent history, the last 200 years, a very large number of major construction projects have been built. The Golden Gate Bridge in San Francisco, Fig. 2-12, the Eiffel Tower in Paris, Fig. 2-13, the Panama Canal, the transcontinental railroad system, the extensive interstate highway system across the United States, Fig. 2-14, the Alaskan pipeline, Fig. 2-15, and the Aswan High Dam, name just a few of these giant construction efforts. All of them have made our lives better and easier.

These more recent projects have been made easier through our increased knowledge of construction. We have learned much more about construction. We have learned to make the entire construction system more efficient. Construction projects that take a year to complete today, took a number of years to finish fifty years ago.

We have learned how to build and use tools, materials, and processes in such a way that we have increased the

Fig. 2-12. The Golden Gate Bridge is one of the most recognized symbols of San Francisco. (Morrison Knudsen Corp.)

Fig. 2-13. The Eiffel Tower stands 984 feet (300 m) high. It was designed for the World's fair of 1889.

Fig. 2-14. The interstate highway system dramatically reduced ground travel time in the United States.

speed at which we can construct projects. Unfortunately, we do not always look far enough into the future to see what effects these construction projects will have on humankind and the environment.

We have become known as builders out of a need to meet the demands of life. As our numbers grow, more and even greater construction demands will have to be met.

SUMMARY

The construction industry is much more complicated and extensive today than it was in ancient times. However, the same concepts that have been used in the building of ancient structures, are in use today and will remain in use in the future.

Great changes have occurred in the construction field involving the speed at which we can build. These improvements are due to more advanced power sources and a more advanced understanding of materials. These advances, to a great extent, have been achieved by individuals specializing in particular crafts. Through specialization, they have advanced their craft to a very high degree.

This specialization has allowed all of our technologies to advance. The great construction efforts of the past and those of the present will undoubtedly be matched or surpassed by construction projects in the future. So long as humans advance, the need to build will exist and more and more construction projects will be completed.

Fig. 2-15. The Alaskan pipeline runs from the Arctic ocean about 800 miles across Alaska's terrain. (Morrison Knudsen Corp.)

KEY WORDS AND TERMS

All of the following words and terms have been used in this chapter. Do you know their meaning?

Agriculture
Cliff dwellings
Great Wall of China
Manufacturing
Pyramids
Shelter
Specialize
Technology

TEST YOUR KNOWLEDGE

Do not write in this book. Please write your answers on a separate sheet of paper.

1. How were the first stone tools acquired?
2. What was a major discovery that allowed humans to remain in one location for extended periods of time?
3. Explain why towns and villages have developed.
4. List several major construction projects that were built thousands of years ago.
5. List several large construction activities that have been built in our recent history.

APPLYING YOUR KNOWLEDGE

1. Develop a chart showing the seven resources required in construction. Demonstrate the difference through drawings and pictures between ancient construction and modern construction in the application of these seven resources.
2. Research the completion dates of major construction projects in your area. Then, develop a timeline that shows these completion dates and project when new major construction projects might be built in the coming years.

3

CHAPTER

THE FUTURE OF CONSTRUCTION

The information provided in this chapter will enable you to:

- *Write a short report on a day in the life of a future construction worker.*
- *Explain the advantages of preassembling some construction components before they reach the job site.*
- *List three major areas in construction that must be further developed to keep up with a rapidly changing society.*

NEW DEVELOPMENTS IN CONSTRUCTION

The construction industry continues to grow and develop. New and better materials are being developed each day. New techniques are being put into practice with each new construction project. One can only imagine what developments are just around the corner that will revolutionize the construction industry. The future for careers and jobs in construction is very bright.

PANELIZED CONSTRUCTION

In the residential construction industry, construction companies are beginning to prefabricate many of the components, Fig. 3-1. To **prefabricate** components means to make them in a factory prior to the construction of the building. Building structures in this fashion is called **panelized construction.**

This process of manufacturing many parts of a home tends to increase the quality of the completed structure. It also makes the structure faster to erect. When this concept is carried further, you can see the complete construction of mobile homes in a factory.

This process of factory building houses and other structures can be applied to larger buildings as well. With larger structures, two or more sections of a home or building are constructed in a factory and then assembled on the site, Fig. 3-2.

Fig. 3-1. These workers are assembling a truss for a house. The completed trusses are then shipped to the construction site.

Fig. 3-2. Almost all of the necessary construction work is complete when these homes leave the factory.

This type of construction is called **modular construction.** It is being used more and more.

Sometimes, various parts of a structure are built in one location and then moved into place on site. Examples are the pipe for a pipeline, Fig. 3-3, sections of railroad track, Fig. 3-4, the trusses for a home, and the beams for a bridge, Fig. 3-5. Kitchen cabinets are generally preconstructed and then set into place in the completed home. Within the next few years, it will be possible to have large sections of homes built in factories and then assembled into a home. This system would then allow the home to be redesigned and adjusted to meet the needs of the family as it changes and grows.

Fig. 3-5. These workers are assembling massive concrete beams for a bridge. (Elk River Concrete Products)

AUTOMATION IN CONSTRUCTION

Currently, nearly all construction projects require a great amount of labor, Fig. 3-6. In manufacturing, robots are being used more frequently, Fig. 3-7. Robots reduce the labor cost and, consequently, the final cost of the products.

Currently in construction, though, robotic aid is not nearly as practical as it is in manufacturing. The large size

Fig. 3-6. Even with today's technology, it still takes many good workers to build a home. (Lowden, Lowden and Co.)

Fig. 3-3. The controlled environment in a plant allows the production of high quality pipe segments.

Fig. 3-4. These segments of railroad track are waiting to be shipped by rail.

Fig. 3-7. Many dangerous manufacturing tasks have been turned over to robots in the last few decades. (Morrison Knudsen Corp.)

Fig. 3-8. Whole communities are springing up in areas that were once vacant.

of many projects, and the fact that they are constructed on site, make it difficult to reduce the amount of labor required to complete the activity. Yet, work is in progress on **automating,** completing with machines, many construction activities. In the future, construction robots will be developed that make building structures faster, easier, and safer.

AREAS OF SPECIFIC CONSTRUCTION GROWTH

Activity will increase in most areas of construction in the future. Most of the increases will be due to growth in population and increases in technology. Some specific areas in construction, though, will be undergoing rapid growth and great changes in the very near future.

TRANSPORTATION CONSTRUCTION

As cities continue to grow, Fig. 3-8, the need for roads, homes, and places to work will continue to grow. A problem that will have to be addressed is the development of a more efficient road system. A system must be built that will allow individuals to go about their business conveniently, yet use up less space, time, and energy than our current system. Fig. 3-9 shows high speed trains common in Europe that travel in excess of 175 miles per hour. These trains move people from place to place rapidly and efficiently.

In the United States, because of its large natural resource of oil, and due to the prominence of the automobile, an extensive system of high-speed rail travel was never developed. Now, though, the world faces a reduced supply of **fossil fuels** (fuels created from decaying plant and animal matter) as well as limited space. Consequently, better systems of travel must be developed. In the future, you may well be involved in the development or construction of such a **rapid transit** system, Fig. 3-10. Whatever transportation method is developed, you can be sure that it will take a great deal of construction activity to complete the project.

Fig. 3-11 shows a major transportation tunnel construction project that took place under the English Channel. The tunnel spans a distance of approximately 26 miles, connecting England and France. To complete this project engineers had to solve many construction problems. The tunnel reduces travel time and the cost of traveling between England and the rest of Europe. The tunnel increases commerce between the neighboring countries.

CIVIL CONSTRUCTION

There are numerous civil construction projects currently underway. In addition, many more civil construction projects

Fig. 3-9. As high-speed trains gain popularity in the United States, they are expected to relieve some of the burden now placed on the air traffic system. (Morrison Knudsen Corp.)

Fig. 3-10. Subway systems such as this one are vital to many large cities. (Morrison Knudsen Corp.)

Fig. 3-11. These are images from the construction of the tunnel under the English Channel. It is one of the world's most impressive engineering accomplishments. (British Consulate General)

will be set up in the near future. Desalinization plants, waste disposal projects, and the interstate highway system name just a few of many government projects.

Desalinization plants

A rapidly growing problem is the shortage of drinking water. Many new **desalinization** plants, plants that remove salt from seawater, may have to be built to meet the needs of our growing population. Extensive pipelines and dams used to transport water many hundreds or thousands of miles may have to be constructed.

Waste disposal

Other construction projects will need to be built to handle the problem of waste disposal. People are producing greater amounts of waste, and in some cases the waste produced is becoming increasingly dangerous. To dispose of the waste safely will require extensive construction projects for new sewer treatment systems and plants. There will also be the development and construction of systems that can collect and recycle solid waste products.

Solid wastes are now, primarily, covered over in landfills. **Landfills** are areas where large quantities of solid waste are buried between layers of earth. At present, many of these landfills are loaded to their full capacity. In the coming years, we will not have the luxury of throwing away all of our trash without second thought. **Recycling,** reusing our resources, of many items such as glass, paper, metals, and wood, will be encouraged, if not required by law. See Fig. 3-12. All of these needs will create additional construction projects and provide quality jobs for well qualified workers.

Interstate highway system

Another example of civil construction is the interstate highway system. This multi-lane system was first designed

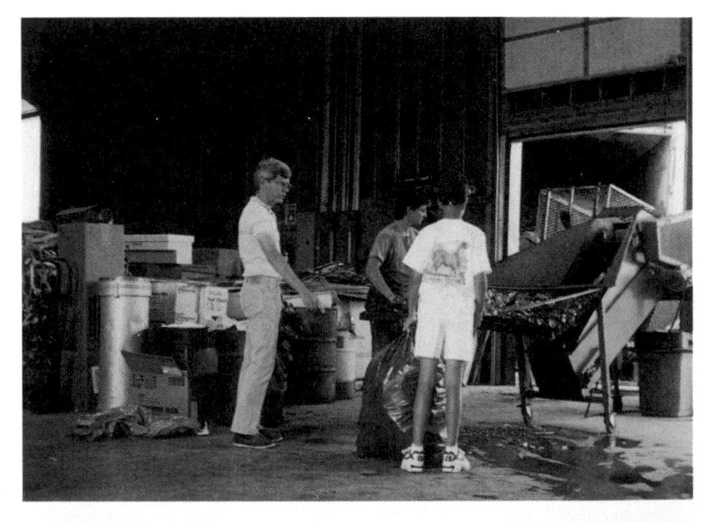

Fig. 3-12. Although the recycling of many common household garbage items is optional in most areas, the recycling of some of these items will be required in the near future.

Fig. 3-13. The Sears Tower in Chicago is 110 stories high. (Sears Roebuck and Co.)

to help reduce accidents and increase the efficiency of highways throughout the United States.

The interstate highway system was designed for the possibility of driving nonstop between any two major cities. The system was designed with limited access to allow quick and safe travel. These accesses made extensive use of cloverleaf interchanges to allow merging traffic to come up to speed before entering the main traffic flow.

This highway system, which was first begun in the early 1960's, is now over thirty years old. In many places, it is much in need of rebuilding. This remodeling will consist of extensive resurfacing of the road and upgrading of tunnels and bridges. It will also require the construction of new interchanges as well as new roads as the population grows. All of this construction will require many well qualified workers with an interest in civil construction.

COMMERCIAL CONSTRUCTION

More large commercial buildings may have to be constructed to house the many workers making up the work force in this age of information. These structures will be major projects and some of the buildings will extend over a hundred stories high, Fig. 3-13. In large cities, where the land is expensive and scarce, this is often the most economical method of arriving at the needed floor space. Again, the need for construction workers for projects of this type will be enormous.

Airports comprise another area in which vast quantities of construction will be needed in the future. Nearly all large airports are now under some continuous form of construction or remodeling. This is necessary in order to keep up with the demand for air transportation. As shown in the artist's rendering of Denver International Airport, Fig. 3-14, larger and

Fig. 3-14. This is an artist's rendering of the Denver International Airport. (Airport Public Affairs, Stapleton International Airport)

more efficient airports must be built to handle the needs of tomorrow's traveler. The airports must also be capable of handling the air transportation of mail and other goods.

Space travel presents a great opportunity for construction work. Launch pads, Fig. 3-15, here on earth and platforms in space, Fig. 3-16, must be built. In addition, construction projects are now being designed for other planets, Fig. 3-17. These structures will likely have to be constructed from the materials found on the planet being explored. This will require construction workers who cannot only understand our present construction system but those who are excellent problem solvers. These workers will be required to work with limited resources to construct structures. The structures will have to protect them from all of the harsh conditions on the various planets or moons visited.

Fig. 3-15. Space shuttle launching from Cape Canaveral in Florida. (Morrison Knudsen Corp.)

Fig. 3-16. An artist's rendering of a space station. Ships can dock without having to enter a planet's atmosphere. (NASA)

ENERGY FACILITIES CONSTRUCTION

Another area that will be undergoing great changes in the future is the area of energy production. The entire energy picture on earth is being carefully studied, and new sources of energy developed. This will require the construction of major energy production facilities, Fig. 3-18, regardless of which types of energy facilities are developed.

Some of these projects may deal with additional coal fired systems. Other projects might be solar or wind powered plants. Still others could be powered by hydrogen or some other energy source.

Coal-fired electrical plants make a good example. These electrical generating plants require extensive construction activities before they are ready to produce electrical energy. First, the supply of coal must be carefully determined. If the coal is available locally, it will only have to be moved a few miles. On the other hand, if the generating plant is a long distance from the coal source, the coal will have to be transported long distances. In either case, some transportation system must be designed and constructed.

Other construction work includes electrical transmission lines to carry electricity from where it is generated to

Fig. 3-17. Missions to Mars by astronauts are not just a fantasy anymore. These missions may take place in the near future. (NASA)

Fig. 3-18. Wind turbines are a common sight on the west coast. They are often grouped together in *wind farms.*

its various distribution points. The coal generating plant must also be built, which is a major construction activity in itself. The plant requires several years to complete. In addition, today's laws require that once the coal is removed from surface mining, the earth must be moved back into a state similar to its original condition. This requires extensive earthwork to be completed by heavy equipment.

COMMUNICATIONS CONSTRUCTION

You often hear that society has entered the communications age. We have the capability of sending information almost anywhere on the planet in minutes. We are able to do this because of the construction of specialized earth stations, Fig. 3-19, to transmit signals to satellites, and the installation of fiber-optic lines, Fig. 3-20. All of these construction projects require construction of a specialized nature.

Our telephone system is a good example of a communications system that needs construction workers. The system consists of telephone lines that are installed to the customer's phone. Beyond this the telephone company must have computer switching offices located in most large cities to handle the switching of phone calls from the caller to the receiver.

Fig. 3-19. This earth station receives and transmits thousands of messages with satellites orbiting the planet.

Fig. 3-20. This construction crew is installing fiber-optic cable underground.

Links between these major switching facilities may be by land line, buried cables, or most recently, fiber-optic cables, microwave stations, and satellites. All of these installations require an extensive labor force of construction workers. Workers are needed not only to build the various systems but to repair and maintain the system once it is up and operating.

RENOVATION, REMODELING, AND REMOVING CONSTRUCTION PROJECTS

Another construction field, which is often forgotten, is the moving, removal, and remodeling of older structures that have outgrown their usefulness. With **renovation,** a building is restored to a former state. With **remodeling,** a building has changes made in its actual structure.

Some day, all of the construction projects that are being erected around you will fall into this area. These structures must be either remodeled, renovated, moved, or demolished. This, again, takes construction workers with special skills and knowledge to do this type of work efficiently and safely.

CONSTRUCTION IMPACTS SOCIETY

The construction industry will play a vital part in the future development of our society. An understanding of construction is a vital part in the education of all citizens. It is important to the individual who is interested in construction as well as those individuals who will only use items constructed by others. The basic understanding of construction and its impact on us will make us all better citizens. In addition, knowledge of construction will make society better qualified to use the remaining resources we have on earth. This knowledge will enable us to provide a better life for all.

SUMMARY

All construction projects are very labor intensive. The reasons for the large amount of labor are that the projects are large and, for the most part, constructed on site. Robotics is just beginning to make an impact on construction projects. Within the next few years, you will start to see more and more robots introduced into the construction field.

Construction in the transportation field will continue to grow. The need to move people and materials safer, faster, and more reliably will have to be dealt with through major transportation construction projects. As the population grows there will be the need to build homes, office buildings, and factories. Some of these structures will be built on undeveloped land while others will require the renovation, remodeling, or the demolition of existing buildings.

Space construction will provide many exciting opportunities for construction workers who have the necessary skills. As exploration of space, nearby planets, and moons grow, the need will emerge for new structures designed and built to house these special workers and their equipment. Career opportunities for well qualified construction workers are very bright. There will be excellent opportunities for those individuals who have basic skills and are interested in solving construction related problems.

KEY WORDS AND TERMS

All of the following words and terms have been used in this chapter. Do you know their meaning?

Automating
Desalinization
Fossil fuel
Landfill
Modular construction
Panelized construction
Prefabricate
Recycle
Remodel
Renovate

TEST YOUR KNOWLEDGE

Do not write in this book. Please write your answers on a separate sheet of paper.

1. Give several reasons why robotics has not yet had a significant impact on the construction field.
2. Explain why there will be an increased need for construction projects such as homes, factories, transportation systems, and communications systems in the future.
3. Explain why tall skyscrapers are built in large cities, and they are not constructed in small towns.
4. Analyze the similarities and differences between building a large structure in space and on earth.
5. List the factors that need to be considered when determining whether to remodel or demolish an existing building in a growing community.

APPLYING YOUR KNOWLEDGE

1. Visit with the local government planning office. Determine what the future plans are for your community that will involve construction activities.
2. Invite an architect to class. Have the architect discuss the need for long-range community planning.
3. Write a letter to NASA requesting information on construction in space.
4. Write a short report on what you think a day in the life of a construction worker will be like in the year 2050.

4
CHAPTER

IMPACTS OF CONSTRUCTION ON SOCIETY

The information presented in this chapter will enable you to:

- *Develop an awareness for the impacts of construction on our everyday lives.*
- *Outline a law that has been passed to help in development of the best interest of society.*
- *List five advantages and five disadvantages for a major construction project in your community.*
- *Identify the impacts on each of the seven resources of a major construction project in your community.*

EFFECTS OF CONSTRUCTION ON EVERYDAY LIFE

Consider for a minute, how construction has affected your life. The home you live in, the sidewalk you walk on, the school where you study, bridges, railroads, pipelines, electrical lines, and the sewer system were all built by construction workers using the resources at hand, Fig. 4-1.

Any time humans build or construct an item that makes a task quicker, easier, cheaper, or more expensive, that item affects our life greatly. Technology is neither good nor bad. How you see the effects of technology, depends upon your point of view.

Fig. 4-1. Large cities contain numerous construction projects. Skyscrapers, bridges, and roadways are some of the most noticeable projects, but there are many structures constructed underground as well.

HISTORICAL IMPACTS OF CONSTRUCTION

Construction has affected the land as well as the development of society in many ways. Think back to when Europeans first settled in North America. At that time, there were few roads, buildings, or houses. Primarily, the people lived in very simple homes, Fig. 4-2. Most of the homes were made of logs or sod. Because there were so

Fig. 4-2. Early homes were of a simpler, and generally smaller, construction than most single-family dwellings built today. (Bureau of Land Management)

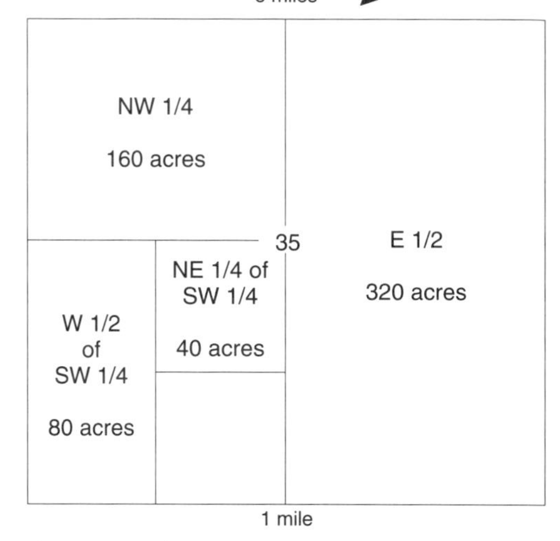

Fig. 4-3. A system of surveying land was established in 1796. This system is still in use today.

few people, the **impact** on the land was small. However, as the population grew, towns developed. More and more people moved closer and closer to each other. The effect on the land became greater.

Disagreements arose as to who owned what land. These disputes brought about a system of measuring and recording the land, Fig. 4-3. This process had a significant impact on the country. Once the land could be described, it could be traded or sold to others. With the development of surveying, Fig. 4-4, came further regulations.

The cities began to grow rapidly. The larger cities had to develop building codes to assure some measure of order in their communities. An example can be seen in the problems that came about when individuals built their houses too close to the neighboring houses. If one house caught on fire, it would burn all of the neighboring homes down.

The Great Chicago Fire of 1871 raced through the city, in part, because of this type of construction. At least 300 people were killed in the fire, and 90,000 people were left homeless. The fire resulted in the passage of many **building codes,** codes that regulate how a building may be constructed. Many

Fig. 4-4. Surveying has become a science. The equipment used is very accurate.

other cities at about this time also enacted building codes that helped to prevent similar major catastrophes. Today, all buildings must follow strict building codes, Fig. 4-5. The codes, though, vary from city to city.

Fig. 4-5. Plumbing systems, mechanical systems, electrical systems, and the building construction itself are all governed by detailed building codes.
(Building Officials & Code Administrators International, Inc.)

With the development of the automobile, more, and better, roads were built. For these new roads to be built, the government had to acquire land. If a landowner did not wish to sell the land, an entire project could be held up, even though the new road would benefit the community. Laws were set up to prevent this from happening. Individuals can be forced to sell their property to the government for a reasonable price. This is referred to as the government having the **power of eminent domain.** The power of eminent domain helps make sure that projects for the good of all citizens can be completed.

There are numerous other examples that can be cited showing the impact of construction on individuals and society. Any construction action will cause a number of reactions. Some actions only affect the surrounding communities. Some actions affect society as a whole. You may be interested in only one change. For example, you wish to pave a parking lot. Yet, the effect of that change makes many changes in the **immediate future** and the **long-term future.** Water drainage patterns in that area will be dramatically altered by the paving of the lot.

IMPACT OF CONSTRUCTION ON THE SEVEN RESOURCES

The impact construction projects will have on the community, Fig. 4-6, as well as the surrounding area is easy to

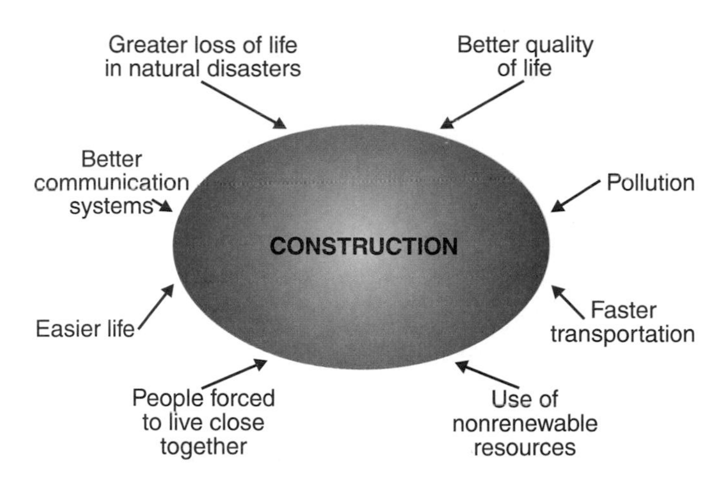

Fig. 4-6. Some of the impacts that construction has on the surrounding community.

underestimate. The seven resources involved in construction are all impacted by the project. If the project is small, naturally, there is less of an impact. However, when there are a large number of small projects, the impact on the area may be as great or greater than one large project.

In all cases, major construction activity will impact the seven resources in the following ways:

- People: A large project will take workers away from other activities in the community. This could cause a labor shortage in some areas. The project will also cause a temporary influx of new members to the community. If the construction project will take several years to complete, construction workers will often move their families close to the construction site. This may impact the local schools, stores, restaurants, entertainment facilities, and traffic.
- Information: Construction projects, particularly in the field of communication, can accelerate or halt the flow of information. A single large project or a number of smaller construction projects will require the addition of more and better communications systems. Additional phones, fax machines, computer links, and mail delivery may be required.
- Materials: The price of materials in the local community will be raised or lowered. Additional materials have to be shipped into the community causing transportation problems or opportunities. Some materials will be difficult to obtain because of shortages caused by the construction. The construction project may also bring in an influx of new businesses.
- Tools and equipment: Are the tools and machinery available locally? They may have to be brought in from outside areas, which could again cause transportation problems. Noise pollution from power equipment may be disturbing. The service and sales of equipment may provide an opportunity for local business to expand.
- Energy: Energy consumption for the construction projects will be fairly high. This could affect the local cost of

electricity, gas, or oil. If the power consumption is too high, there also could be power shortages or power outages in the community. On the other hand, the increased power demand might cause the local power company to upgrade its utility lines to the community. This would leave residents with better service.

- Capital: A construction project will affect the amount of money local banks have to loan to other customers in the community. The addition of more construction could also encourage other lending agencies to move to the area and set up new banks or savings and loan offices.
- Time: During construction, a project will cause time delays in traffic or time delays in accomplishing some tasks. However, many construction projects, when completed, will ease up delays that were experienced before the construction began.

IMPACT OF CONSTRUCTION ON THE COMMUNITY

Large construction projects have the most noticeable effects on the surrounding community. Suppose that a new coal-fired electrical generating plant, Fig. 4-7, was to be constructed in your community. What would be some of the most common impacts the plant would have on your community? Would the construction of the plant be beneficial to the community? If the decision is made to build the plant, then there must be a transportation system that would be able to get the coal to the plant, Fig. 4-8. Is this system in place or would it have to be constructed also?

PLANNING FOR CONSTRUCTION

The actual construction of the plant involves a large number of workers, Fig. 4-9. The construction of the plant would take place over a period of several years. If these workers have children of school age, the school system will need more classrooms for these new students, Fig. 4-10. The community may have to construct new schools, or at least provide temporary school facilities, for the children.

With the added amount of workers in the community other community services would have to be expanded. Is the current road system adequate to handle the added traffic of workers and their families? Are there enough stores to supply the growing community with goods and services? These questions must be considered before construction begins.

STARTING CONSTRUCTION

Once construction begins, additional problems will develop. Some of these problems will involve the community government. Other problems will only affect certain individuals. How will the added noise produced by the construction activity be handled? What will be done with the added amount of waste produced by the construction? Are the telephone lines adequate to carry the added load?

COMPLETING CONSTRUCTION

Once the facility has been completed, many of the large number of workers who were involved with the project will leave the community. Then, the area will need to deal with different problems. Some of these problems are:

- What is to be done with extra public services and utilities that have been built to handle larger numbers, when they now must be paid by a community with smaller numbers?
- What effects will the pollution from the plant have on the lives of people in the area, Fig. 4-11?

Fig. 4-7. How many advantages can you see in having this power plant erected near your community? How many disadvantages can you see? (Montana Power)

Fig. 4-8. A dirt road is fine for small amounts of country traffic, but it would be an unfit transportation system for supplying coal to a large power plant. (Bureau of Land Management)

Fig. 4-9. The construction of a power plant is an immense job. (Montana Power)

Fig. 4-10. Mobile homes are often used as temporary schools to handle sudden and dramatic increases in enrollment.

Fig. 4-11. Power plants release heat as well as chemical pollutants into the surrounding area.

- How could a water pollution problem be controlled if some part of the plant does not function properly?
- How long can the plant stay in operation before it must be either closed or modified?
- Will the plant attract additional industry?
- Will the plant force some industries to close or move to other communities?
- What might the plant do to the environment over the next 20 or 50 years?

These are some of the questions and concerns over the advantages and disadvantages that a community must deal with when decisions are made about community growth. Often, the decisions have not been studied enough. Many times people would have been better off had more careful planning taken place.

The impacts of construction should not be simply viewed as all good or all bad. Whether the impacts are good

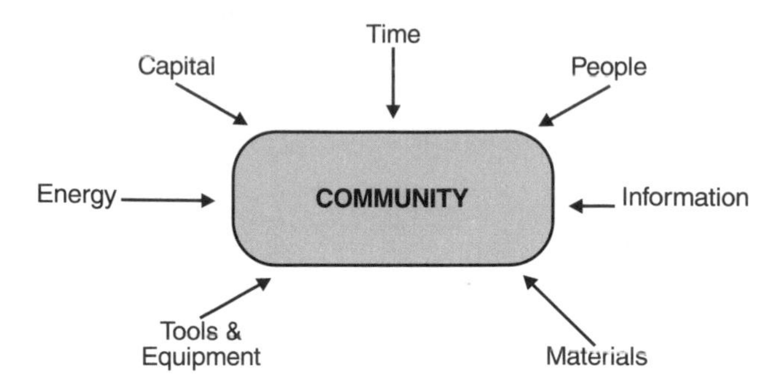

Fig. 4-12. Construction affects the seven resources, which in turn affect the community.

or bad will depend on what interests you have in the construction activity. The important action is to be able to think far enough into the future to see all of the impacts that the construction project will have. Once a careful study is completed, a decision can be made by weighing the advantages and disadvantages of a project.

As society moves forward, we are more able to understand the impacts our technological activities have on the planet earth. This enables us to design systems that better allow us to live in harmony with our **environment,** the land, vegetation, and animals around us.

CONSTRUCTION IN SPACE AND ON OTHER PLANETS

As we begin to construct space stations and develop colonies on planets with limited resources, we will be forced to be examine closely how our construction activities are planned and carried out, Fig. 4-12. Careful planning will be required for survival. Perhaps we have been spoiled here on earth where all of our resources are very plentiful. Many resources have not been used as wisely as they should.

SUMMARY

Construction activities are a very large part of our society. Any time there is a construction activity, there will be impacts to the environment and society. Each impact will depend upon how extensive the project is. How a particular construction project impacts individuals or the environment depends greatly on how it is viewed. What is very desirable for some people, may be undesirable for others.

In order to assess the future impact of a particular construction project, a careful study must be conducted and information gathered from a variety of sources. Without adequate information it is difficult to determine what the long-range impacts of a construction project will be.

In recent years, we have become more aware of the fragile nature of the planet. The earth is being viewed as a whole system and, therefore, people are becoming more conscious of the limits of our resources. One way to determine the potential impact of construction activities is to view the activity in terms of its impact on the seven resources.

KEY WORDS AND TERMS

All of the following words and terms have been used in the chapter. Do you know their meaning?

Building code
Environment
Immediate future
Impact
Long-term future
Power of eminent domain

TEST YOUR KNOWLEDGE

Do not write in this book. Please write your answers on a separate sheet of paper.

1. What historical event promoted the use of building codes?
2. List the seven resources used in construction.
3. Describe several advantages and disadvantages of the construction of a new school in your community.
4. How might space travel make us more aware of how we use our resources?
5. The construction of a high speed rail system is taking place near your community. Write a paragraph that analyzes the possible impacts that this might have.

APPLYING YOUR KNOWLEDGE

1. Take a field trip to your local courthouse. Ask to see the land records. Find out who owns some particular pieces of property in your community.
2. Make a simple map of the school yard and buildings.
3. Review a building code book. Ask a city official to discuss the need for building codes with the class.
4. Obtain a copy of a local environmental impact statement for a construction project near your community. After reviewing this document, determine for yourself the merits of the proposed construction.
5. Make a diagram demonstrating the impact of a major construction activity on your community. Include not only the impacts on the environment, but also the impacts on the seven resources.

Modern technology creates powerful new tools for the construction industry.

section 2

The Relationship of Construction to Other Technology Systems

Technology can be divided in many ways and into a large number of smaller systems. In this text, you study technology systems in four major areas. These systems are: construction systems, manufacturing systems, communications systems, and transportation systems. Although the central focus of this text is construction systems, it is important to understand that construction systems do not operate alone. They are dependent on support from the other three systems of technology.

Anytime a project is constructed it must rely on communications technology, manufacturing technology, and transportation technology. Without these other technology systems no building or structure could be built. In this section, you will examine this relationship between construction and the other three technology systems, and you will see how all four systems are interrelated.

Large manufacturing plants are monuments to modern construction technology.

5

CHAPTER

MANUFACTURING SYSTEMS

The information provided in this chapter will enable you to:

- *Understand the similarities and the differences between manufacturing and construction systems.*
- *Explain the interdependence between construction systems and manufacturing systems.*
- *Explain the differences between custom manufacturing and mass production.*
- *Tell why many raw materials must be processed before reaching a construction site.*

INTERDEPENDENCE OF TECHNOLOGY SYSTEMS

The field of technology consists of a variety of communications systems, construction systems, manufacturing systems, and transportation systems. These systems work together to produce an item, service, or structure that meets our needs. Those systems that do not meet our needs are soon eliminated.

The four technology clusters, Fig. 5-1, rely on each other to provide needed information, equipment, structures, or transportation. Each technology area contributes one set of vital pieces to the rest. If you remove any one of these areas from the total system, the others will operate at a lower level of efficiency or not at all. There is a vital dependence of each technology cluster on the others, or all of the technology clusters can be said to be **interdependent.**

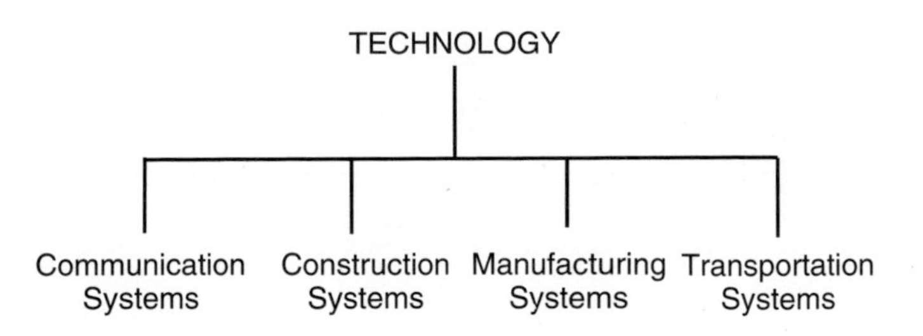

Fig. 5-1. Technology can be divided into four separate, but interdependent, systems.

DEPENDENCE OF MANUFACTURING ON CONSTRUCTION

All manufacturing systems rely on the transportation of materials from their source to the manufacturing plant and then to the customers. Without an extensive system of roads, railroads, waterways, and airports the manufacturing system could not move its products. The transportation system allows this movement of materials. Without a construction system to build these various means of transportation, the manufacturing system would not be able to operate.

Construction of a plant or factory is directly dependent upon the construction system. Without construction activities the manufacturing plant could not be built or remodeled. In order for a manufacturing company to stay current with today's advancing technology, they often have to remodel or add new equipment to the factory. Without well-trained construction workers, a factory will not meet the needs of their customers.

Factories and manufacturing plants rely on various forms of energy in order to produce products. This energy (oil, electricity, and natural gas) must be transported to the factory by various means. The building of these pipelines and electrical lines are part of the construction industry. Also, the construction of electrical generating plants, hydroelectric dams, and oil refineries are all the end products of construction activities.

CONVERTING RAW MATERIALS INTO USABLE FORM

There are few materials that can be used in construction directly from nature. Some form of processing must be done. The **raw material,** the crude form of the material, must be transformed into standard shapes, sizes, and purity, Fig. 5-2. Then, efficient use of the products can be made. These processing activities are accomplished in **manufacturing systems.**

Manufacturing can be defined as the processes of changing materials into more usable forms in a manufacturing plant or factory. This covers the converting of raw materials such as iron ore, trees, or crude oil into standard stock products. Manufacturing also includes the further converting or refining of standard stock into products that meet the needs of other industries or society.

The inputs to a manufacturing system consist of the raw materials. An example of one of these raw materials would be trees. Trees are converted into building products. The tree is first cut and hauled to a sawmill. There it is cut and processed into standard stock. This stock, Fig. 5-3, whether it is framing lumber, plywood, particle board, shingles, or finished trim, must meet certain specifications. This assures the **contractor,** the person who is under contract to erect the structure, of a standard size and quality of products. This allows the contractor to meet the requirements of a construction project. Without standards that are agreed on by the manufacturer and the contractor, the efficiency of any project would be greatly reduced.

Fig. 5-2. Iron is heated in giant furnaces and alloyed with other metals to produce steel. The molten metal is then formed into bars of a standard shape and size.

TYPES OF MANUFACTURING

Manufacturing can be broken down into two major methods, Fig. 5-4. These methods are custom manufacturing and mass production.

CUSTOM MANUFACTURING

In **custom manufacturing** one, or very few, of a particular item is manufactured. Custom manufacturing allows the production of items that only a small number of people would want, or items designed for special purposes.

Fig. 5-3. This wood has been trimmed into standard stock and is waiting to be shipped.

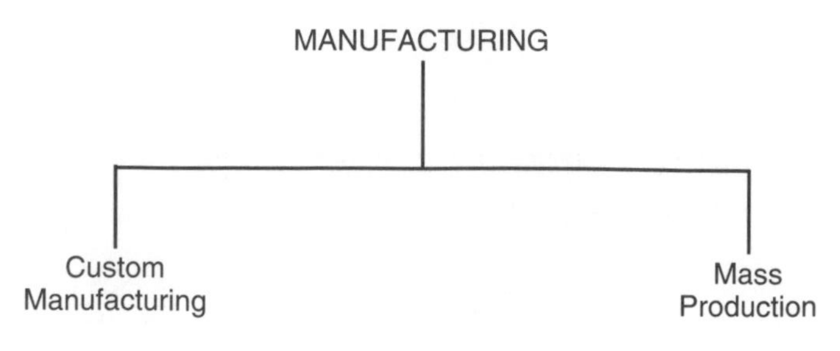

Fig. 5-4. Manufacturing can be subdivided into custom manufacturing and mass production.

This type of manufacturing is very labor intensive. It requires a great deal of skill by the worker.

Custom manufacturing is also quite expensive, because each part of the item manufactured is custom-built. A **custom-built** item is built to individual specifications. A good example of custom manufacturing is a car designed to race in the Indianapolis 500. Each of these cars cost hundreds of thousands of dollars. They are designed and handmade for a special purpose.

Fig. 5-5. This factory mass-produces cement for many purposes.

MASS PRODUCTION

With **mass production** many like items are produced. While the car built for the Indianapolis 500 cost hundreds of thousands of dollars to produce, an automobile mass produced in a plant may cost less than $10,000. In this case, the car is one of thousands that are manufactured on an **assembly line.**

With mass production, the amount of skilled labor involved in each car is greatly reduced. Since each car does not require custom parts, the cost of each auto is also greatly reduced. Without mass production, people would not be able to afford many of the goods and items that they have today.

MANUFACTURED PRODUCTS IN CONSTRUCTION

Examples of manufactured products used in the construction field are almost endless. The individual parts of a building's entire electrical system are manufactured in plants. The same is true for plumbing systems, heating and air-conditioning systems, and the parts of the interior finishing. Often, even the cabinets are manufactured in a plant and then installed in a building.

Road construction also uses manufactured products. The culverts, the bridge members, and even the materials that go into a batch of concrete are manufactured or converted into a usable form by a manufacturing facility, Fig. 5-5. Any steel that is used must first be extracted from the earth or recycled from scrap. These processes put the steel into usable forms. Then it can be properly used in the construction industry.

ASSEMBLY LINE MANUFACTURING FOR CONSTRUCTION

As discussed in Chapter 3, manufacturing is being used more in the creation of building components in a factory. Entire homes may be built on an assembly line. These homes are generally built in modules or sections, as shown in Fig. 5-6. The pieces are then transported to the construction site and assembled with other modules, completing the building structure.

These assembly line methods are used in more than just home construction. In road construction, an entire section of a bridge may be manufactured off site. Sections of large tanks are pre-shaped before they are shipped to the construction site. Many common components in building construction are fabricated in plants. These components include trusses, wall sections, cabinets, bathrooms, and windows, Fig. 5-7.

ADVANTAGES OF MANUFACTURING COMPONENTS

Manufacturing components off-site allows better control over the conditions involved in producing the part. A factory can be set up with special equipment and environments, Fig. 5-8. Control of workers and materials can be carefully monitored. This increases both the efficiency of the workers as well as the quality of the completed product. This gives the construction contractor an assurance of quality products with which to construct the project. Also, with standard materials, the contractor can better determine the time and cost to complete a job.

The interdependence of construction with manufacturing is easy to see. With careful coordination between these two major technology systems, the efficiency of both can be greatly increased. This works to steadily improve our standard of living.

SUMMARY

Few items that are used in the construction field are found in a usable form in nature. Raw materials must be manufactured into a form that can be used on the construction site. The manufacturing of these items into

Fig. 5-6. Shown is a modular home in several stages of factory construction. When the modules are complete, they are transported to the construction site for assembly. (Bud Smith)

Fig. 5-7. Cabinets, trusses, and even entire walls are now commonly produced in factories.

construction materials is required so that the progress of the construction can be efficient.

Nearly all components of the construction industry rely first on the manufacturing of standard materials. Once these materials are delivered to the construction site, the contractor can then assemble them into a structure that meets the needs of the customer.

The two major methods of manufacturing are custom manufacturing and mass production. Custom manufacturing requires a great deal of individual attention and labor. Mass production is more efficient. The large number of a particular item or product produced allows mass production to take advantage of assembly line techniques.

KEY WORDS AND TERMS

All of the following words and terms have been used in this chapter. Do you know their meaning?

Assembly line
Contractor
Custom-built
Custom manufacturing
Interdependent
Manufacturing
Manufacturing systems
Mass production
Raw material

Fig. 5-8. The workers on the left construct a wall in the controlled environment of a plant. On the right, the workers must contend with the cold and snow while completing their work. (Bud Smith)

TEST YOUR KNOWLEDGE

Do not write in this book. Please write your answers on a separate sheet of paper.

1. List two construction materials that are manufactured into standard stock before they arrive on the construction site.
2. List five items that are often built in factories and then transported to a construction site.
3. Explain some of the ways that manufacturing and construction industries are dependent upon each other.
4. What are the two major types of manufacturing, and how do they differ?
5. What are the advantages and disadvantages of each type of manufacturing?

APPLYING YOUR KNOWLEDGE

1. Visit a lumber yard. Determine how many standard stock items are available for the construction of small buildings.
2. Compare the price of several automobiles, some custom-built or modified and some that are mass produced. Discuss your findings with others in your class.
3. Make a diagram with a construction site in the center. Around the construction site, list all of the manufactured components you can think of that go into the construction project.
4. Trace the manufacturing processes used to convert raw materials into usable forms. List some of the careers that would be associated with these manufacturing processes.

Slag is made up of the impurities that float to the surface while processing steel. Here, slag is being poured off in a waste area to cool.

COMMUNICATIONS SYSTEMS

The information provided in this chapter will enable you to:
- *Illustrate the relationship between construction systems and communications systems.*
- *List the communications links between different workers from the beginning of a project to its completion.*
- *Explain the importance of the communications links between all those who work on a construction project.*

COMMUNICATIONS SYSTEMS IN CONSTRUCTION

Communications is critical to construction, manufacturing, and transportation technologies. Without communications taking place between individuals, no work can be completed. No one would be able to direct the work of others. Without direction, individuals are not able to coordinate their efforts with other workers. A communications system is critical in accomplishing any construction activity requiring more than one individual, Fig. 6-1.

COMMUNICATIONS IN THE BEGINNING OF A PROJECT

Before any construction project begins, there is first an individual or societal need. This need has to be communicated to others. All individuals must be able to communicate their ideas with each other. The **originator,** the person who conceives the idea, and all others in this information loop, have to adjust those ideas to best meet the needs of the group and society, Fig. 6-2.

All of the communications involved with the project ideas take place long before any design work for the construction project is started. As the initial idea takes shape, architects and engineers communicate to refine the plan and design of the project. Sketches and drawings of the project are developed. These are then checked with manufacturers to determine if the necessary materials are available. Again, all of these activities require workers to be in communication with each other, Fig. 6-3.

Finalizing plans involves a great deal of communications, Fig. 6-4. Suppliers, engineers, architects, geologists, and government officials all must be in contact with one another. At the same time, more communications are taking place between city planners, real estate personnel, and lending

Fig. 6-1. Proper communication is important in construction work. Workers communicate through many mediums. They may speak face-to-face, over the telephone, or through a computer.

Fig. 6-2. This model communicates the designers ideas to the other people involved in the project. The model is not a final decision. Many changes will be made before the structure is complete. (Tom Wood)

agencies to assure the proposed plan will meet all of the local building codes and requirements. **Financing,** the money for the project, must also be made available.

COMMUNICATIONS IN THE CONSTRUCTION PHASE

As the construction project begins, the communications between the architect, contractor, subcontractors, suppliers, and workers become even more critical. The **architect,** the person who designs the project, communicates to the contractor primarily through detailed plans and specifications. The contractor, then, communicates these ideas to the job superintendent. The superintendent, in turn, provides instructions to the supervisor. The supervisor communicates with the workers. These communications are made using a variety of media–written, spoken, and electronic, Fig. 6-5.

In the past, the most common way to communicate to another person was face-to-face. With the advent of a solid postal system, it was reliable to communicate with others by writing letters. Though, this still takes days to receive a reply. When the telephone came into common use, communication between two people long distances apart became more highly efficient.

However, until more recently drawings and diagrams still had to be sent through the mail. Today, this information can be transferred by a facsimile machine, or FAX. This allows drawings and diagrams to be transferred in a matter of seconds making the total communications systems even more efficient. With the progress of the com-

Fig. 6-3. These construction workers need information to do their job. Information is conveyed to them through drawings.

Fig. 6-4. During the final planning of a large construction project, continuous streams of information are maintained with lending agencies, real estate personnel, and the city planners. (IBM)

Fig. 6-5. Supervisors can communicate information very precisely and dramatically with computers. (IBM)

puter and computer links, entire sets of drawings can be transferred between computers in a few minutes. This ability to communicate large amounts of various data almost instantly makes the construction system more efficient and better able to meet societies needs and desires.

All of these **communications links** must be efficient. Each work crew is assigned specific tasks to perform based on the plans for the entire project. Problems could come about if someone does not understand what has to be done and when it is to be completed.

The entire field of scheduling work on a project is a study in communications. **Scheduling** a construction project involves determining in what order tasks must be done and when they need to be completed. In construction, certain jobs must be completed before other parts of the job can begin, Fig. 6-6. If materials or labor are not provided at the correct time, or if they are not provided in the proper amount, the entire job may be held up. This delay is a result of a **communications breakdown.**

If contractors do not communicate clearly to their subordinates, their projects will not be completed within the original time estimate. In addition, with poor communications, the quality of the completed project will be reduced. Quality contractors are excellent communicators and great managers of their resources.

Communication is not only important for the owner or upper management of a construction company. All the workers on a project must be able to communicate effectively. The entire construction company must work as a team, or efficiency will be reduced.

One prime example of communications between members of a construction crew is shown in Fig. 6-7. In this case, materials are being off-loaded from a truck to the construction site. The workers must communicate with the crane operator, but noise and distance at the site prohibit voice communications. The workers rely entirely on standardized hand signals. Signals are flashed to let the crane operator know when to pick up the load and transport it to the construction site. There, other workers communicate with the crane operator to place the load in the right location. This is, again, all done with hand signals.

COMMUNICATIONS IN THE COMPLETION OF THE PROJECT

During the closing stages of a project, communications are used for final inspections, final payments, and project operating procedures. For the **final inspection,** the owner, contractor, and lending agency tour the project. They determine if the completed work meets the original specifications, Fig. 6-8. Naturally, inspections take place while

Fig. 6-7. A set of hand signals allows this worker to communicate clearly to the crane operator. Noise on the construction site interferes with verbal communication at long distances.

Fig. 6-6. Construction had to be halted until this shipment of steel arrived on the site.

Fig. 6-8. Safety is important on all construction sites. During the final inspection, inspectors check for flaws in mechanical systems that may endanger a project and its employees. (Chevron Corp.)

the project is being constructed, but the final inspection has special significance. This inspection reviews all previous inspections to assure that the standards agreed upon at the start of the project have been met.

Before the contractor leaves a project, the owner must be provided with specific operating instructions for the various systems installed in the project. This information may be communicated through something as simple as leaving manuals about the equipment installed in a house for the home owner to review. In a complicated construction project, such as a large hydroelectric dam, detailed instructions and **training sessions** for management and employees would be conducted, Fig. 6-9. In the case of the dam, there are volumes of operating procedure manuals that must be well understood by the workers. These also fall under the area of communications systems.

If any miscommunication occurred between the various parties, the **judicial system,** Fig. 6-10, may be brought into place to settle the dispute. Generally, individuals and companies do not want to have judges or juries determine how some document should be interpreted. However, this system must be used to solve contract problems. These problems arise when two parties did not have the same understanding of the policies, procedures, or quality of work that was to be completed.

These are only the major parts of the communications systems that come into action for any construction project. In addition to these systems, there are a vast number of smaller communications systems that are involved. The combination of all of these communications helps bring any construction project from the initial idea to the finished project.

Fig. 6-9. An instructors job is to communicate information. Instructors are trained to communicate. (IBM)

Fig. 6-10. Taking a dispute to court is a last resort. The first step is to discuss your case with a lawyer.

SUMMARY

Communications is a critical element in the completion of any construction project. Without communications systems in place and properly working, the efficiency of any project is greatly reduced.

The construction industry relies on communications systems from the period long before the project is physically started to well after the project is completed. First, an idea must be formulated in someone's mind. Then, this idea is communicated to others. This idea must meet some individual or societal need. Preliminary information must be communicated to all interested individuals and groups before any construction plans are started.

The plans themselves are one form of communication. During the actual construction, each worker must receive directions from other construction workers based on the construction plans, if work is to progress on schedule.

Once the project is completed, operating instructions must be communicated to the owner/operator of the project. This is necessary for correct operation and maintenance of the structure. These are but a few of the communications networks that allow us to construct projects efficiently and on schedule.

KEY WORDS AND TERMS

All of the following words and terms have been used in this chapter. Do you know their meaning?

Architect
Communications breakdown
Communications link
Final inspection
Financing
Judicial system
Originator
Scheduling
Training session

TEST YOUR KNOWLEDGE

Do not write in this book. Please write your answers on a separate sheet of paper.

1. List ten communications links between individuals and/or groups to other individuals or groups that you think are common in the construction field.
2. Name several types of communications systems used to relay information between individuals.
3. What are some of the advantages and disadvantages of each type of communication system you named in question #2?
4. Analyze the possible impacts if the plans for a construction project had a major error relating to the location of supporting members.
5. Obtain a set of plans for a small construction project and list all of the specific information the plans provide the builder.

APPLYING YOUR KNOWLEDGE

1. Visit your local planning office and ask how communications is important to the entire planning process of a community.
2. Make a very simple sketch using only a few lines. Select a partner. Then:
   a. Describe to your partner how to duplicate your drawing. Your partner is not allowed to see your drawing or allowed to ask you any questions. Compare the finished drawing with your own.
   b. Repeat the activity, but now allow your partner to ask questions.
   c. Allow your partner to see your drawing. Compare the efficiency (the time spent in combination with the accuracy of the drawing) of each method. Which method works the best?
3. Make an organizational chart to show lines of communications on a construction project.
4. Visit a construction site and ask the contractor to show you plans for the project. See if you can relate the current construction activity with that portion of the construction plans.

Satellite dishes can receive a world of information.

7

CHAPTER

TRANSPORTATION SYSTEMS

The information provided in this chapter will enable you to:

- *Identify technologies that are primarily involved with transportation.*
- *Illustrate the relationship between the transportation and construction systems.*
- *Diagram the transportation path a product travels from raw material to the construction site.*
- *Explain the interrelationships between the four major systems of technology.*

TRANSPORTATION SYSTEMS IN CONSTRUCTION

A major part of the construction enterprise deals with the transportation cluster. Roads, bridges, sidewalks, tunnels, railroads, and pipelines are all parts of the transportation system that must be constructed on site. A great number of workers in the construction industry are involved in the construction of **transportation systems**.

These transportation systems are just as much a necessity for construction as construction is for transportation. Without transportation, no construction project could be started or finished.

In today's society, it is unlikely that all of the tools, equipment, labor, and materials required to complete any construction project will be found very near the construction site. All, or some, of these items will have to be **transported** to the site, Fig. 7-1. The bricks for houses or the pipe for pipelines must be transported from where they are manufactured to where they are needed. Some items will be moved only a short distance. Other items will have to be transported many thousands of miles. In fact, some may have to be brought in from another country.

Movement of workers, goods, and materials is required for any present day construction activity. Consequently, without a transportation system, construction is not possible. All construction projects require careful consideration of the available transportation systems. This assures that supplies, tools, equipment, and labor are in the proper place at the proper time.

TRANSPORTATION BEFORE CONSTRUCTION BEGINS

In the early *idea* stages of a construction project, it is important for individuals to get together to discuss their ideas, Fig. 7-2. For two people to get together, one or both must use some transportation system.

One person may walk to see the other. This is a human transportation system using paths or sidewalks. If the distance is greater, a person might ride a bicycle or drive a car. Bike paths and roadways are needed for this. For even

Fig. 7-1. Items must be transported to all construction sites. Trucks haul lumber, concrete, and bricks to the site. In the case of modular housing, materials are still transported to the site. The materials have simply been assembled prior to shipping.

Fig. 7-2. Two construction workers are examining the plans to a home. When people share ideas, the entire project benefits.

greater distances, one party may fly or take the train. These systems require the construction of airports or railroads.

TRANSPORTATION TO THE CONSTRUCTION SITE

As plans for a construction project are completed and the actual construction begins, a great deal of transportation of material must take place. In most construction projects, as shown in Fig. 7-3, dirt must be moved about the site or transported from one location to another. If the project is a road or dam, the amount of earth materials that must be moved may be massive. If a house is being built, just enough dirt must be moved to allow for a sound foundation to be placed.

Fig. 7-3. Large projects may require a fleet of dump trucks to bring the necessary quantity of earth to the site. The trucks are loaded with cranes, bulldozers, or special conveyors. (Washington Construction Corp.)

Once the dirt is moved, building materials must be transported to the site, Fig. 7-4. These materials may be lumber, concrete, steel, special fill material, asphalt, or any number of other products that have been manufactured at some other location.

Fig. 7-4. The dirt has been cleared from this site. Concrete structural members are now been transported into place.

Often, the materials may be moved on several transportation systems before arriving at the site. Lumber is a good example, as shown in Fig. 7-5. After the trees have been chopped down and cut into logs, they must be transported from the forest. Workers first use a simple transportation system to move the logs to a central place. In some cases, the logs are towed by tractors. In other instances, they are pulled by a large **winch,** a machine with a cable that is attached to a powerful drum and motor. The winch drags the logs to the central location.

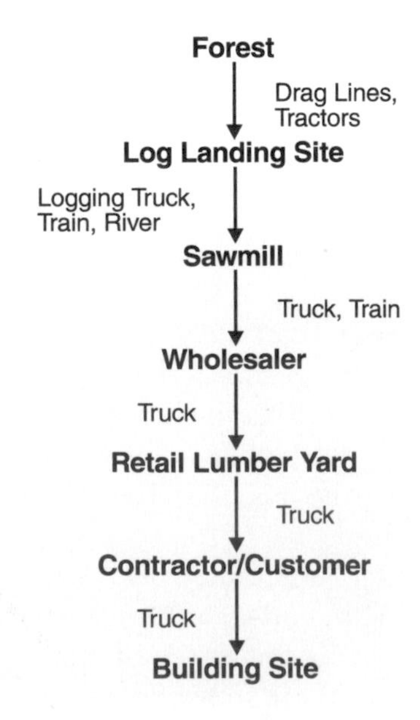

Fig. 7-5. On its trek from a tree in the forest to a beam in a house, a board of wood is transported many times.

The logs then use a second transportation system that takes them to the **sawmill** to be cut into boards. The most common way of moving the logs is by truck. A second method used to carry logs is by train and the railroad system. If the forest is near a river, the most **economical** (cost efficient) method might be to float the logs to a nearby sawmill.

After the logs have been cut in the sawmill, the wood must be transported once again. Most often, the lumber is sold to a **wholesale distributor** and transported to the distributor's storage facilities. Later, the lumber is sold to a lumber dealer and it must be transported to a **lumberyard,** were the wood rests until it is sold. From the lumberyard, it is sold to a contractor and delivered to the job site. Once on the job site, it may be transported several times while it is being used in the construction of the project. The majority of this movement between locations is by truck. In the lumberyard and on the job site, a lift truck is used to move the lumber from place to place.

Similar systems of transportation for other materials can also be traced. In most cases, materials are transported a sizable number of times in their journey from raw materials to the finished construction project.

TRANSPORTATION AT THE SITE

Transportation of materials at the construction site can be accomplished by a number of means. Materials can be hand carried from trucks to a storage area. This is by far the most labor intensive and is only used on small projects or where no other means are possible. Cranes, Fig. 7-6, forklifts, Fig. 7-7, and concrete pumps, are all common methods of efficiently moving large volumes of construction materials on modern job sites.

Fig. 7-6. Cranes are on-site transportation devices. Large steel beams are lifted into place.

Large cranes are able to lift hundreds of tons of materials in a very short period of time. They are capable of locating the material precisely where it will be needed. Concrete pumps are able to deliver the correct volume of concrete to exact locations. This allows workers to place the concrete directly into the forms. Fork lifts are the workhorses for moving more conventional materials such as lumber, bricks, steel, and drywall to new locations. All of these systems are important to the efficient transportation of materials at the job site.

The efficient transportation of goods, materials, and people requires a safe, well-constructed transportation system. In the case of lumber, sound roadways and railroads are extremely important. If the transportation system is not well-designed, it is difficult to assure that information, equipment, supplies, or workers will be able to arrive at the construction site at the correct time.

TECHNOLOGY IS INTERRELATED

As we become more dependent on worldwide resources, not just those of a single country, transportation technology will play a more and more important part in making a quality life for all. The construction system is a key element in the development of society, both now and in the future.

By this time, it should be clear that the four major clusters of technology–construction, manufacturing, communications, and transportation–must all work together. If any one of the systems is operating incorrectly, the other systems will suffer. By correctly using the interdependence of the systems, society may operate efficiently and profitably.

Fig. 7-7. This forklift is raising stacks of cinder blocks up to the crew on the scaffolding.

SUMMARY

Transportation systems consist of more than just trucks, buses, trains, and planes. They also include roads, waterways, and pipelines. An effective and efficient transportation system is essential to any construction activity. Without a transportation system, no large construction project could be completed or even started. Before a construction project begins, designers must meet to review a variety of ideas. Often, they must use some form of transportation system to reach the meeting site.

Once the design is final, transportation facilities must be utilized to transport the drawings to potential contractors. After a contractor is selected, the construction company must begin to transport workers, materials, supplies, and equipment to the job site.

On the job site, there is again a great deal of transportation activity taking place to get the various construction components to their final location. Once the project is completed, the equipment and excess materials must be transported to a storage area or on to the next job site.

The successful contractor must be aware of all of the available methods of transportation and their costs. With this information about transportation systems, the contractor will be able to best estimate transportation costs for the project. There are a very large number of transportation systems that interrelate to the construction system to help meet the construction project time schedule and bid price.

KEY WORDS AND TERMS

All of the following words and terms have been used in this chapter. Do you know their meaning?

Economical
Lumberyard
Sawmill
Transport
Transportation system
Wholesale distributor
Winch

TEST YOUR KNOWLEDGE

Do not write in this book. Please write your answers on a separate sheet of paper.

1. List three different transportation systems used to move logs to a sawmill.
2. What are some of the advantages and disadvantages of each of the transportation systems you listed in the above question?
3. Analyze some of the possible impacts if one of the major transportation systems was not able to deliver specified materials on time to a construction site.
4. Make a list of construction materials that need to be transported for assembly on a construction site.

APPLYING YOUR KNOWLEDGE

1. Design a bulletin board with a construction activity at the center. Around the activity at the center, list all of the transportation systems that make the construction project possible.
2. Visit a construction site and observe the various types of transportation equipment used in the construction project.
3. Develop a diagram of a material commonly used in construction projects. Trace all of the transportation systems that are used to place that material in its final location on the construction site.
4. Make a list of all of the transportation systems you use in one day.
5. Select a single item which might be required on a construction project. Use that item to demonstrate how the technologies of communication, construction, manufacturing, and transportation work together to produce a higher standard of living for all of us.

For transportation between one or two floors, the escalator is a common choice. For transportation between three floors or more, elevators are more efficient.

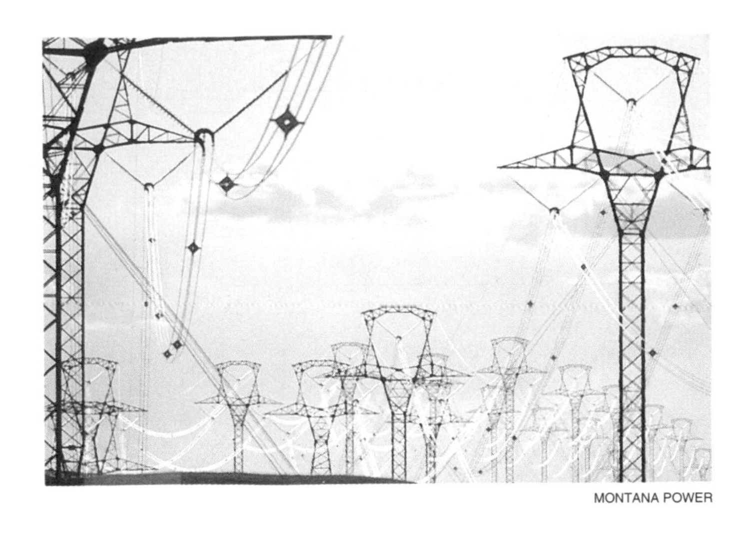

MONTANA POWER

MONTANA POWER

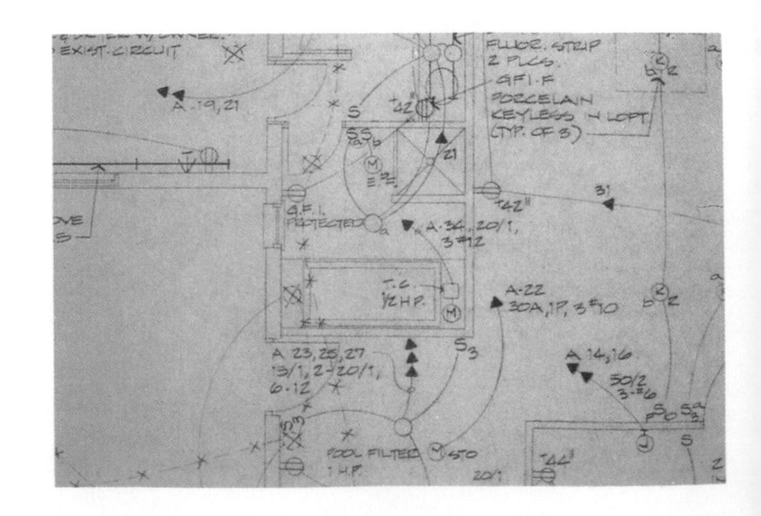

section 3

Elements of Construction

In this section, each chapter will deal with one of the seven major resources used in construction. These resources are: people, information, time, materials, capital, tools and equipment, and energy.

Think about the great construction efforts of the past. You have seen or read about the great pyramids, the Eiffel Tower, and the Golden Gate Bridge. Building these structures required the same seven resources. Furthermore, the resources that were used to complete these projects of the past are the resources used today.

Some things have changed. Instead of using stone as a construction material, high strength plastics, steel, or alloys are used. Yesterday's projects used muscle power. Today's projects use hydraulics, diesel, and electrical power. Then, builders used simple hand tools. Today, builders are using sophisticated power tools. In the past, calculations could have taken hours or days. Information today is stored in computers. They allow us to calculate complicated formulas in a fraction of a second.

While the categories of resources remain the same, there are major changes within each. We have already noted that there is a greater variety of materials and equipment. There has also been a speeding up of the transportation and assembly of materials. This makes the modern construction enterprise vastly different from what it was even at the turn of the century.

Construction enterprises will continue to change to meet changing human need. Who knows what it will be like when humans inhabit space and other planets? You can be sure that the same resources will be required when building on the moon or other planets. No matter how far or fast technology advances, you must first have an understanding of the basic concepts. These will prepare you for a better understanding of specific skills. Some of you will use those skills for rewarding careers.

Spending a few minutes with a set of construction drawings can save a contractor from doing hours of corrections.

8

CHAPTER

PEOPLE IN CONSTRUCTION

The information provided in this chapter will enable you to:

- *List and explain nine general safety concepts.*
- *Recognize safe and unsafe working practices in the construction enterprise.*
- *Explain the importance of management for efficient use of construction personnel.*
- *List the three functions of management.*
- *Recognize the importance of quality workers on any project.*
- *Explain what a labor union is and why they arose.*
- *List the four directions of progression that careers may take.*
- *Explain retirement.*

The most important resources for all construction companies are good managers and workers, Fig. 8-1. Without a well-trained work force, the companies simply cannot build anything. They, also, cannot make a profit!

In this chapter, you will learn about the importance of people. You will see why it is critical for a construction company to hire well-trained individuals. You will also learn the importance of workers doing quality work. In addition, managers must be able to manage all of the company resources in an efficient manner. Without an understanding between the worker, the manager, and the employer, labor problems may arise. Misunderstandings can cause situations that are a disadvantage to all.

CONSTRUCTION SAFETY

Safety is of primary importance for any construction enterprise. Construction safety is a combination of government regulations, company practices, and worker attitudes. No contractor will stay in business very long if he or she does not follow strict safety practices, Fig. 8-2.

REASONS FOR SAFETY

There are many reasons that safety practices are enforced on construction projects. State and federal laws require many safety practices. On the business side, lost time, even from minor accidents, means the loss of profit to the contractor and lost wages to the worker. The workers, naturally, are concerned for their health as well.

At times a worker may be tempted to ignore safety practices. To finish a job more quickly, people might wish to avoid the inconvenience of proper safety equipment or proper safety procedures. But, a single mishap that costs the sight in an eye, or causes some other serious injury, is a heavy price. Safety conscious companies do not tolerate workers who endanger themselves or other workers by not

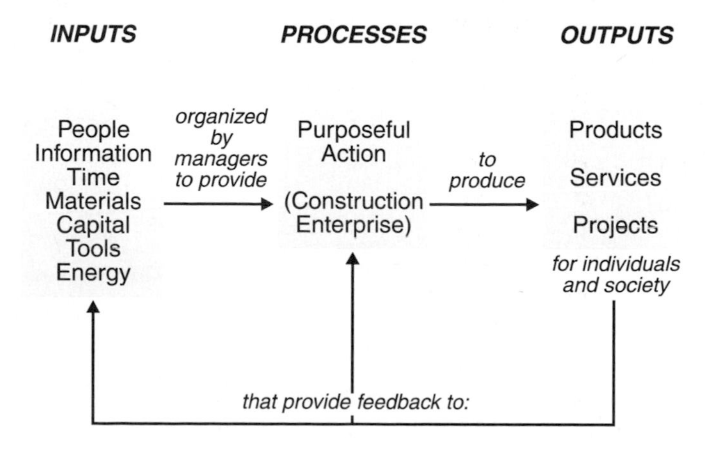

Fig. 8-1. Workers and managers are necessary inputs in the Universal Systems Model. Good work and strong organization produce superior products.

taking the proper safety precautions. The unsafe worker will likely be hunting for a new job shortly.

Fig. 8-2. Hard hats are required gear for everyone in many construction sites. How many safety devices can you see on the worker in the foreground?

Fig. 8-3. The exact locations of all underground pipes and cables are a must when doing excavation work. If the backhoe were to accidentally strike the gas line, the operator would be in for quite a surprise. (Montana Power)

SAFETY CONCEPTS

Because safety is so important, all new equipment comes with specific safety rules. The operator of the equipment should read these rules before working with the tool or equipment. Fig. 8-3 shows another example of a written safety warning.

It is also important that a worker understand general safety concepts. These are to be followed in all cases, regardless of the tool or equipment that is being used. The **general safety concepts,** listed on the following pages, cover the major areas of safety. Through careful study of these safety concepts, you can become a safe and efficient worker.

Have the proper mental attitude

The operator's **mental attitude** is the single most important safety consideration, Fig. 8-4. If you are not in the proper frame of mind, or if you do not respect the tools or equipment you are using, the potential for an accident increases greatly. You should never engage in horseplay in the work area. Never distract others while they are operating equipment. Keep all equipment guards in place. In addition, operate equipment *only* after proper instruction.

Fig. 8-4. People who enjoy their work make fewer mistakes.

Be aware of your positioning

Positioning refers to the operator's location in relation to the piece of equipment to be operated. Is the operator out of harm's way? Many pieces of industrial equipment have areas marked where operators should and should not be while the equipment is running, Fig. 8-5.

Positioning also refers to the position of bystanders and helpers near the piece of equipment being operated, Fig. 8-6. Workers must be aware of everybody in their work area.

Fig. 8-5. The yellow paint on the floor serves as a warning to all who pass by.

Fig. 8-6. Construction horses and cones have been set up to keep people and vehicles from wandering into the path of the bulldozer. (John Deere)

Check the condition of the tool or equipment

Maintenance, or regular upkeep, of equipment is vital. Otherwise, the equipment will not operate in a safe and efficient manner. Tools not in proper working order should be promptly repaired or **reconditioned.** Many pieces of equipment can take away much more than valuable time from projects when they brake down at inopportune times. The equipment can take lives.

Maintain the safety zone for cutting machinery

The operator should maintain a safe distance, or a **safety zone,** from the cutting area of the tool, Fig. 8-7. This distance will vary from tool to tool. The distance will be stated in the operator's manual of the specific tool.

Practice good housekeeping

Observe general cleanup and housekeeping procedures in the laboratory and on the construction site. **Housekeeping**

Fig. 8-7. Power cutting tools can be dangerous. Experience, patience, and the proper protective gear cut down the risks for this construction worker.

includes maintaining a clean working area around equipment as well as on the construction site, Fig. 8-8. Tools should be put away when not in use. Waste products should be disposed quickly.

Carefully set up and adjust equipment

Operators should be sure all equipment used is set up and adjusted properly. Adjustment should be done while the equipment is disconnected from the power source. If you are not sure of the proper setup or adjustment on a tool or piece of equipment, carefully study the **operator's manual** or get instruction from a knowledgeable person, Fig. 8-9.

Be aware of your work environment

Operators should be aware of safety constraints unique to the work areas. **Safety constraints** would include the wearing of proper protective clothing, safety glasses, hearing protection, or protection from potential electrical problems, to name a few. See Fig. 8-10.

Anticipate possible results

A good worker will take a moment to study the potential problems that might occur when a particular action is taken. For example, how would one protect against kickback from a table saw? By **anticipating** results and reactions, many accidents can be prevented.

Report all accidents

No matter how small or insignificant, report every accident to the individual in charge, Fig. 8-11. The full reper-

Fig. 8-8. Keeping a work area clean and materials in order saves time and money for a contractor. (Bud Smith)

Fig. 8-10. This man is wearing heavy denim clothing as well as insulating leather gloves and boots. To protect his head and eyes he has a hard hat, safety glasses, and a welding mask.

Fig. 8-9. A diver is being instructed on the proper application of heavy diving gear. This suit can withstand tremendous pressures. (Morrison Knudsen Corp.)

Fig. 8-11. Even minor accidents should be reported to your supervisor.

cussions of an accident are not always obvious. Some equipment damage and some personal injuries take time to be seen.

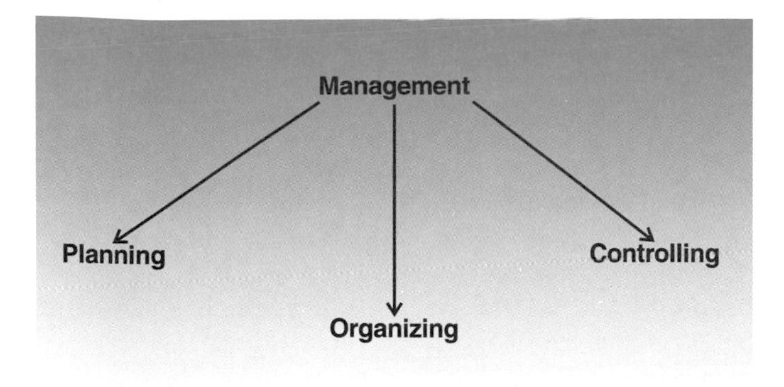

Fig. 8-12. Planning, organizing, and controlling are the three tasks of management.

Fig. 8-13. This crew has gathered together to discuss how the piping will handled in this room.

Fig. 8-14. Large construction projects require tremendous amounts of materials. These materials have been organized in such a way that they are easily located when they are needed. (Morrison Knudsen Corp.)

CONSTRUCTION MANAGEMENT

For any construction activity to function efficiently there must be individuals who:

- Plan the project.
- Organize the labor, tools, and materials.
- Control the progress and completion of the project.

These three activities are the **functions of management,** Fig. 8-12.

All construction activities contain these three functions. On simple projects, they may not be clearly defined. The three functions may occur individually or all at the same time. Each function is not rigorously defined.

The success or failure of a company depends, to a large extent, on how well the company is managed. Management must control the seven major resources important to all technological activities.

PLANNING

Planning is a very important part of management. **Planning** is the process of setting up goals, policies, and steps for an activity. It is used throughout a construction enterprise, Fig. 8-13.

First, planning determines whether or not to make a bid on a job. Then, if the company is successful in getting the job, more planning must be done. Plans must be made to get building permits, acquire materials, find subcontractors, and hire labor. Beyond each single job, the company must also do long-range planning. With **long-range planning,** the company sets goals for the company's future. These goals need to keep the company active and prosperous in the construction field.

ORGANIZING

Organizing is the assigning of different tasks and resources to meet the planned goals of the project or company. Management must organize its work force to work effectively. Each person on the force must know the worth of his or her work and how that work fits into the total effort of the company.

Materials and equipment must also be organized. It is the job of management to decide what materials, tools, and machines will be needed to complete a task. Then management must make sure that all resources meet at the proper time and place. Timing is important so the activity can be completed in the most efficient manner, Fig. 8-14. If workers show up to knock down a building on Monday, but the wrecking machinery does not arrive until Thursday, valuable time and money are lost.

CONTROLLING

Controlling is the third function of management. **Controlling** means directing how and when tasks are done. Controlling includes the activities of:

- Supervising workers.
- Coordinating their work.
- Assuring a constant supply of materials.

Control makes certain that the construction project is completed on time. The construction supervisor is most often the key to this phase of management, Fig. 8-15. This individual regularly meets with the project superintendent, the architect, and the owner. These meetings review various parts of the project during the project's construction. The meetings are used to assure that the project goes as planned.

Fig. 8-15. The supervisor (right) is coming to examine some of the work being performed by people under his direction. The supervisor checks that the work is being done correctly and on schedule. (Bud Smith)

ENTREPRENEURSHIP

Entrepreneurs are persons who strike out on their own to establish businesses. They take risks to make money for themselves. Most companies operating today were started by an entrepreneur. Sometimes the individual had a good original idea. Other entrepreneurs just had a strong interest in an already established product or service area.

When entrepreneurs make a profit in their enterprise, they continue to operate. If sales go very well, the business may expand. However, if the idea was not sound and does not turn a profit, the would-be entrepreneur does not prosper. The entrepreneur will then be forced to do something else.

There are large numbers of entrepreneurs in the construction field. Included are those individuals who feel they have a better way to operate a business or provide a service. Often these individuals start out as small subcontractors. Then, because of quality work and competitive prices, they become larger contractors. They, in turn, hire subcontractors to work for them.

Good construction workers are in demand. They are a valued resource for any construction company. Locating a job in construction generally requires many hours of searching through want adds, working with placement agencies, or directly seeking out employment with a construction company. In some cases, it may require taking tests and or joining **labor organizations.** Advancement in the construction field may require additional training. The need for quality construction workers in the future is great, and a career in this field can be very rewarding.

RECRUITING AND HIRING

Recruiting construction workers means actively seeking new employees. It is an important early step in the hiring phase of employee relations. Many construction workers are hired each season.

Before any recruiting strategy can be set up, management must first assemble data concerning the type of work newly hired employees would be expected to accomplish. This is a **job analysis.** From this job analysis, a job description is written. **Job descriptions** include statements relating to general duties and specific tasks the worker must perform. From these given tasks, management may then preselect a labor source most likely to yield a prospect with the desired skills.

Once the job is analyzed and a likely labor source is identified, a number of recruiting methods may be used to locate a prospective employee. Some common recruiting methods include: advertising in trade magazines and newspapers; working with school, public, or commercial employment agencies; doing extensive field recruiting; making an internal search within the present employment force; going to trade meetings; recruiting from a related company; securing direct applications from prospects.

WANT ADS

Placing an advertisement in appropriate trade publications is a relatively inexpensive, yet effective manner to communicate with a select portion of the population. A more popular recruiting method among construction firms is to use local newspapers to advertise for help. Newspaper **want ads,** Fig. 8-16, cover a large, general labor market.

PLACEMENT AGENCIES

School, public, and commercial **placement agencies** are popular means of locating certain kinds of construction

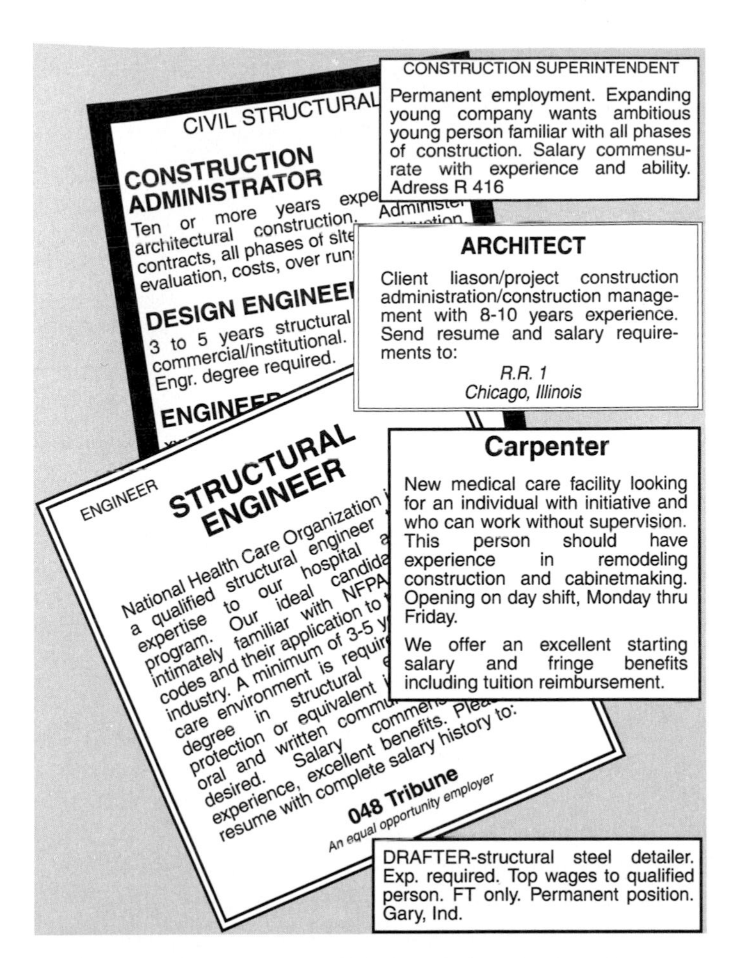

Fig. 8-16. Want ads carry brief job descriptions designed to attract a strong response from prospective workers.

workers. One frequently voiced complaint is that referrals (prospects provided) do not match the requirements of the company. Usually, this is a problem of lack of communication as an inadequate or incomplete job description is submitted.

FIELD RECRUITING

Field recruiting is done by a representative of the construction firm. With **field recruiting,** a recruiter covers college campuses in search of appropriate graduates. A recruiter may also visit military separation centers in search of men and women with construction experience.

INTERNAL SEARCH

Probably one of the most overlooked sources of construction workers is within the contractor's organization itself, Fig. 8-17. This is called an **internal search.** The internal search for a specific type of worker has one important advantage. It creates and sustains high company morale and effort. If present employees know they are

Fig. 8-17. Internal promotion keeps morale high. Good workers know they have the opportunity to advance.

likely to move up through promotion to an existing vacancy, they are likely to put forth greater effort.

Hiring outside personnel without considering internal promotions can have the opposite effect on morale. Employees are more likely to have a "care less" attitude if positions are always filled from outside of existing personnel.

PROFESSIONAL MEETINGS AND CONVENTIONS

Professional trade meetings and conventions sometimes are used for recruiting purposes. They usually provide a ready pool of prospective employees. The obvious shortcoming here is that participants of the meetings probably are satisfied with their present positions. They are top prospects, but much less likely to shift jobs.

DIRECT APPLICATION

Direct application may not qualify as an active recruiting practice, yet a construction company may exert control over it. Common methods of control are an application form, Fig. 8-18, an appropriate written test, and a personal interview.

When releasing application forms, an employer should receive the following basic information:

- Whether the applicant has the mental aptitude to perform the work.
- If the applicant is willing to do the work.
- If, after additional training, the applicant has the aptitude to do the work without close supervision.
- If the applicant is able to work in harmony with fellow employees.

In addition, the employer may want to know why the applicant left a previous job. A sampling of the applicant's

Date ____/____/____

EMPLOYMENT APPLICATION

Name: ______________________________

Address: ______________________________
Street
____
City State Zip Code

How long have you lived at present address? ____________
If hired can you furnish proof of age? _____ Telephone ____________

Employment Record for 5 Years Starting with Last Job

Employed by ______________ From _____ To ________
Address ______________________________
Salary ________ Job description ______________
____

Employed by ______________ From _____ To ________
Address ______________________________
Salary ________ Job description ______________
____

Employed by ______________ From _____ To ________
Address ______________________________
Salary ________ Job description ______________
____

*Use the back of sheet if more space is required.

Type of work applying for: ______________
Expected starting salary: ______________

REFERENCES
____
____
____

Comments:

____
Signature

Fig. 8-18. Application forms give employers a brief overview of a candidate and his or her past employment.

attitudes is also helpful. An application form covering most construction jobs would be difficult to produce. Yet, it is in the company's best interest to do so.

SELECTING A WORKER

Selecting the best individual for a given job is not an easy task. It is especially difficult if many of the applicants have similar skills. In this situation, the applicant's desire to work and ability to get along with fellow workers will have great bearing on the final selection.

TESTS

Test results can be helpful to the construction firm. By comparing test scores of the applicants to actual job performance, a company can predict how well the applicant will perform. Sometimes, however, job performance depends on qualities that are hard to identify. Tests are expensive and difficult to construct and administer. Even with these shortcomings, tests are valuable when used in conjunction with other methods of selection.

PERSONAL INTERVIEW

Possibly the most important employee selection tool is the **personal interview.** Much can be learned by an experienced interviewer. Often the decision to hire or pass over an applicant is made during the interview. The applicant may give undesirable answers, or lie in an attempt to please the interviewer. A trained interviewer is expected to detect this type of deception.

WORKING CONDITIONS

Our work surroundings are very important to us. If we are too hot or too cold we are uncomfortable. If we find ourselves among other workers who are unfriendly or unpredictable we are unhappy. In either situation, we must adapt to the working condition, Fig. 8-19, or change jobs.

Ever since people started to leave their homes to work, they have had to contend with these two problems. The emotional effect of work surroundings are such that they may influence our life habits. Many are the instances in which a person's success or turmoil at work created a like situation at home.

So important are our work conditions that improving them has created new jobs and opened up whole areas of study. A large company may invest heavily in industrial psychology, safety engineering, and systems analysis to improve morale, safety, and efficiency.

We say, then, that working conditions are an important area of concern to both the worker and to management. We call working conditions an environment and distinguish between two elements. Actual work conditions, such as dust or temperature, are called **physical environment.** The interactions of people at work is called **social environment.** Both management and unions become involved in bettering these environments.

PHYSICAL ENVIRONMENT

If our workplace is in an enclosed area, the company can control things like the dust in the air or the temperature. Rooms can be air conditioned and miners may have air conditioned suits to cope with hot mining conditions. Not much can be done about dust and temperatures on a construction site. However, other aspects, such as safety clothing and safe equipment, can be provided, Fig. 8-20.

Fig. 8-19. Construction workers on the site of a building are directly exposed to the weather. Construction workers must be prepared to accept what nature gives them.

Strikes or work stoppages are sometimes called by labor because of physical conditions. Labor and management work together to improve such things as safety equipment, lighting, dust, and noise control. Certain conditions cannot be eliminated or made less disagreeable. These must be endured as part of the work condition.

SOCIAL ENVIRONMENT

The social working environment refers to the actions and interaction of workers with the people who work around them. If the worker is happy with these people, the work will generally be of better quality. Social environment includes:

- The general attitude workers have toward each other.
- What after-hours get-togethers they have.

The workplace should have a centrally located place where the workers can congregate and interact in their leisure time. This place might be a cafeteria or lounge to be used during breaks and lunch hours. These places are important to the morale of the workers. It is also helpful to have after-work activities. Companies may sponsor bowling leagues, softball teams, Little League teams, picnics, outings, and other company functions. See Fig. 8-21.

ECONOMIC REWARDS

Economic rewards involve more than the money the construction worker receives for performing certain tasks. Fringe benefits such as life and hospital insurance, social security, workmen's compensation, and paid vacation, Fig. 8-22, are other important economic rewards.

Construction trades offer especially rewarding career opportunities for those who are not planning to go on to college but who are willing to spend several years learning

Fig. 8-20. Boots, gloves, hard hats, and welding masks are provided by construction firms to protect their employees. (Montana Power)

Fig. 8-21. Company sponsored events, such as picnics, are good for employee morale and fun for the whole family.

with pay. Well-trained construction workers can find jobs in any part of the country. Generally, their hourly wage rates are much higher than those for most other manual workers. Also, those who have above-average business ability have a great opportunity to establish their own business.

Fig. 8-22. Paid vacation gives workers time to spend with their families without putting them in a financial bind.

COLLECTIVE BARGAINING

One person alone has little bargaining power. When people with common interests join together, they can jointly elect one person to speak for them. This person has more authority than any other single member in securing pay raises or benefits for all.

When this leader seeks benefits for the group from the employer, it is called **collective bargaining.** The group of organized workers who wish to bargain collectively is called a **labor union.**

Working together through a union, all employees can speak with one voice at the bargaining table just as management does. Unions negotiate for benefits such as better wages, better working conditions, and a comfortable retirement program.

LABOR UNION HISTORY

During the early industrial revolution, labor did not become organized for several reasons:

- Free movement of families from one area to another and rapid growth of the frontier found workers drifting in and out of industrial occupations.
- The agricultural nature of society kept public support away from organized labor.
- Laborers accepted the idea that they were like servants.

Labor did not become a significant part of the American industrial scene until mass production placed many employees under one management.

Some local attempts at strikes and work stoppage began as early as 1800. However, not until the middle of the

1800's did these groups gain numbers. They were local groups very loosely organized.

Generally, the courts did not favor labor unions. Also, during this period, immigration to the United States was heavy. Employers were able to hire cheap labor. Although many attempts at organization were made during this period, unions generally were short lived.

In 1869, a group called the **Noble Order of the Knights of Labor** organized in Philadelphia. They grew to over a million members. However, after unsuccessful attempts at striking, they lost many members.

In the 1870's a cigarmaker, Samuel Gompers, began to organize local unions. In 1881 he and a few leftovers of the Knights of Labor organized a national union called the Federation of Organized Trades and Labor Unions of the United States and Canada. This appeared to be too loose an organization and five years later, Gompers seized total control. This was the beginning of the **American Federation of Labor (AFL)**.

At the very best, labor/management relations in those days were strained. Labor organized strikes, boycotts, pickets, and civil disruption. Management fought back with injunctions, lockouts, blacklists, and more civil disruption. Labor unions became more militant.

This pattern continued well into the 1900's. The labor unions were doing more than just creating civil disorder. Unions were pushing for legislation to give them recognition and support.

The depression of the 1930's hurt labor unions. No one escaped the effects of the economic downswing. Labor had been against the government-sponsored work programs. However, they were unable to provide any relief of their own for out-of-work people. The labor unions found they needed help.

The United States government responded with legislation. Though short-term in scope, measures such as the National Industrial Recovery Act, and its successor, the National Labor Relations Act had long-range effects. These pieces of legislation gave legal recognition to the fact that employers had responsibilities to laborers. When these were declared unconstitutional, Congress responded with the Walsh-Healey Act and the Fair Labor Standards Bill, which provided maximum work hours and minimum wage standards.

Many benefits have been made possible by labor unions. They have fought for and gotten shorter hours, higher pay, and better working conditions, Fig. 8-23. Often, improvement in working conditions has also made a construction operation more efficient.

Fig. 8-23. Taking time out for lunch after a good morning's work was not always the right of an employee.

LABOR AGREEMENTS

What management and organized labor are committed to do for each other is governed by the terms and conditions of the **labor agreement.** Labor agreements have varied widely from one industry to another since Samuel Gompers first organized the AFL. A labor agreement is the total relationship between a union and management. Agreements are of two types, simple and supplementary.

SIMPLE AGREEMENT

Simple agreements set down a few elementary rights and duties of the employer and employee. Simple agreements are confined to a few fundamental and procedural steps for disposing of problems as they arise from day to day. This document seeks to prescribe a general method of handling grievances.

SUPPLEMENTAL AGREEMENTS

Supplemental agreements are agreements that are added on to previous agreements. Supplemental agreements may be classified as either major or minor.

Major supplemental agreements

Major supplemental agreements are those issues which either labor or management consider vital to their operation. The issues discussed usually center around policies and rules that have been accepted for a long period of time. One party often is trying to get the other party to change or modify the contract. Both parties will take a militant position all the way to the arbitration.

The party that loses the decision will keep that issue in mind until collective bargaining begins for a new contract. The issue can be used in bargaining for a wage increase, or perhaps a comprehensive health and welfare plan.

Minor supplemental agreements

Minor supplemental agreements are the result of grievances that are settled locally and without arbitration. In this way, the words of a contract come to have a specific meaning in a variety of circumstances. Records are kept of all grievances handled and how they were settled. These records constitute a system of precedents, acts that justify similar acts, which fill in the general provisions of a simple agreement.

In some labor/industrial management relations, minor supplemental agreements are created another way. They are reached through the practice of exchanging letters setting forth agreements on problems that arise during the contract period. These letters deal with issues that are not spelled out in the contract. They are useful to both parties involved.

GRIEVANCES

A **grievance** is a complaint by a worker that he or she has not been treated fairly in some aspect of work or pay. Usually, the grievance is in an area specifically covered by the collective agreement.

Seniority grievances are common. **Seniority** is established by the number of years spent with the employer. Workers may feel their seniority has not been reckoned with properly in cases of promotions or transfers.

When unions bargain with small employers, the grievance system is usually very simple. The workers' complaints are delivered verbally to their immediate supervisor. If satisfaction is not given, they go to the union president or its business agent. That person talks to the owner or manager of the firm.

Such informal procedure is clearly unworkable in large companies. Therefore, large formal systems of handling grievances have been worked out over the years. In large companies, the worker would go first to the **shop steward.** If the grievance appears to be valid, the steward will help the worker put it in writing.

If the steward cannot settle the grievance with the supervisor, it will travel upward through a well-defined series of steps. Each step involves increasingly higher levels of authority within the union and company management.

In a large firm, the second step might be a conference between an officer of the local union and management's personnel department. A third step might be a meeting between the grievance committee of the local union and a committee from management.

The grievance procedure helps to keep unions in close touch with their members. It is an important function that maintains life and interest in the local union.

MEDIATION

Mediation places an impartial person into the role of adviser in collective bargaining. Mediators have no real authority. The parties make all the decisions by agreement. Mediators must rely on persuasion. They may suggest or they may recommend, but the individual parties may choose not to accept the advice.

ARBITRATION

Suppose labor and management get into an argument over the contract and cannot come to an agreement. What then?

Both parties submit the facts before a third party who has nothing to gain no matter which way the argument is decided. This is called **arbitration.** There is a difference between arbitration and mediation. Arbitrators have the responsibility and authority to decide one or more disputed issues. The decision is binding on all parties. Both labor and management agree in advance to live by the decision made by the arbitrator. There can be no appeal to the decision later.

Two kinds of disputes are handled by arbitration.

- A difference of opinion over what a collective bargaining agreement means. This is called a **rights dispute.**
- An argument over the terms and conditions of a collective bargaining agreement. This is called an **interest dispute.**

Like a judge, the arbitrator listens to arguments and reviews evidence offered by both sides. Then the arbitrator decides on the basis of the evidence what is fair. Arbitrators are, in effect, both judge and jury. However, it is not part of our system of courts. Rather, it is designed to keep the argument *out* of the courts. The legal process is slow and expensive.

Labor arbitration is not collective bargaining. Though, it may become a part of it. Its purpose is to arrive at a decision. Interpretation of a collective bargain contract may be arbitrated.

The arbitrator's fee is usually divided equally between the disputing parties. Services of an attorney are not necessary but may be helpful at times. The process is quick. Arbitration may last from two days to two months.

Voluntary labor arbitration provides a process for an orderly solution to disputes. It also gives a solid basis for good labor/management relations. Nearly all issues to be arbitrated go through a grievance procedure set up by both labor and management.

STRIKES

In construction, as well as in other industry and business areas, it takes people to operate the various pieces of equipment and perform the various tasks. Unlike machines, people have a voice in their destiny. Consequently, people organize to bring about a change.

A strike is a means of doing this, Fig. 8-24. The term **strike** means a concerted withdrawal from work by a group of workers employed in the same economic enterprise. Its purpose is to force the employer to be aware of their demands. The work stoppage is understood to be temporary and subject to laws.

WHEN WORKERS STRIKE

Strikes occur most often in those nations that have a large industrial base and a relatively free, democratic political system. The right to strike is normally guaranteed in democracies. Strikes occur more often when employment is rising and when economic conditions are favorable. At such times, employees are less concerned about economic security and employers are more sensitive to work stoppage.

The public is usually unaware of reasons behind a particular strike. A few common reasons are listed here.

- An employee group strikes to impose its will on an employer in the broad area of wages, hours, and working conditions.
- Many strikes result from conflicting interpretation of the terms of a collective bargaining agreement.
- **Jurisdictional strikes** may be caused by a disagreement between two unions over the assignment of work or jobs.
- In the recognition or **organizational strike,** employees try to force the employer to deal with the specific union as the collective bargaining representative.
- A strike may be called to change the collective bargaining to national bargaining.
- A strike may be called to change the effective date on contracts so that they expire on dates more favorable to labor.

Fig. 8-24. Workers strike for many reasons. These mechanics are asking for higher wages as well as ensured contributions for their health insurance and pension fund.

There are other reasons for strikes, but most strikes fall somewhere in these categories. Most strikes come fairly quickly to a point of settlement.

SETTLEMENTS

Most strikes are settled fairly soon by mediation, fact finding, or arbitration. The resulting agreement between the parties is called the **settlement.** Governmental or private third parties enable management and labor to narrow the issues and bridge differences while saving face with their constituencies (those they represent).

People are capable of making mistakes. People can rationally make decisions to live by. Strikes, mediation, and arbitration are tools whereby people can come to a mutual understanding and control change.

CONTRACTS

Labor contracts are the result of mediation, arbitration, and collective bargaining. They are statements of the provisions agreed on by both parties. Labor and management have come to an agreement on hours, wages, and working conditions. Labor contracts usually are written for a given period of time. At the end of the contract period, collective bargaining is repeated and a new contract is bargained for.

CAREER PROGRESS

If employees have worked for a company long enough and exhibit desirable work habits, they have a good chance for advancement. As new positions or job classifications open up in the company, their record will be reviewed. Job classifications in the middle or upper part of the job scale are quite often filled by transferring people from lesser jobs.

In the construction industry, as in other industries, there are four directions that a career may take, Fig. 8-25:

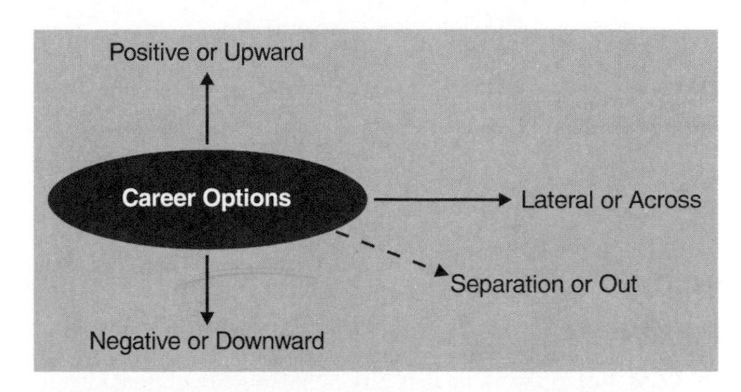

Fig. 8-25. Career progress can move in one of four directions.

- Positive progression–upward.
- Negative progression–downward.
- Lateral progression–across.
- Separation–out.

POSITIVE PROGRESSION

We usually think of career progression as positive, upward. With **positive progression,** employees move up to positions with more responsibility and generally a larger salary. Promotion is often a reward for years of good service to a firm. An extended period of time is not always needed to be promoted. Competent employees may find themselves given more responsibility rapidly, Fig. 8-26.

NEGATIVE PROGRESSION

Negative progression or **demotion** is being moved to a job with less responsibility or to one that requires less knowledge or skill. This may or may not mean less pay.

If a company has a layoff, there is a good deal of demoting especially in a company that acknowledges seniority. Employees with the least seniority are laid off first. This type of demotion is usually temporary until the company's work load or production is back to normal again.

Another type of demoting occurs when a person is found to be incapable of handling a certain job. The employee is moved back to a lesser job while another person is tried in the more responsible position.

LATERAL PROGRESSION

Lateral progression is usually called a transfer. To **transfer** is to move to another job of the same level and pay. This may happen for several reasons.

Fig. 8-26. A good work ethic is one of the most important factors in positive progression. (Amoco Corp.)

- Employee may not have a good attitude or may have trouble getting along with others. If reliable in other ways, the worker may be transferred to another job or to another department.
- Personality clashes between supervisor and worker. Management may want to keep this situation from getting worse or affecting a whole department.
- Lack of advancement opportunity. Employee may have advanced as far as possible without realizing true worth to the company. In this case, the worker may choose to transfer into a job of equal pay but with a chance for further advancement and higher pay.

SEPARATION

Separating is the process of discharging, relocating, laying off, or retiring of personnel. Discharging means firing or removing an employee from the work force.

Many reasons may be given for discharging: lack of skill, an undesirable attitude, or simply inability to get along with coworkers. In any case, the construction crew can function better without that member.

Laying off is the temporary release of an employee during a cutback in the number of workers. Laying off may be the result of a work slowdown, strikes of other building trades, bad weather, or possibly material shortages.

RETIREMENT

Retirement is the last, and a very important, form of separation. Retirement has become a very important part

Fig. 8-27. Many people use their retirement as an opportunity to travel around the country.

of the American scene. We take retirement at a time in life that many look forward to. Retirement gives workers time to do the things they never had the time to do, Fig. 8-27.

However, retirement is a relatively new concept. Never before could individuals quit earning their livelihood with so many productive years ahead of them. Today, company retirement plans, Social Security, life savings, Medicare, and investments make retirement possible from an economic point of view.

SUMMARY

People are the primary resource for all construction companies. Without a labor force no construction project can begin or proceed. Workers are not interested in working for a construction company that does not provide a safe working environment. The workers must be able to work efficiently and safely in order to complete a construction project on time.

General safety can be summarized in nine major concepts. These concepts are: having the proper attitude, having the proper working position, keeping the equipment in proper condition, maintaining safety zones, practicing housekeeping, carefully setting up equipment, being aware of your environmental, anticipation of results, and reporting all accidents. By practicing the above concepts workers can recognize unsafe practices as well as prevent accidents.

Management consists of the three main functions of planning, organizing, and controlling. Planning is the process of setting up goals, policies, and procedures. Organizing consists of assigning various tasks and resources to meet the planned objectives. Controlling is directing how the various tasks will be completed.

Labor unions were formed to give laborers one voice in dealing with management. When disputes arise, the opposing parties may try mediation or arbitration to settle the problem. Sometimes workers will strike to try and get their demands met.

Careers paths have four directions: positive, negative, lateral, and separation. Positive progression generally means more responsibility and higher pay. A form of separation that most workers eventually face is retirement.

KEY WORDS AND TERMS

All of the following words and terms have been used in this chapter. Do you know their meaning?

American Federation of Labor (AFL)
Anticipating
Arbitration
Collective bargaining
Controlling
Demotion
Economic rewards
Entrepreneur
Field recruiting
Functions of management
General safety concepts
Grievance
Housekeeping
Interest dispute
Internal search
Job analysis
Job description
Jurisdictional strike
Labor agreement
Labor organizations
Labor union
Lateral progression
Long-range planning
Maintenance
Mediation
Mental attitude
Negative progression
Noble Order of the Knights of Labor
Operator's manual
Organizational strike
Organizing
Personal interview
Physical environment
Placement agency
Planning
Positioning
Positive progression
Reconditioned
Recruiting
Retirement
Rights dispute
Safety constraint
Safety zone
Seniority
Separating
Settlement
Shop steward
Simple agreement
Social environment
Strike
Supplemental agreement
Transfer
Want ad

TEST YOUR KNOWLEDGE

Do not write in this book. Please write your answers on a separate sheet of paper.

1. List nine safety concepts that should be understood by all safe workers.

2. Explain why it is important for a contractor to have employees follow safety rules.
3. What are the three functions of management?
4. Define entrepreneur.
5. State three methods you might use to find a job with a construction company.
6. Why were labor unions formed?
7. Which person has power in settling labor disputes, the mediator or the arbitrator?
8. If a welder transfers to another construction project at the same pay it is called _____ progression.

APPLYING YOUR KNOWLEDGE

1. Visit a construction site and observe what safety practices are followed. Identify where the nine safety concepts apply.
2. Write a letter to a local contractor. Ask them to visit your class and discuss topics such as safety and how to get a job in the construction field.
3. Ask the local job service or contractor about the wages a construction worker earns. Discuss the different types of jobs such as a laborer, a carpenter, and a welder.
4. Find out and list the qualifications for workers described in item three.
5. Research a local labor dispute involving a disagreement between labor and management. Divide into two groups, one representing labor and one representing management. Discuss the problems and arrive at a solution to the disagreement.

Worker safety is important to all employers.

9

CHAPTER

INFORMATION IN CONSTRUCTION

The information provided in this chapter will enable you to:

- *List and discuss the three basic factors of design.*
- *List the steps to problem solving.*
- *Define the terms designing, engineering, drawings, specifications, and estimating.*
- *Distinguish between different types of drawings.*
- *State the value of specifications.*
- *Explain the reasons for codes.*
- *Discuss the need for inspections.*

When information is unavailable or inaccurate, a construction project has little chance of being properly started or successfully completed. In this chapter, you will learn some of the important types of information that a contractor needs in order to be a successful builder, Fig. 9-1. Designing, engineering, and planning aspects of a construction project all require and provide information about construction. Additional information required in construction deals with specifications, codes, and final inspections. This information assures that the construction project will meet certain standards.

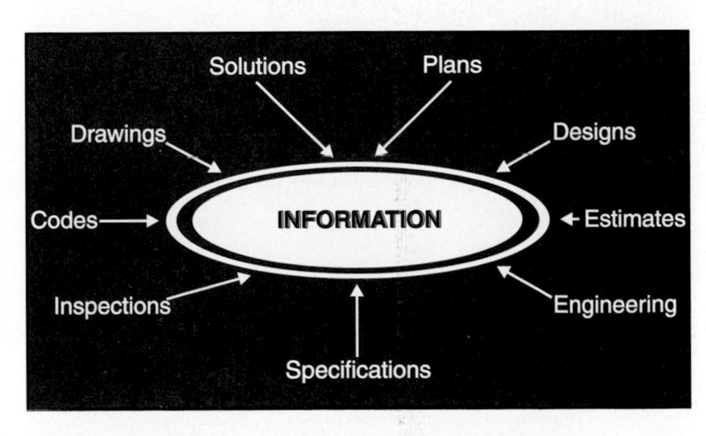

Fig. 9-1. Contractors use information in many forms.

CONSTRUCTION DESIGNING

The construction design process requires a great deal of information to allow a quality structure to be completed within budget. **Designing** is an organized sequence of steps that leads to a final decision. This design procedure, which an architect or engineer goes through, can be analyzed with the use of a systems model shown in Fig. 9-2.

PROJECT GOAL

The beginning of any construction project is the establishment of a **goal,** or the defining of a problem to be solved. A goal may be as simple as building a structure to protect us from the elements. The goal could be as complicated as determining the most efficient method of constructing a space station. Once a goal or problem has been carefully defined, the design or problem solving approach can begin.

INPUT PHASE

The next step is gathering input. During the **input phase** of the systems model, the client, the architect, and other

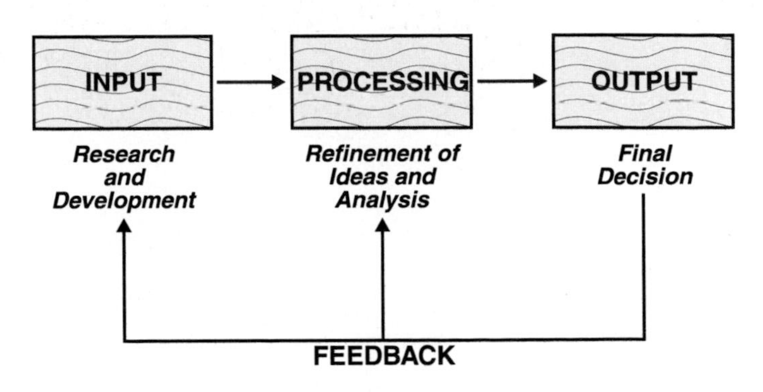

Fig. 9-2. Design procedure sequence.

individuals provide ideas for possible solutions to the problem. The more ideas (input) gathered, the greater chance the solution will meet the original needs of the client.

Research and development

The design process moves forward as various persons investigate structural elements of the design. This is called **research and development.** Each structural element and subsystem is evaluated in its own right. A **subsystem** is a separate but necessary part of a structure. In a building, for example, the electrical wiring, the heating and plumbing are subsystems.

This work may involve the mechanical engineer, electrical engineer, civil engineer, structural engineer, and the landscape architect. Each of these individuals makes a judgment about the tentative solutions that are offered, Fig. 9-3.

PROCESSING PHASE

After an adequate amount of information has been gathered, the information must be processed into some usable form. This is called the **processing phase.** This phase includes careful review of all ideas generated during the input phase.

Refinement of ideas and analysis

More information is compiled in support of various solutions. These facts are needed to prove the value of one design over all others. This is known as the **refinement of ideas.** Architectural artists produce very real looking sketches called **artist's renderings** to give the client a better idea of what the designs will look like when built.

Analysis is the part of the design process that has the architect working out how strong the related components need to be to support the design. This phase of the design process is more closely related to engineering than any other step. It becomes difficult to separate architecture and engineering. Engineers often aid architects in the electrical, drainage, and structural subsystems. See Fig. 9-4. To engineer a building or structure involves two things. The engineer must first analyze the materials and then determine the structural requirements for supporting the weight of the building.

Construction technology has given us many new facts about metals, woods, and other materials. We are more aware of what stresses materials are capable of withstanding. Materials receive stress not only from the structure's own weight but also from outside forces like the wind, the climate, and earthquakes.

Engineers may never guess at their work. They must establish, without any doubt, the load-bearing capabilities of certain materials. The foremost function of construction materials is to develop strength, rigidity, and durability. These qualities must be equal to the service for which they are intended. Structural requirements are defined largely in such terms as *load, stress, weight,* and *distance.*

OUTPUT PHASE

Using the refinement of ideas and analysis phase, the architect or engineer will have arrived at several feasible designs to take to the client for a decision. This leads to output phase. The **output phase** of the process is the decision that is based on the best information available at the time.

At this point, the client and the designer come to some agreement on the design. This must be done before the design process can go on. Remember, no work has been done at the site. Three possible decisions can be made:

Fig. 9-3. During the early stages of a project, systems engineers gather to discuss their ideas. (Federal-Mogul)

Fig. 9-4. This engineer (standing right) is checking some figures before work continues.

- Accept one design and proceed working out the individual details.
- Combine features (ideas) from each design.
- Reject all the tentative designs and start over again.

The last choice is not necessarily a bad decision. Repeating the process a second time may bring forward many more good ideas.

FEEDBACK

A **feedback** loop provides a means of adjusting the design so it may better fit our original expectations. Even the best efforts and intentions to design a construction project may not fully satisfy the original goal. Feedback from all parties involved in design and construction of the project help to make adjustments so the completed project will best match the expectations of the owner/operator, Fig. 9-5. Feedback from the use of a construction project can assist in future changes in the structure.

An example of feedback can be seen in the construction of a new highway. The highway is designed to reduce accidents and move traffic more efficiently into a city. The original design expected normal traffic rates of 2000 cars per hour. After public hearings feedback on the proposal was received. It was determined that during peak times (mornings and evening hours) 4500 autos per hour would be on the road. This feedback provides the designers with additional information that may necessitate the redesign of the proposed road.

The earlier that feedback is received, the simpler it is to make changes. The changes are also much less costly. Unfortunately, a great deal of feedback is received after a project is well underway. In some cases, important feedback arrives when projects are nearing or at completion. At this time, changes may be difficult to make. Though, this feedback is still very important. The feedback may prove useful for future projects. Feedback should be continuous throughout the project. It should come into play during all phases of design and construction.

FINAL DECISION AND IMPLEMENTATION

When the final decision is made to accept a design, the architect or engineer begins to make final drawings. Dimensions and specifications are prepared. These indicate sizes and type of material. This information is necessary before the project can be constructed.

DESIGN CONSIDERATIONS

The design considerations affect every step of the construction designing process. Regardless of the type of structure being developed–a bridge, a tower, a church, or a residence, there are basic factors common to all designs, Fig. 9-6. These are:

- Function.
- Form.
- Cost.
- Materials.

All architectural designs are a product of each of these factors. It is hard to generalize as to which is the most important factor. Though, in individual cases it becomes easy to pick one factor that is stronger than the others.

FUNCTION

Function is considered, in many cases, by designers to have priority in design. That is to say, the purpose or intended use of the structure should have first consideration by the designer.

Function is the ability of the structure to meet the need it is built to fulfill. A foundation, Fig. 9-7, is functional if it meets the need of the total structure. If the foundation does not meet this need, the design is bad. It does not matter how economical or attractive it may be.

Fig. 9-5. Models are assembled for many projects to assist in the feedback process. The three-dimensional view of this ballpark allows additional insight into problems that may arise.

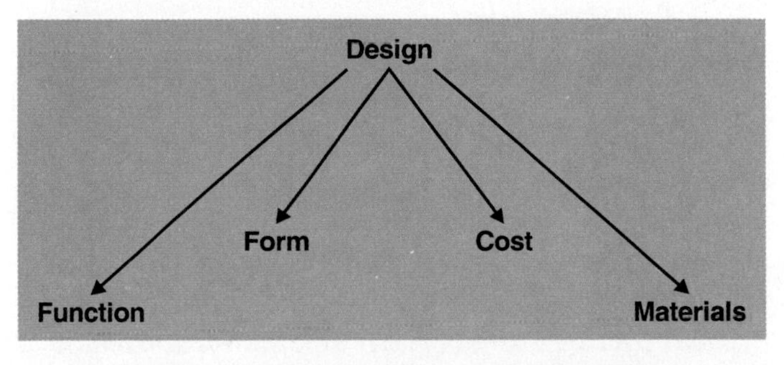

Fig. 9-6. There are four important design considerations.

Fig. 9-7. There are many different foundation designs. Different foundations perform varying functions for each unique construction project. (Morrison Knudsen Corp.)

Fig. 9-8. Need to solve a heat build-up problem led to the basic form of this building. (City of Tempe)

The functional home contains all necessary services for family activities. It does this efficiently without waste. Yet it is economical to build and maintain.

FORM

Form is the shape or outline of the building. It has nothing to do with color or material texture of the structure. For instance, the building in Fig. 9-8 is shaped like an inverted pyramid. It was given this form to prevent heat gain from solar energy coming through the glass. By overhanging each floor level, an insulated heat transfer barrier is created. Only 18 percent of the sun's heat energy passes through into the building. This reduces air conditioning costs. Heat loss or gain should be considered in the design of any structure that houses people.

COST

Cost is another important design consideration. Architects must make each project safe and attractive, yet most construction projects have a limited amount of funds to be used. Decisions must be made on the most cost-effective materials and designs. Both form and function choices are weighed against their price before any final decision is made.

SURVEYING

The job of **surveyors** is to find exact locations to build structures such as roads, dams, or buildings. Special instruments that give extreme accuracy are used. Basic surveying equipment includes:

- A **transit,** which establishes angles and straight lines.
- A **stadia rod,** which is placed over a point being established by the transit.
- A **tripod,** which supports the transit while a line or point is being established.

Survey parties are usually made up of two or more workers. A chief plans and directs the work. The chief or an assistant on a large party will record findings and operate the transit. The instrument is set up over a known point while a rod person holds the rod over another point. Fig. 9-9 shows a surveyor performing a typical survey operation.

Measurements are made with a steel tape called a chain or with laser equipment. Data collected during the survey are recorded in a field notebook or on a portable computer.

Fig. 9-9. Surveying is one of the most important activities in preparing a site for construction.

Surveying of large areas may be done from the air. Aerial photographs will show features of the area being surveyed. Other expensive equipment is needed to finish the survey.

PLANNING FOR CONSTRUCTION

The planning of the project comes next in the chain of activities needed to build a structure. The planning cannot begin until one design has been chosen over the others. A structural plan is engineered from the preliminary sketch, and all of the ideas regarding the plan have been recorded.

The plans now need final drawings. These are called **working drawings** and will include all the necessary views, dimensions, and details. Working drawings can only be prepared after the design has been established and all engineering notes have been made. Fig. 9-10 shows the order in which work has been done.

Construction involves contracts and subcontracts. Many individuals as well as individual construction companies must work together. With so many legal aspects and people involved, there is a danger that someone will misunderstand the plan. Drawing is very important to construction. It gives the architect/engineer a means to communicate ideas accurately to the builder.

WORKING DRAWINGS

A set of working drawings for a small project such as a residence has the following basic parts:

- Floor plan.
- Foundation plan.
- Elevations.
- Framing plan.
- Cabinet details.
- Wall sections.
- Details.
- A plot plan.

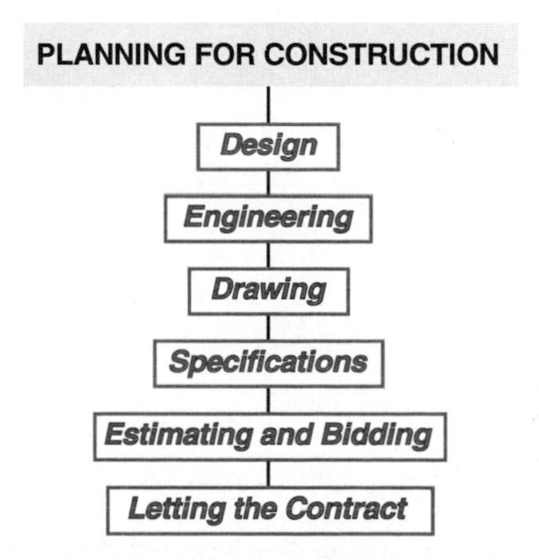

Fig. 9-10. Sequence for construction planning.

Fig. 9-11 shows a floor plan and an elevation plan. In simple projects all of the physical details are shown in one set of drawings. In large projects, there are many more drawings. Each drawing shows much more than the physical form of the project and its dimensions. Various installations such as the mechanical, electrical, structural, and plumbing drawings may be included.

In large projects, the drawings are more complex. The project needs much more explaining. A set of drawings may include a site plan. The **site plan** contains records such as the soil test boring schedule, excavation limits, paving requirements, river control, or detour construction. When necessary, the working drawings also include:

- Heating and cooling systems.
- Drainage and plumbing.
- Stairs.
- Elevators or escalators.
- Lighting and electrical requirements.
- Structural details.

Layout drawings

Through working drawings, the physical layout is clearly shown. The layout drawing of a highway interchange shown in Fig. 9-12 includes the overpasses, entry and exit ramps, location of service roads, and survey station points.

Elevations

Elevation drawings are important to establish the vertical location, Fig. 9-13. A builder, studying the bridge elevation will understand pile location and height. Elevations give a general idea of how the finished construction will look.

Structural drawings

Structural drawings show details of the structural parts of the project. Fig. 9-14 shows a cross frame for a bridge. This frame is made of steel. The drawing gives information on:

- Cross section configuration (outline or shape) of members.
- Size of steel members.
- Type of welds needed to fasten these members.
- Dimensions.

Civil drawings

Civil drawings have to do with the land on which, for example, a road is built. Civil drawings or contour maps show such information as cut and fill necessary for a roadbed. See Fig. 9-15. Contour maps are based on survey data.

Before a contour map can be drawn, elevations must be measured in the field for various key points controlling the establishment of the contour lines. Various methods are used:

- Grid system is used where all intersection elevations are grid lines.

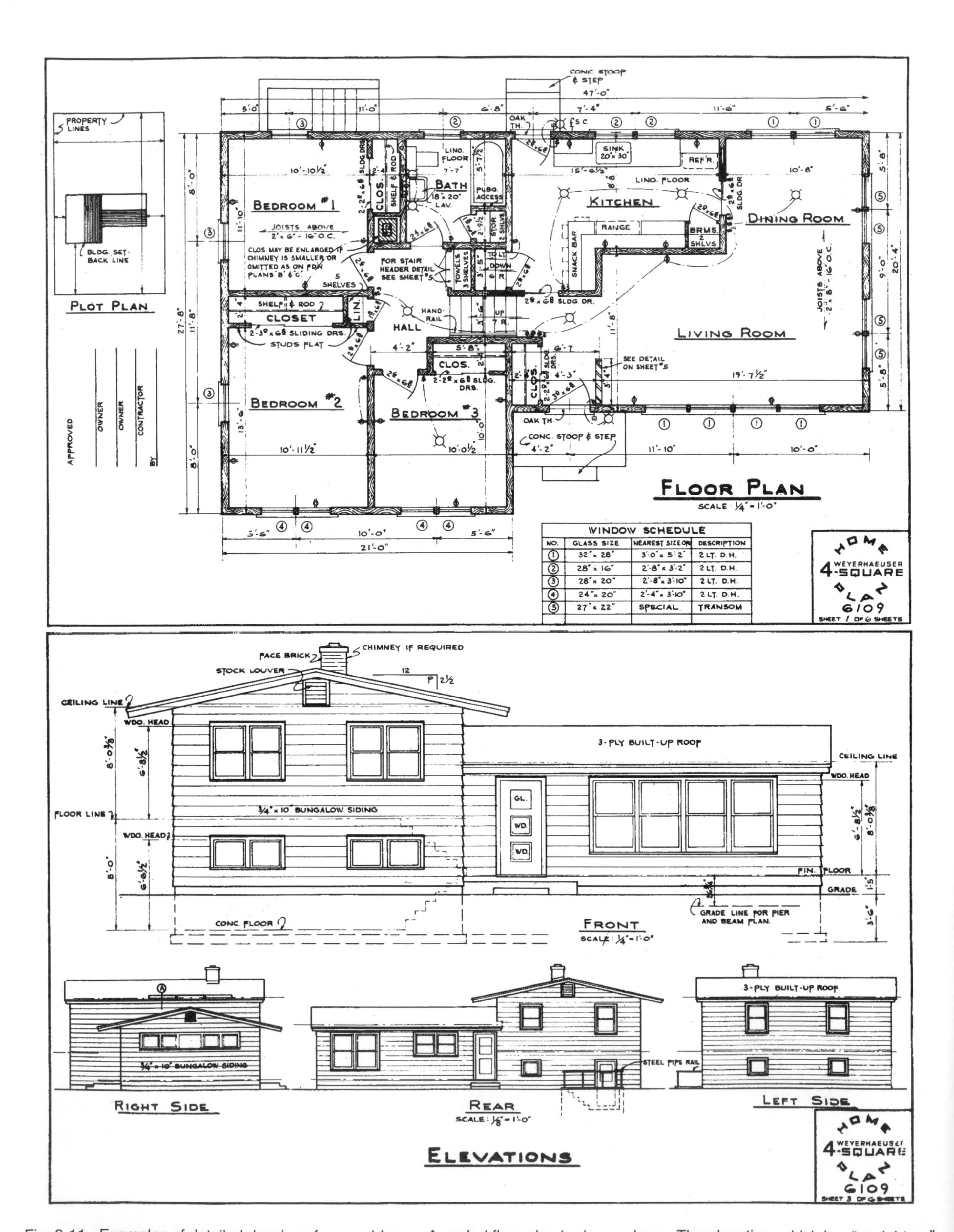

NO.	GLASS SIZE	NEAREST SIZE OF	DESCRIPTION
1	32" x 28"	3'-0" x 5'-2"	2 LT. D.H.
2	28" x 16"	2'-8" x 3'-2"	2 LT. D.H.
3	28" x 20"	2'-8" x 3'-10"	2 LT. D.H.
4	24" x 20"	2'-4" x 3'-10"	2 LT. D.H.
5	27" x 22"	SPECIAL	TRANSOM

Fig. 9-11. Examples of detailed drawings for a residence. A scaled floor plan is shown above. The elevation, which is a "straight-on" drawing of all sides of the construction, is shown below. In addition to the floor plan and elevation, a full set of drawings include a foundation plan, framing plan and detail drawings and section drawings. (Weyerhaeuser)

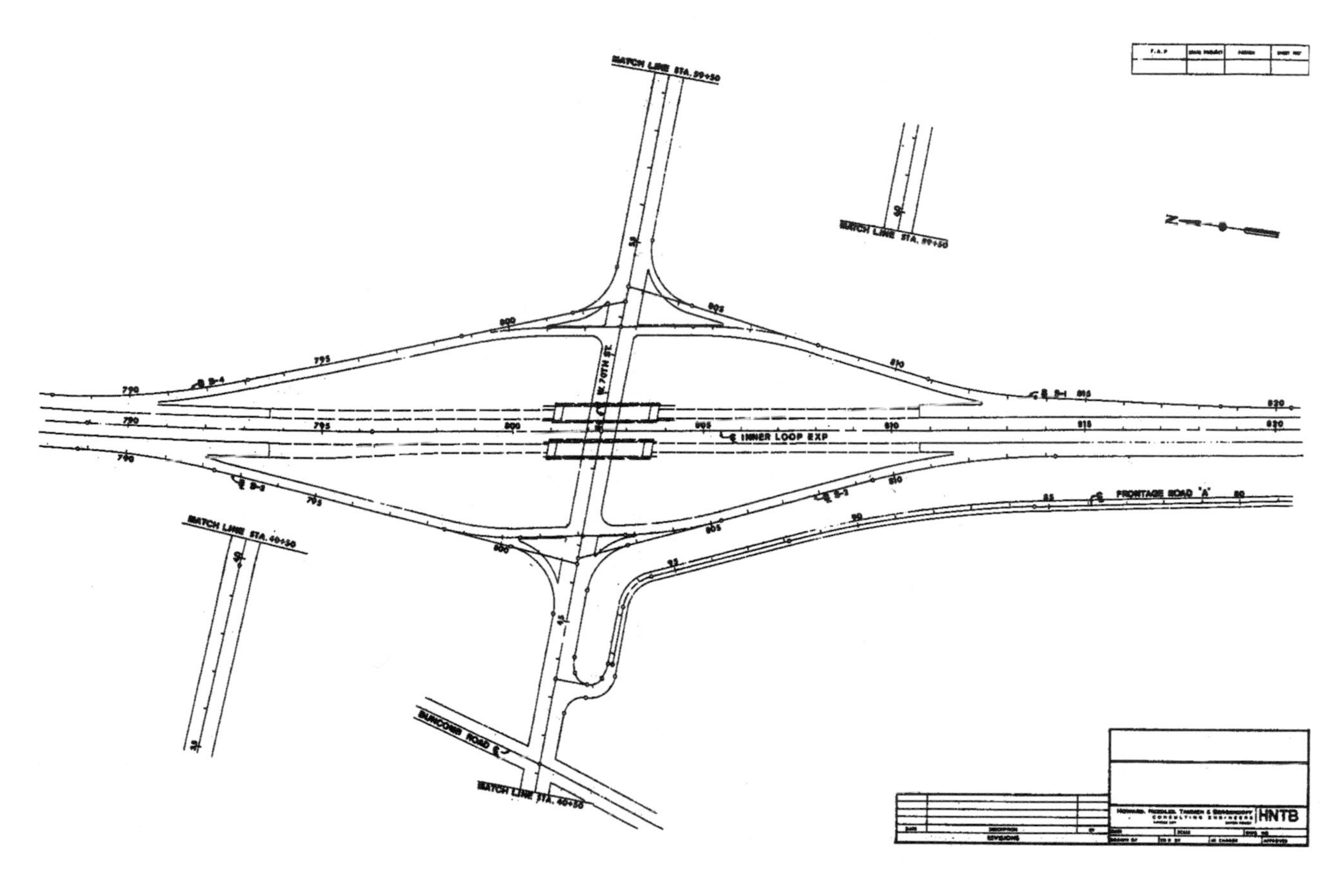

Fig. 9-12. Layout of a highway interchange. (Howard, Needles, Tammen & Bergendoff)

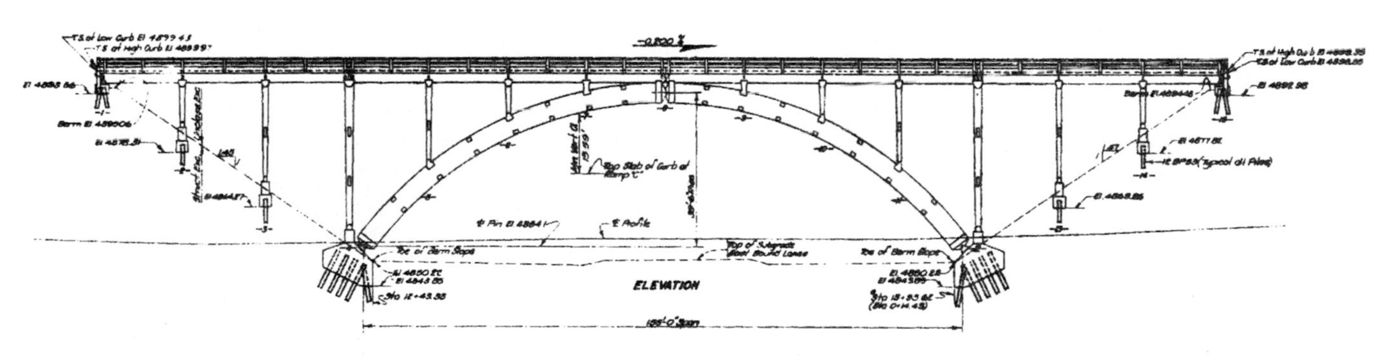

Fig. 9-13. Elevation drawings reveal a view that looks somewhat like the finished project. They also provide some basic information. (American Wood Preservers Institute)

- Points located by transit and stadia rod with corresponding elevations calculated by plane table.
- Aerial photograph surveys.

All three methods require experience in surveying. This is an important function of civil engineering.

Architectural drawings

Architectural drawings show many details of the physical form of the project. Architectural drawings include elevations, floor plans, foundation plans, sections, and details. Look back to Fig. 9-11. Elevation and floor plans are shown.

Electrical and mechanical drawings

Electrical and mechanical drawings show the plumbing, Fig. 9-16, heating/cooling systems, electrical requirements, ventilating systems, and lighting. Usually the mechanical drawings appear in a set of working drawings as sections or detailed views. Any type of mechanical installation must be planned in detail to ensure that space is saved for the installation.

SCALED DRAWINGS

It is not possible to prepare working drawings full size. Instead, they are drawn to **scale.** This means that drawings

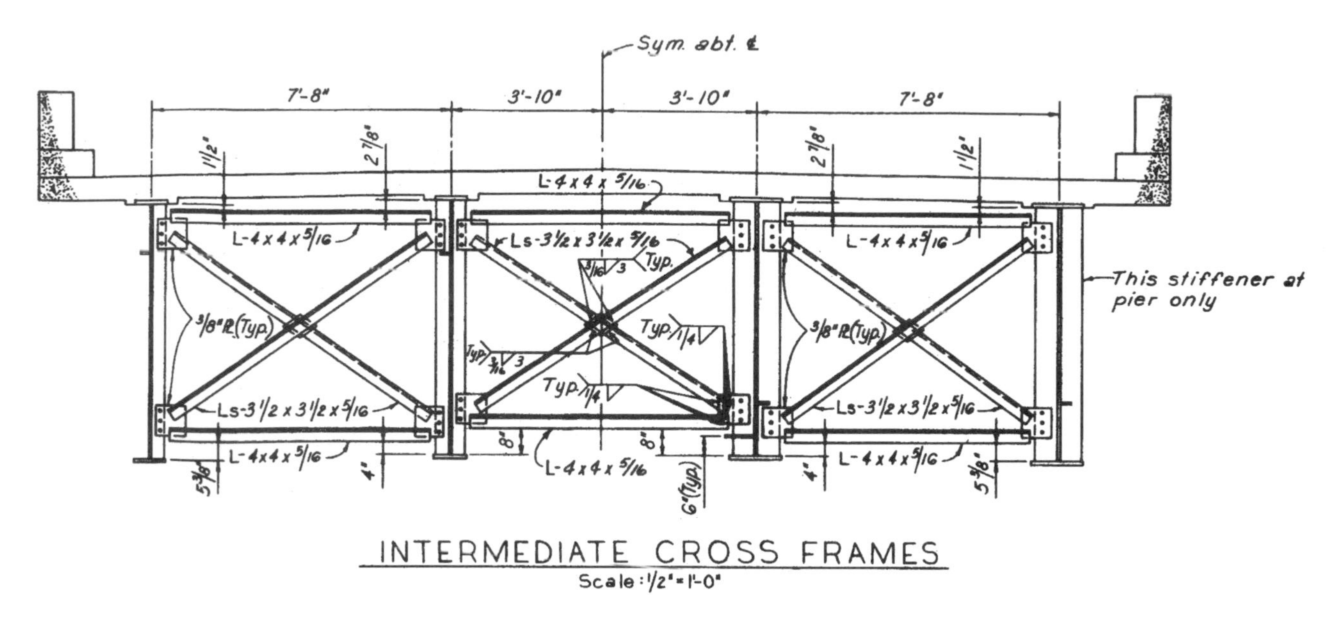

Fig. 9-14. Structural drawing of a bridge cross frame. (Corps of Engineers, Kansas City District)

have been reduced proportionately from actual size so that they will fit on a drawing sheet. Small units of measure stand for large units.

The floor plan for an average size residence is prepared using a small unit–usually 1/4 in. (6 mm) will represent one foot of the completed project. This reduced scale makes the drawing of the residence 48 times smaller. This allows it to fit on a 17 in. x 22 in. (43 cm x 58 cm) sheet of paper. Most architects use computer-aided drafting (CAD) tools and software to produce their drawings.

Dimensioning a working drawing

In **dimensioning** a working drawing, the first concern is to show the information in a way most helpful to the workers building the structure. This concern has led to development of some general rules of dimensioning:

- Everything is dimensioned to actual size even though the drawing is in scale.
- Dimensions are placed on the view that describes the feature best.
- Dimensions are always placed where they can be read from the bottom or right side of the drawing.
- When the metric system is used, distances are recorded in meters. The English system uses feet and inches.

Each drawing should be completely dimensioned. If it will help the builder, dimensions may be repeated. At no time should anyone have to add or subtract to find a size.

Architectural symbols

Since working drawings are made to a very small scale, it is not possible to draw many features or actual parts of the structure as they appear to the eye. Therefore, many parts are represented by standardized symbols.

Architectural symbols are standardized to represent walls, wall materials, and methods of construction. See Fig. 9-17. The different styles of windows each have a symbolic representation.

Electrical symbols, Fig. 9-18, are another good example of simple signs representing actual parts of a structure. These electrical symbols may contain specific information related to the installation. Fig. 9-19 shows use of electrical symbols in a sectional view of a stairwell.

Where steel framework is welded together, the architectural drafter uses welding symbols, Fig. 9-20. These symbols tell the welder what type of welded joint is wanted. Look again at Fig. 9-14. Can you locate the welding symbols and explain each one? There are many other welding symbols used by the construction industry. You will learn about them if you take more advanced courses in this field. Symbols for plumbing and piping are shown in Fig. 9-21.

MODELS

Use of building models is widespread. They are made of a variety of materials–paper, illustration board, and wood are but a few. In model building, all features must be kept in true scale. The advantage of a building model, Fig. 9-22, is showing the appearance of a completed building from any station point. When compared with a pictorial drawing, which shows a structure from a single station point, its advantage is easy to see.

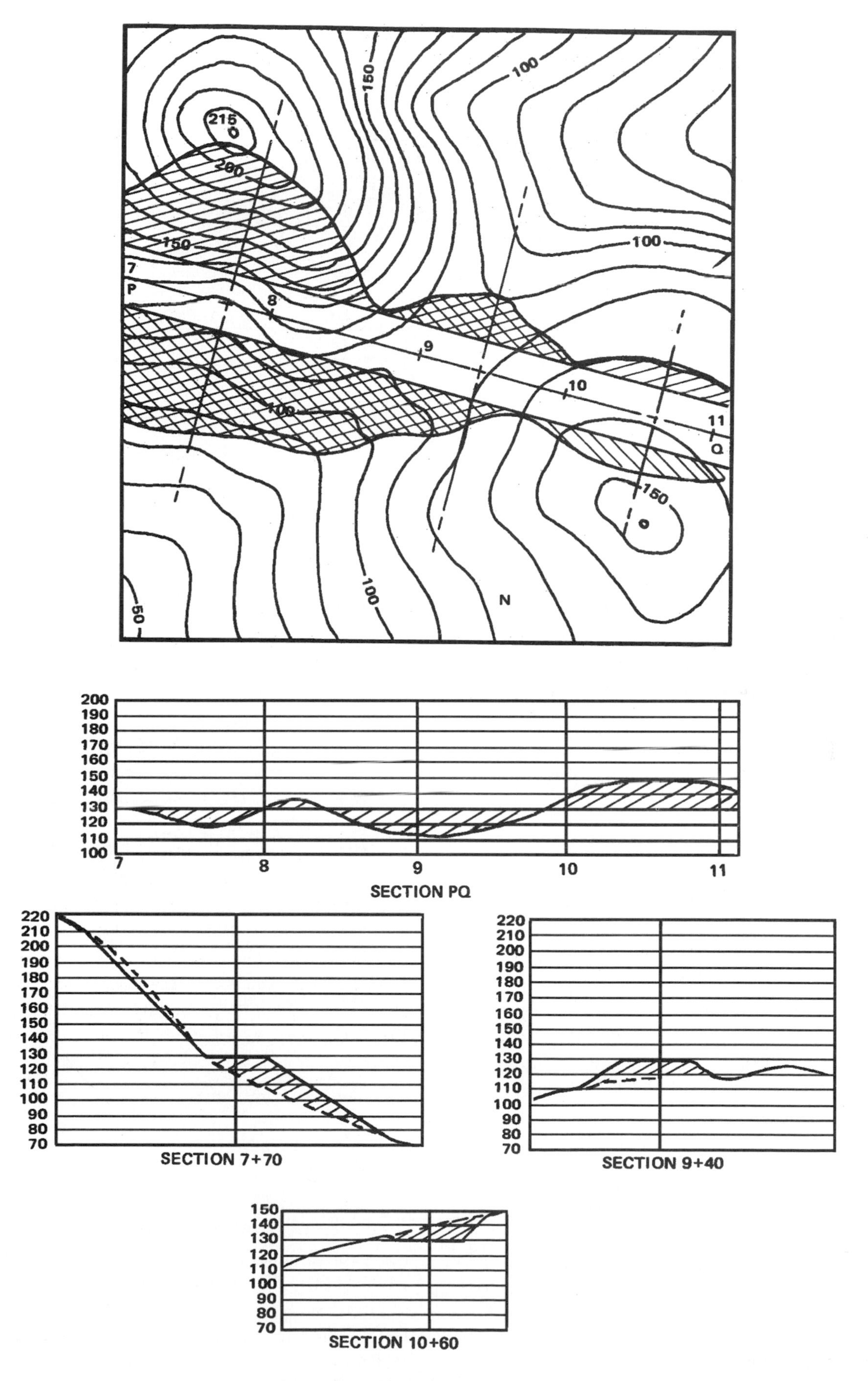

Fig. 9-15. Cut and fill for a road bed. Note that illustration at top is like a contour map. Other illustrations are cross-sectional views. Elevations above sea level are shown at the left of these cross sections.

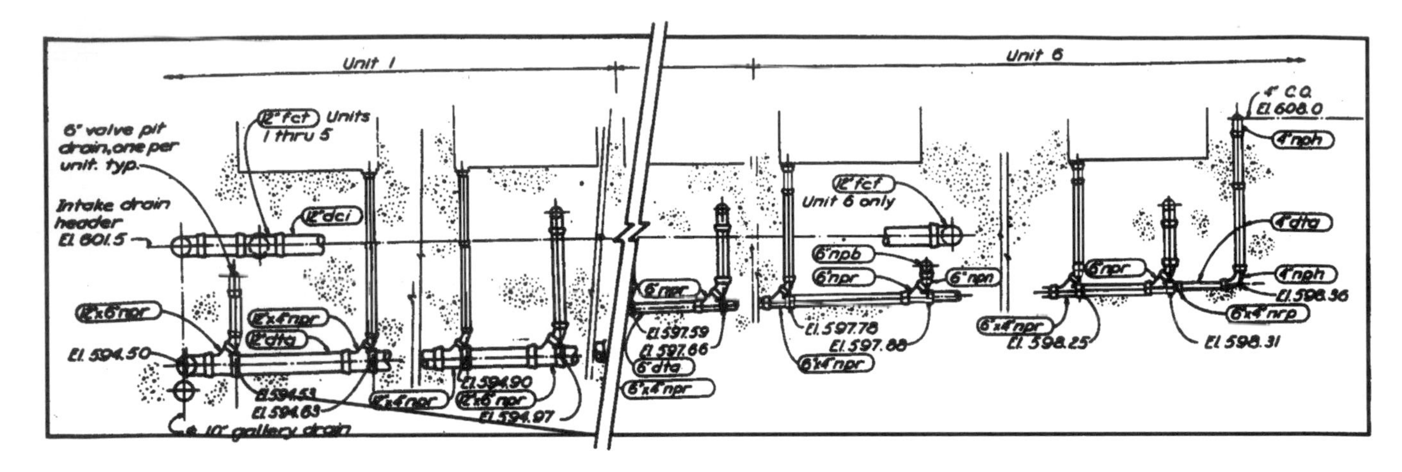

Fig. 9-16. Detail of a plumbing system.

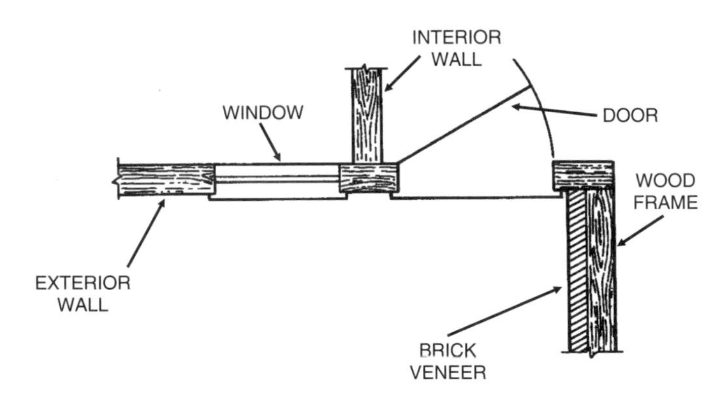

Fig. 9-17. Symbols for wall sections are easily understood.

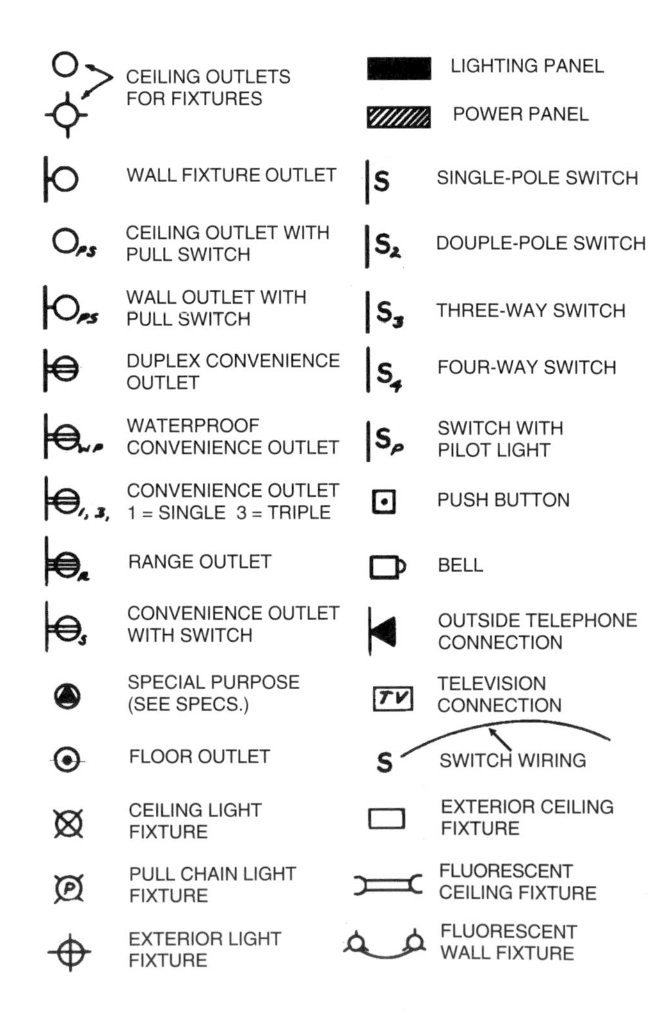

Fig. 9-18. Electrical symbols used in construction.

SPECIFICATIONS

Working drawings are complete in expressing shape and size of a construction project. It is, however, impossible for drawings to describe the quality of materials and work expected. This is even more true on large public projects, such as a stadium. This information cannot be placed on the working drawing. Therefore, certain information is given through specifications. See Fig. 9-23.

Construction **specifications** are written statements which inform the builder what is to be built, what materials are to be used and how the job is expected to be done. They are needed to prevent misunderstanding between the contractor/builder and the client/owner.

Those preparing specifications must know all aspects of construction. They must understand all materials, their classification, strengths, and structural capabilities. They must be able to present clearly all the information necessary to identify construction techniques and processes.

Provisional categories

Written specifications have three categories:

- Legal provisions.
- General provisions.
- Technical provisions.

Legal provisions consist of legal documents involved in the advertisement for bids. A bid form is shown in Fig. 9-24.

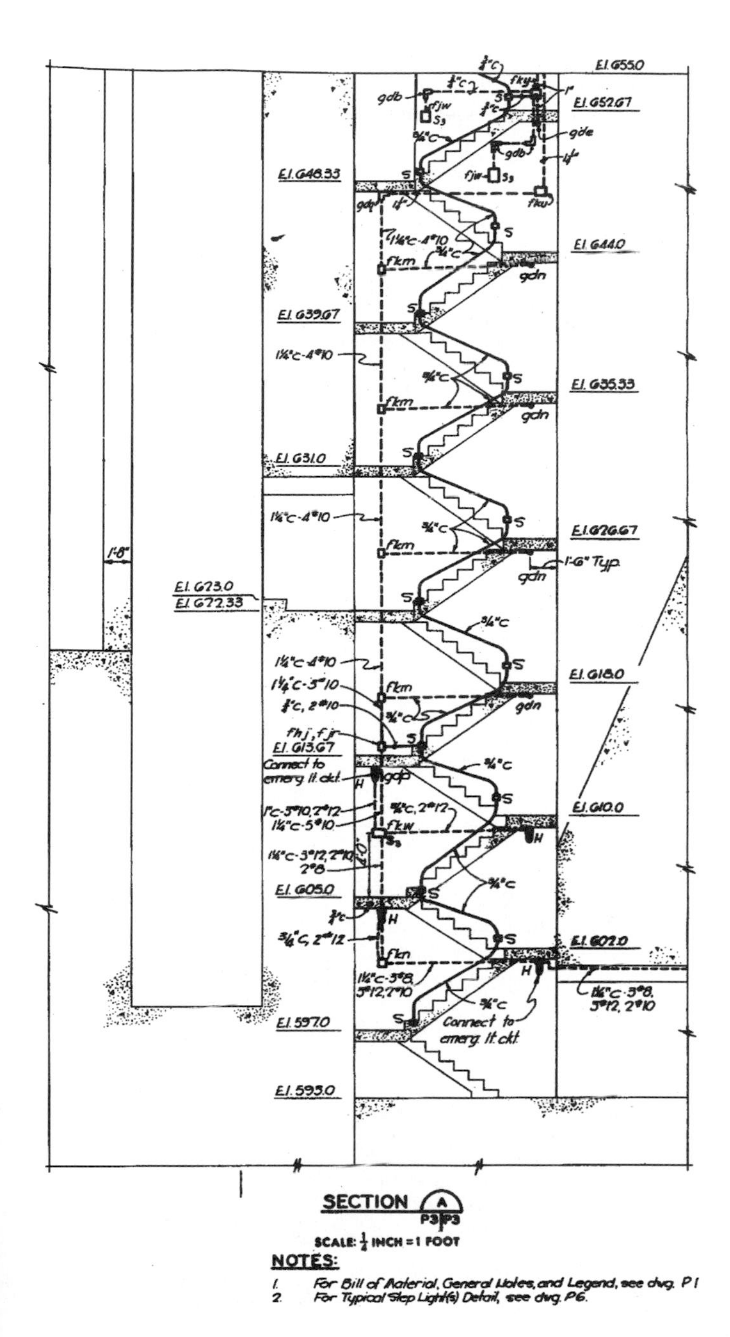

Fig. 9-19. A stairwell with location of conduit, fixtures, and switches. Note technical data on conductor sizes.

Instructions to the bidders, bond forms, and owner/contractor agreements make up the legal provision section.

General provisions state the conditions and responsibilities of all people involved in the construction project. The architect/designer, owner, contractor, and all subcontractors are made to understand the conditions under which they are to function while doing their jobs.

Model sets of these general conditions are published by the American Institute of Architects. The general provisions are understood to be a necessary part of the specifications. The general provisions include items that are not covered in any part of the trades section or in any other division of the specifications.

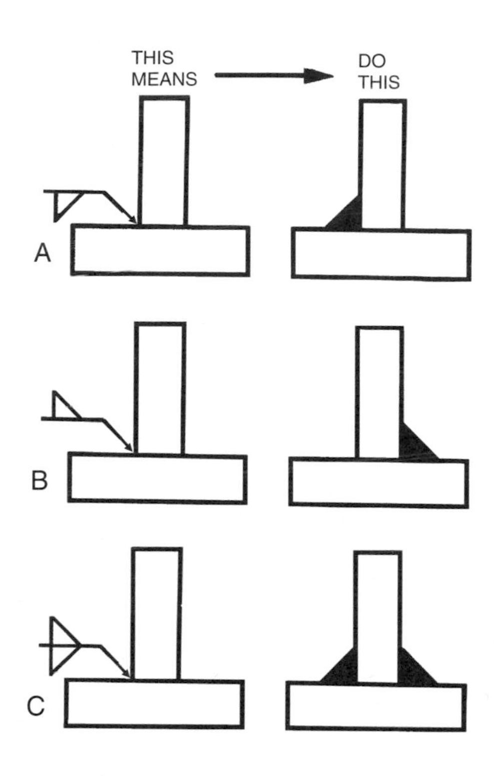

Fig. 9-20. Standard welding symbols are a way for drafters to show what is wanted. Study the symbols above. What they mean is drawn on the right. A–Symbol drawn on the underside of the arrow means to weld on the same side as the arrow. B–Symbol on top of the arrow means to weld on the opposite side. C–Symbol on both sides of the arrow means to weld both sides.

Technical provisions, Fig. 9-25, list the kinds of materials and processes to be used in the construction. Generally, the technical provision section includes architectural, civil, structural, plumbing, electrical, heating/cooling, and mechanical specifications. Good technical specifications reduce construction problems on the site, Fig. 9-26.

Each section of the technical provisions of the construction project is organized into:

- The general scope of the work.
- Description of the work.
- Materials to be used.
- Time of installation.
- Installation procedure.
- Testing and measurements.
- Special conditions.

BUILDING CODES

Building codes are laws or ordinances that apply to the materials selected and used in building projects, Fig. 9-27. They are based on local standards which, to a great extent, govern the quality and characteristic properties of the building materials. These codes also identify procedures and techniques used in construction.

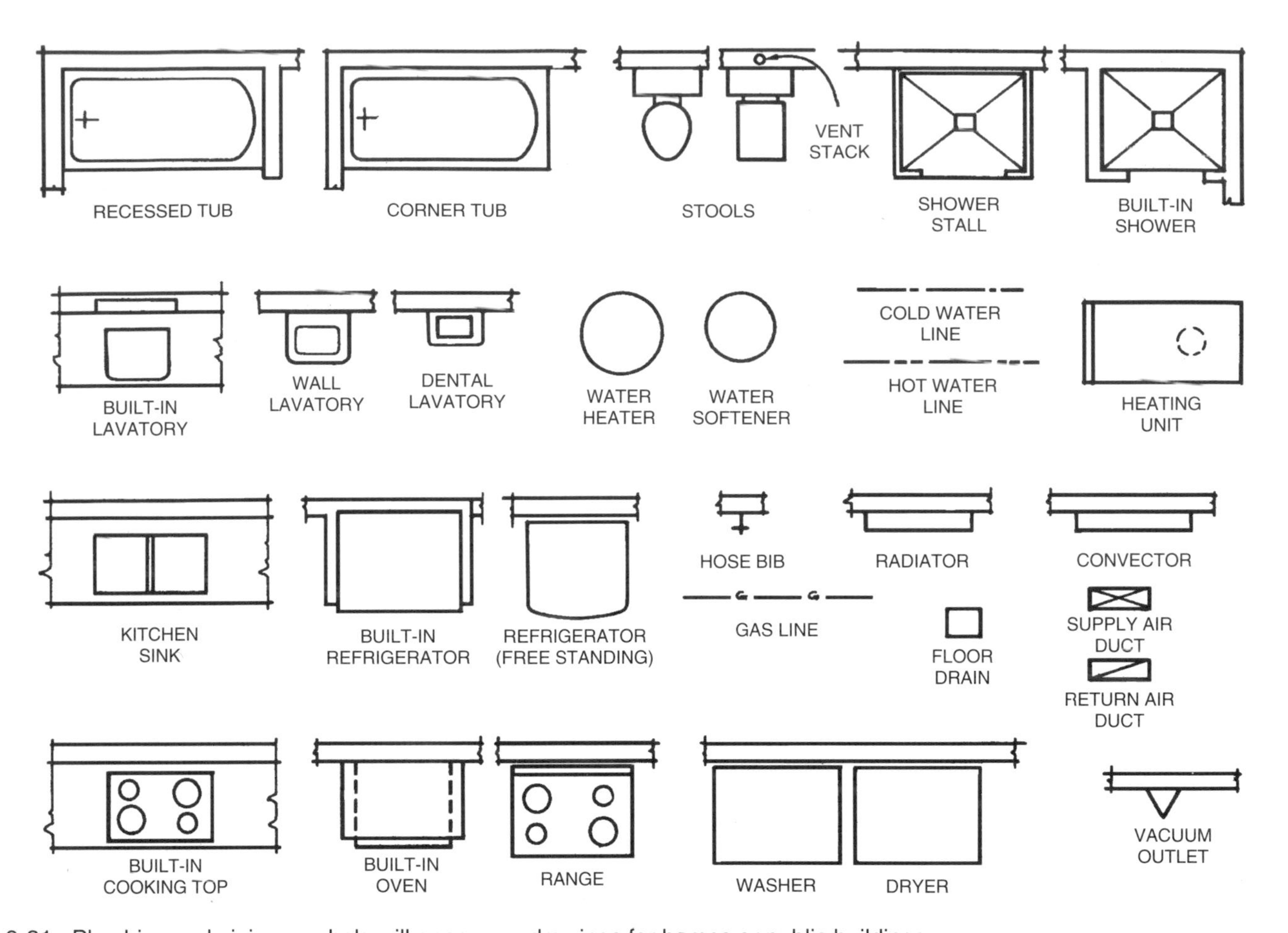

Fig. 9-21. Plumbing and piping symbols will appear on drawings for homes or public buildings.

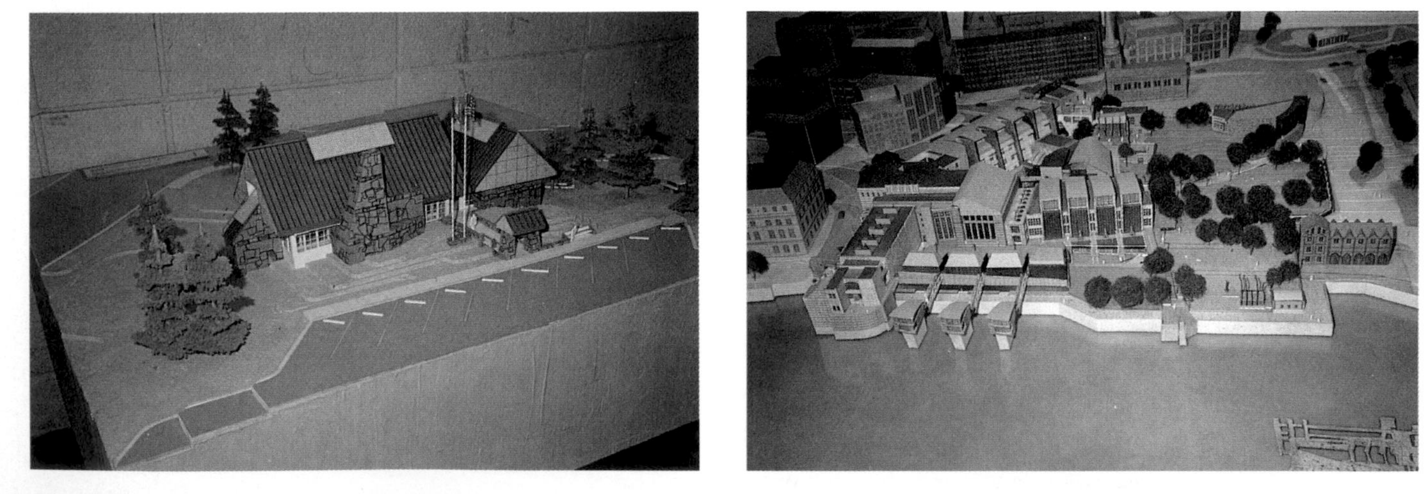

Fig. 9-22. Models give people a three-dimensional view of the project to be built. Models may be created of just a single building or they may be made to represent a large parcel of land.

Building codes are established for the purpose of providing minimum standards to safeguard life, health, and property. They protect the public welfare by regulating and controlling the design, construction, technique, quality of materials, use and occupancy, location, and maintenance of all structures within a political jurisdiction (city, county, or state).

The first known building code was prepared by Hammurabi, King of Ancient Babylon in 2250 BC. Its six short statements cited the wages that the architect/builder was to receive. This first code also spelled out the penalty that would be incurred for using inferior materials and employing faulty construction techniques. Simply stated:

DIVISION 6 ROOFING

General Conditions:

This contractor shall read the General Conditions and Supplementary General Conditions which are a part of these Specifications.

Scope of Work:

Furnish all materials and labor necessary to complete the entire roofing as shown on the drawings or hereinafter specified.

Roofing:

Flat roofing shall be 4 ply tar and gravel Spec. 102 of the Chicago Roofing Contractors Assn. Except single pour on gravel.

Carry new roofing onto existing roof a minimum of 3'-0" and roof new saddles on existing.

Rigid Insulation:

Cover all flat roof surfaces over steel deck with 2" of rigid insulation.

Insulation board shall be Fesco or equal.

Insulation shall be installed according to Spec. 102 of the Chicago Roofing Contractors Assn. Form saddles on existing roof of rigid insulation so pitch is to new roof drain.

Fig. 9-23. Specifications for projects large or small protect the interest of everyone involved: architect, builder, and owner.

STANDARD FORM 20
JANUARY 1961 EDITION
GENERAL SERVICES ADMINISTRATION
FED PROC. REG. (41 CFR) 1.16 401

REFERENCE

Invitation No.
DACW41-71-B-0010

INVITATION FOR BIDS
(CONSTRUCTION CONTRACT)

DATE

13 August 19

NAME AND LOCATION OF PROJECT

Construction of Harry S. Truman
Dam, Stage III
Harry S. Truman Reservoir, Missouri

DEPARTMENT OR AGENCY

Department of the Army
Corps of Engineers

BY *(Issuing office)*

Kansas City District

Sealed bids in one copy for the work described herein will be received until

1:00 p.m., local time at the place of bid opening, 24 September 19 , in Room No. 140, Federal Building, 601 E. 12th Street, Kansas City, Missouri 64106, and at that time publicly opened. Hand-carried bids delivered immediately prior to bid opening shall be delivered to the above-designated room.

Information regarding bidding material, bid guarantee, and bonds

1. Bid Bonds. Each bidder shall submit with his bid a Bid Bond (Standard Form 24) with good and sufficient surety or sureties acceptable to the Government, or other security as provided in paragraph 4 of Instructions to Bidders (Standard Form 22) in the form of twenty percent (20%) of the bid price or $3,000,000 whichever is lesser. The bid bond penalty may be expressed in terms of a percentage of the bid price or may be expressed in dollars and cents. (ASPR 16-401.2)

2. Performance and Payment Bonds. Within ten (10) days after the prescribed forms are presented to the bidder to whom award is made for signature, a written contract on the form prescribed by the specifications shall be executed and two bonds, each with good and sufficient surety or sureties acceptable to the Government, furnished namely a performance bond (Standard Form 25) and a payment bond (Standard Form 25A). The penal sums of such bonds will be as follows:

Fig. 9-24. First page of an invitation for bids. (Corps of Engineers, Kansas City District)

3.5 Special workmanship requirements:

3.5.1 Welding procedure qualification shall be in accordance with the qualification requirements of AWS Standard Specification D2.0 and shall be submitted for approval in accordance with the requirements for shop drawings of the SPECIAL PROVISIONS. The procedures shall be such as to minimize residual stresses and distortion of the finished members of the structure. Heat treatment, if required, shall be included in the procedures specification. Should it be found that changes in any previously approved welding procedure are desirable, the Contracting Officer will direct or authorize the Contractor to make such changes. Approval of any procedure, however, will not relieve the contractor of the sole responsibility for producing a finished structure meeting all requirements of these specifications.

3.5.2 Stress-relief annealing: Where stress-relief annealing is specified or indicated, it shall be in accordance with the requirements of Article 412(a) of the AWS Standard Specification D2.0 unless otherwise authorized or directed.

3.6 Welding of aluminum: Aluminum members specified to be welded shall be welded by the inert gas shielded metal arc-welding method following the recommendations of the manufacturer of the materials and the instructions outlined in the current edition of the American Welding Society, "Welding Handbook."

3.7 Brazing shall be in accordance with the recommendations and instructions outlined in the current edition of the American Welding Society, "Brazing Manual."

4. Flame Cutting: Low-carbon structural steel may be cut by machine-guided or hand-guided torches instead of by shears or saws. Flame cutting of material other than low-carbon structural steel shall be subject to approval and where proposed shall be definitely indicated on shop drawings submitted to the Contracting Officer. Where a torch is mechanically guided, no chipping or grinding will be required except as necessary to remove slag and sharp edges. Where a torch is hand guided, all cuts shall be chipped, ground, or machined to sound

Fig. 9-25. A typical specification of the technical provisions providing for work quality requirements.

Fig. 9-26. When working on a large project, problems are bound to arise. Specifications help to ensure project uniformity. This keeps problems to a minimum. (Morrison Knudsen Corp.)

"If the building collapsed and killed the owner, the architect/builder was to be put to death. If the owner's son should be killed by the collapsing structure, then the architect/builder's son shall be put to death."

From this clear and simple performance code have progressed nearly 18,000 local codes in the United States alone.

MODEL CODES

There are no national or universal codes for governing how buildings are to be constructed. However, at present, there are three model building codes. These are sponsored by three different organizations. Copies of these codes can be purchased. The codes and their sponsors are:

- *Basic Building Code* by Building Officials and Code Administrators International.
- *Southern Standard Building Code* by Southern Building Code Congress.
- *Uniform Building Code* by International Conference of Building Officials.

These codes are simply *models* that may be accepted through legislative action of a local municipality, county, or state.

Some governmental agencies elect to write their own set of standards. These standards are adopted by many cities or counties, with modifications the authorities feel are necessary when writing a building code for their locality. In this way, they are certain that the material and construction techniques are valid and comply with proven standards.

The Office of Building Standards and Codes Services, a division of the United States Department of Commerce, aids the building industry, government, and building users by assisting in the improvement of the nation's building regulatory system. Actions by the Building Standards and Codes Services reflect the needs of the building industry and society. Technical assistance in revising the codes and standards is provided by this office.

The Office of Building Standards and Code Services serves as Secretariate to the National Conference of States on Building Codes and Standards. The Conference, in turn, provides a forum for the interchange of building regulatory techniques. It provides a stimulus to needed research on technical questions, such as energy conservation in construction.

Fig. 9-27. There are stringent building codes for industrial sites. These codes protect the people who will be working in the complex as well as the surrounding community.

CODE SPONSORS

The four sponsoring groups for the establishment of building codes provide a testing service whereby each material or product of construction is tested in many ways, Fig. 9-28. From these tests, written research reports are prepared and made available to the architect and builder to aid in their selection of building components.

Research reports are available containing information on fireproofing, waterproofing, strength and stress values, and many other topics of concern to the designer or builder. See Fig. 9-29. It is to the benefit of each manufacturer of building components to have each of their building products tested. The manufacturers are responsible for submitting their building product to the sponsoring organization for testing. The sponsor, in return, conducts the evaluation test and prepares the research report.

Special interest organizations, such as the American Gas Association, provide certification of gas-operated equipment, Fig. 9-30.

Fig. 9-28. A fire test of an insulated steel column places the column in the center of a furnace, which is closed and sealed. The column is exposed to fire and its temperature is determined by thermocouples attached to the column before the insulation is applied. During fire exposure, the temperature of the column must not exceed certain limits. (Underwriters Laboratory)

BUILDING OFFICIALS & CODE ADMINISTRATORS INTERNATIONAL, INC.
1313 East 60th Street • Chicago, Illinois 60637 • 312 / 324-3400

Research Report No. ______________________

APPLICATION FOR BOCA RESEARCH AND EVALUATION

Date Received ______________________
mo. day year
(above for BOCA use)

Date ______________________

A. Applicant ______________________
Address ______________________ Telephone ______________________
City ______________ State ______________ Zip Code ________

B. Product ______________________
Trade Name ______________________

C. Performance of product for which evaluation is requested:
☐ Structural ______________ ☐ Fire/Flame resistance ______________
☐ ______________ ☐ Weather resistance ______________
☐ ______________ ☐ ______________
☐ ______________
☐ ______________
☐ ______________

...to ... comply the Research ... cant will ... sult in automatic suspension of ... valuat... of a product does not imply any warranty or guarantee by BOCA ... offic... any responsibility in regard to patent infringements and applicant agrees to ... employees, officers and members harmless from any litigation arising from the use or operation of defend and indemnify same against any loss, expense, liability or damage, including reasonable attorney's and e.) that the terms on reverse side hereof are part of this agreement.

______________________ *(Firm Name)*

Signed and Sworn to Before Me, this ________
day of ______________ *19* ________

______________________ *(Signature of Proprietor, Partner or Authorized Officer and title)*

______________________ *(Notary Public)*

Attest: ______________________ *(Secretary)*

Corporate Seal

Fig. 9-29. It is the manufacturer's responsibility to have company products evaluated by an independent organization.

Electrical and plumbing components are tested and approved by special interest organizations. If evaluation is positive (the product is not dangerous to the health and safety of the user), the components are recommended for public use. All tests are based on the condition that the components have been properly installed.

Building codes for each locality may be obtained from the building inspector's office or from the office of the governmental official having jurisdiction over building codes.

Many municipal building codes do not permit a residence to be built on a lot smaller than a given size. Also, many subdivisions have additional restrictions regarding the size of residences to be built within its jurisdiction. In these cases, the two separate codes work together. For example, they might require that each lot must be in excess of 3000 sq. ft., and the residence to be built must have at least 1000 sq. ft. of living space. This requirement would prevent someone from building a 500 sq. ft. residence on a 2300 sq. ft. lot, which might adversely affect the appearance and value of other houses in the neighborhood.

Fig. 9-30. Special interest groups give certificates of approval for equipment that meets requirements fixed by their codes.

INSPECTIONS

Inspection of a structure under construction involves a thorough investigation and careful evaluation of the technique, materials, and work quality of the contractor. Inspections are a management function, involving the examination of construction projects at various intervals, Fig. 9-31. A progressive inspection timetable should be a part of the construction work schedule, planned and conducted by people with knowledge and authority. Construction management uses the inspections as a method of controlling, monitoring, reporting, and correcting the construction process.

As a result of an inspection report, the finished project is either accepted or rejected by the owner. Rejected work is corrected and a final inspection is made to determine if the quality of materials used and the work are equal or superior to those stated in the specifications. Inspection schedules start as early as the planning stages and continue until the project is completed.

INSPECTOR

People who serve as inspectors for the construction project fall into three general groups:

- Those who serve the architect or engineer.
- Those employed by the contractor or subcontractor.
- Those who conduct inspections for outside agencies.

During the planning stages when drawings and specifications are prepared, the architect or engineer carefully reviews the work to be done. If the project is private, the plans are submitted to the city, county, or state building department. Here the building engineer reviews the proposed project carefully to see if the construction plan is in keeping with all building codes.

The owner may inspect the project as it progresses, or employ an inspector to make periodic inspections. Often, the architect who designed the building is hired to inspect the work. In all cases, inspections are made to ensure that the specifications are being followed to the letter of the contract by the contractor.

The construction superintendent is, in effect, an inspector employed by the contractor and subcontractor, Fig. 9-32. The superintendent makes certain that the floor joists and beams are level, the walls are plumb, and in general inspects all details of a project to ensure that the crews continue to turn out high quality work.

The individual worker on the project is an inspector, too. Anyone who takes pride in the finished product is an inspector, Fig. 9-33.

Agencies and organizations other than the architect and contractor also provide inspection service. Insurance companies may send their inspector to the construction site to monitor the progress and see if the work is properly and safely done. If an insurance company is to provide insurance after the structure is complete, the company will insist that proper materials are used and correct installation procedures are followed.

Inspections are made by various governmental agencies. U.S. Government inspections are concerned with health and safety precautions. The city building inspec-

Fig. 9-32. This superintendent is looking over the results of tests on several concrete samples. The materials must be up to the contractor's specifications.

Fig. 9-31. This inspection is providing important feedback on the new roadway.

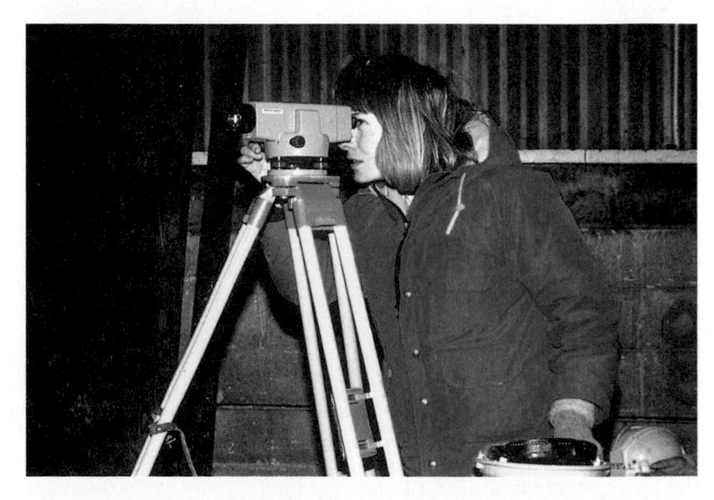

Fig. 9-33. Checking your work is important. Here a level is being used to examine the project. (Morrison Knudsen Corp.)

tor inspects the structure to see if local building codes are being met. For example, a local building code specifically calls for certain plumbing practices and materials; the inspector checks the subcontractor's work for code violations.

The loan agency, bank, or other lending institution usually sends an inspector to the constructing site to observe the progress being made on the project.

The manufacturer of construction materials and components makes periodic inspections before shipping those materials to the construction site, Fig. 9-34. Even before the manufacture of certain building materials and components, destructive tests are conducted to better understand the materials from which each is manufactured. From the results of these tests, the inspectors can predict the limitations of each material or component.

Each inspector makes inspections for a specific purpose. The worker inspects for pride or satisfaction; the lending agency inspects to protect its loan; the contractor inspects to protect its employees and the owner; the city inspector inspects to make sure all construction practices comply with the local building codes. The inspector, in each case, must be completely familiar with the activity or area of construction being inspected, Fig. 9-35.

INSPECTION

Inspection covers nearly every phase of construction:

- Materials.
- Techniques of construction.
- Work quality.

Materials to be inspected include soil tests, size and composition of aggregates, cement, glass, tile, lumber, metals, roofing material, flooring, and every other material that goes into the construction job, Fig. 9-36.

Fig. 9-34. Manufacturers of materials for construction must be able to guarantee that their materials are of a high quality. The tensile strength of steel is examined here. (Spokane Steel)

Inspection calls for testing and evaluating structural techniques. It also includes the testing and evaluation of all mechanical equipment such as heating and cooling systems, plumbing systems, and electrical systems.

Work quality is inspected in concrete finishing, reinforcement bar placement, framing carpentry, electrical and plumbing installation, and all other work done on the structure. Every phase of construction comes under the watchful eye of an inspector.

QUALITY CONTROL

Quality control, to most people, means the manufacture of a product of high quality. To the manufacturer of construction components, quality control means producing components of consistent quality. To a contractor or subcontractor, it means building the best structure for the cost.

Actually, the process of inspection is a means of locating inferior products, material, and work. The inspection, therefore, serves the purpose of quality control by saving time and further effort by discovering the fault early. Inspection is not only a remedial measure, it also is preventive.

Fig. 9-35. A construction supervisor is performing a slump test on concrete.

The inspector's job is to find out why the job is substandard or inferior. In the construction field, it takes effort on everybody's part to ensure that the job is completed in a professional manner.

SUMMARY

All construction projects first begin with information from the client. The client needs to have specific goals and objectives to be met with the proposed construction. The successful construction company relies on accurate and detailed information in order to meet the needs of a proposed project. Without quality information, no construction project can begin.

This information is used by designers, architects, and engineers to arrive at possible solutions to meet those goals of the owner. This input phase gathers information from a large variety of sources to arrive at several solutions to the construction project.

During the processing phase, the information gathered is analyzed and used to produce specific designs. Next, the output phase comes into play to produce a single design. This design is then reviewed to assure that it will meet the needs of the client. The client provides feedback to the designers to continue to refine the design so the proposed structure best fits all requirements.

The quality and accuracy of the information gathered during all phases of construction is critical to the success of any project. Without quality information, construction will not be successful.

Detailed working drawings are important to the success of a construction project. Working drawings include layout, elevation, structural, civil, architectural, electrical, and mechanical drawings. All of these drawings are drawn to scale. Models and specifications are also important tools for construction.

Every community has a set of building codes that must be followed on all construction projects. Building codes are designed to protect life and property. Inspections take place throughout the construction process to make sure that codes are followed.

Fig. 9-36. Lumber is visually inspected for knots, holes, and warping. (Miller Electric Manufacturing Co.)

KEY WORDS AND TERMS

All of the following words and terms have been used in this chapter. Do you know their meaning?

Analysis
Architectural drawing
Architectural symbol
Artist's rendering
Building Codes
Civil drawing
Contour map
Cost
Designing
Dimensioning
Electrical and mechanical drawing
Elevation drawing
Feedback
Form
Function
General provisions
Goal
Input phase
Inspection
Legal provision
Output phase
Processing phase
Quality control
Refinement of ideas
Research and development
Scale
Site plan
Specification
Stadia rod
Structural drawing
Subsystem
Surveyor
Technical provision
Transit
Tripod
Working drawing

TEST YOUR KNOWLEDGE

Do not write in this book. Please write your answers on a separate sheet of paper.

1. List the basic factors of design.
2. Diagram the four components of problem solving.
3. The _____ _____ contains records of the soil test boring schedule, excavation limits, and paving requirements.
4. List five sets of drawings that fall under the broad category of working drawings.
5. What are the three provisional categories of specifications?
6. During which stages of the construction process do inspections take place?

APPLYING YOUR KNOWLEDGE

1. Work with several other students and design a birdhouse that meets the needs of the bluebird. Each individual should come up with several designs based on research. Discuss each proposed design and then select one of the designs. If time permits build one birdhouse.
2. Visit an architect's or engineer's office. Discuss with them the design and engineering processes they use in their careers.
3. Make a listing of all of the various types of information you would like to have in order to construct a sidewalk at your school.
4. Select a single construction material such as wood or metal. Then, list the various information categories related to that material.
5. Design an inspection form that can be used to check if your technology lab is clean and properly organized at the end of each class.

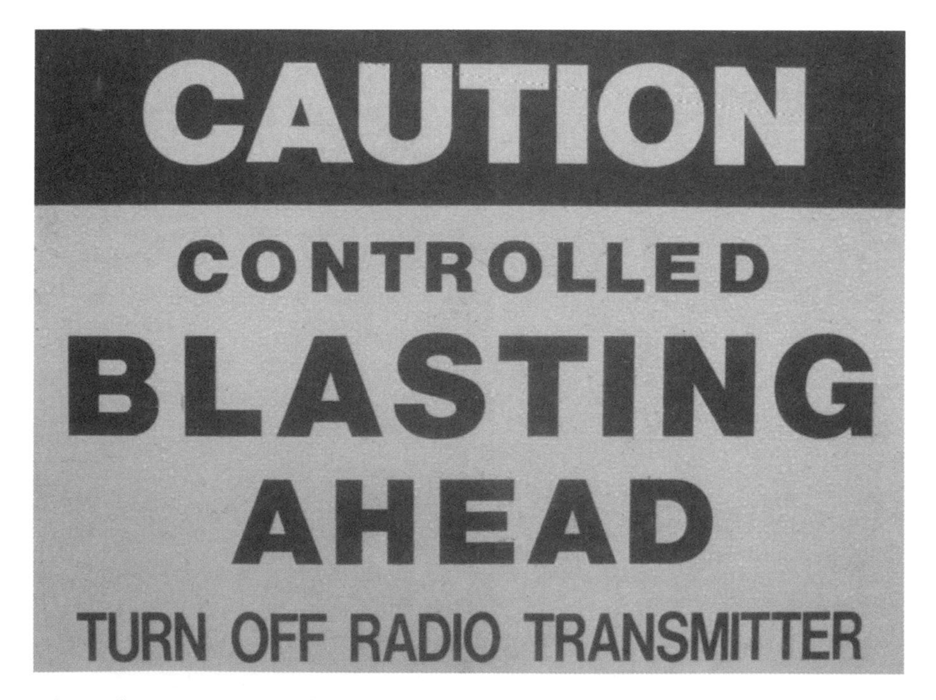

Information on a construction site needs to be clearly presented and to the point.

10

CHAPTER

TIME MANAGEMENT IN CONSTRUCTION

The information provided in this chapter will enable you to:

- *Define scheduling as it relates to the construction industry.*
- *Use the critical path method to estimate how long it will take to complete a project.*
- *Use the bar chart method to estimate how long it will take to complete a project.*
- *Use the progress chart method to estimate how long it will take to complete a project.*

Fig. 10-1. Construction projects are very labor intensive. Costs can be excessive if workers have to wait on materials.

You must always manage the limited time you have. Time regulates when you get up, when you eat, when you attend classes, when you work, when you play, and when you sleep. In work-a-day life, the regulations of time are even more critical. Many employees have been fired because they did not show up to work on time. Their poor judgment of time held up the progress of other workers.

In any career, you must be able to make realistic estimates of the amount of time it will take to complete a job or task. If you can learn to manage this resource efficiently, you will very likely become successful.

TIME MANAGEMENT

Management of time is critical to the construction industry. Labor costs to complete a construction project represent the major expense of any structure, Fig. 10-1. If contractors cannot manage their time resource efficiently, their company will not stay in business very long. Companies that manage their time effectively will remain competitive.

In the construction business, managers must be able to rely on their employees to show up to work on time. This allows jobs to be completed in the proper sequence. If this does not happen, the contractor may lose money on the project. Then, the contractor might not make the profit needed to stay in business.

Time is one of the seven resources that must be properly managed in order for a company to complete its work. All of these seven resources are dependent on the others. In this chapter, you will learn about time management as it relates to estimating and scheduling construction activities.

SCHEDULING

The completion of large construction projects can take from a number of days to many years. The contractor must bid on the project well before any construction begins. Therefore, the contractor must be able to **estimate** how much time it will take to complete the project. This is to allow the contractor to calculate a price that will produce a reasonable profit. Fortunes can be made or lost by being able to accurately estimate the length of time it will take to complete a job.

Time estimating is also an important function in the construction industry for determining when materials should be delivered to the job site, Fig. 10-2. If scheduling the arrival of building materials is not planned properly, the materials may arrive too late. Then, workers may be without materials and equipment. If the materials and equipment arrive too soon, they may be damaged, or just in the way, during early parts of the construction project. In addition, the company's money is tied up in materials that are not needed until much later in the project. This does not allow that money to be used to purchase other materials needed earlier in the project. **Scheduling** is having the right materials, people, and equipment on the job site at the proper time.

On almost all construction projects, the installation of a material or an assembly depends on the installation of previous materials or products. The contractor must have a good understanding of how the particular project will be constructed. This allows the proper scheduling of labor, materials, and equipment.

ANALYZING THE PROJECT

Regardless of the project's size, planning the work is a necessity. There is no use in receiving the roofing material on the construction site before the foundation is poured. The mechanical equipment should not arrive on the site before it is to be installed. Weather or construction damage could impair the use of that equipment in the meantime. By the same token, if the job is delayed because material, equipment, or workers do not arrive on the site at the scheduled time, the job will be slowed. Cost to the contractor and the owner will thereby be higher.

Fig. 10-2. The timing of arriving materials is important. When these large sheets of metal arrived, a number of workers had to be available to safely unload them.

CONSTRUCTION SCHEDULER

Proper scheduling of resources for a construction project requires careful study and a complete understanding of the project to be built. This is the work of a construction scheduler. The **construction scheduler** understands all the important components of a construction project and is able to estimate the length of time each component will take to complete. By adding the length of time each part will take to complete, the length of time to complete the entire project can be determined. This is assuming that each activity must be completed before the next activity can begin.

In many cases in the construction industry, several parts of the project can be worked on simultaneously. On a road construction project, for example, all of the clearing does not have to be complete before any dirt work can begin. If the project is a number of miles long, clearing work, beginning dirt work, gravel crushing and hauling, and finished road surface work can all be worked on at the same time. On a small building, however, the footings must be completed before the foundation wall can be constructed. Schedulers must have this understanding so they can accurately determine how long it will take to construct a project.

SCHEDULING METHODS

Scheduling a project is possible through understanding the individual units of work required. A project is made up of major tasks like excavation, building the foundation, erecting the frame, applying exterior walls, erecting the roof, installing electrical and mechanical systems, and finishing the interior. The scheduler estimates how long it will take to complete each of these tasks. He or she works out a time table and prepares a work schedule. A schedule can be organized by one of several methods.

- The critical path method.
- The bar chart.
- The progress chart.

CRITICAL PATH METHOD

The **critical path method (CPM),** Fig. 10-3, is the most common and economical way to plan operations on a complicated project. It provides a means of assessing the effect of all variations, changes, extra work, or deductions upon the time of completion and upon the cost of the work. The critical path method is an open-ended process that

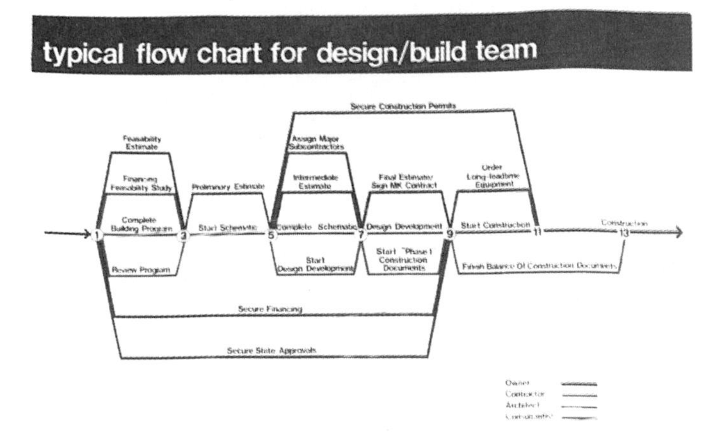

Fig. 10-3. The critical path method is the most common planning method used in construction. This critical path method charts the planning work that must be done before construction begins. (Morrison Knudsen Corp.)

permits different degrees of involvement by management to suit their various needs and objectives.

Generally, the critical path method consists of a pattern of circles and connecting lines, Fig. 10-4. The lines represent a task to be completed and are labeled with the task. Also, the line may contain a series of numbers indicating an estimate of time it will take to complete the task. The first number in the series represents the shortest possible amount of time it could take to complete the task. The last number represents the longest amount of time it could take to complete the work. The number in the middle is the best estimate of time needed to complete the activity.

The circles represent the starting or stopping point for the various events. All activities must enter the circle before the next activity can begin. When setting up the chart, the scheduler must list the order of construction activities. The scheduler must determine which of the activities can be carried on at the same time and which activities must be completed before others can begin. This is shown in the critical path chart.

To determine the length of time that it will take to complete a construction project, add up all of the time required to complete each activity. Then, a total time schedule for the job can be determined.

In applying CPM to construction planning, it is necessary to have accurate estimates of time and costs for each operation. The breakdown of the project into its individual operations may be simple or detailed as desired. The essential requirement is that the direct cost for each operation be estimated separately.

BAR CHART

The **bar chart** is often used on very large projects such as a dam. Here, large quantities of materials have to be stockpiled several years in advance of their actual use. With the use of a bar chart, it is easy to identify the time needed to complete each task.

The bar chart method helps prevent delays. The contractor makes a complete analysis of the work to be done. Then a bar chart is prepared as shown in Fig. 10-5. This chart was made up for major concrete work on the Harry S. Truman dam. In a project of such great size, time is scheduled in much longer periods than would be necessary in a smaller project. The bar chart is capable of providing sufficient lead time to stockpile materials in advance of need. A double bar, Fig 10-5, shows whether the work is construction or the installation of equipment and preassembled units.

The bar chart is really a time chart. The chart allows the contractor to break major activities into subgroups and estimate the time needed for each subtask as part of the major task. Fig. 10-6 shows how this breakdown might appear.

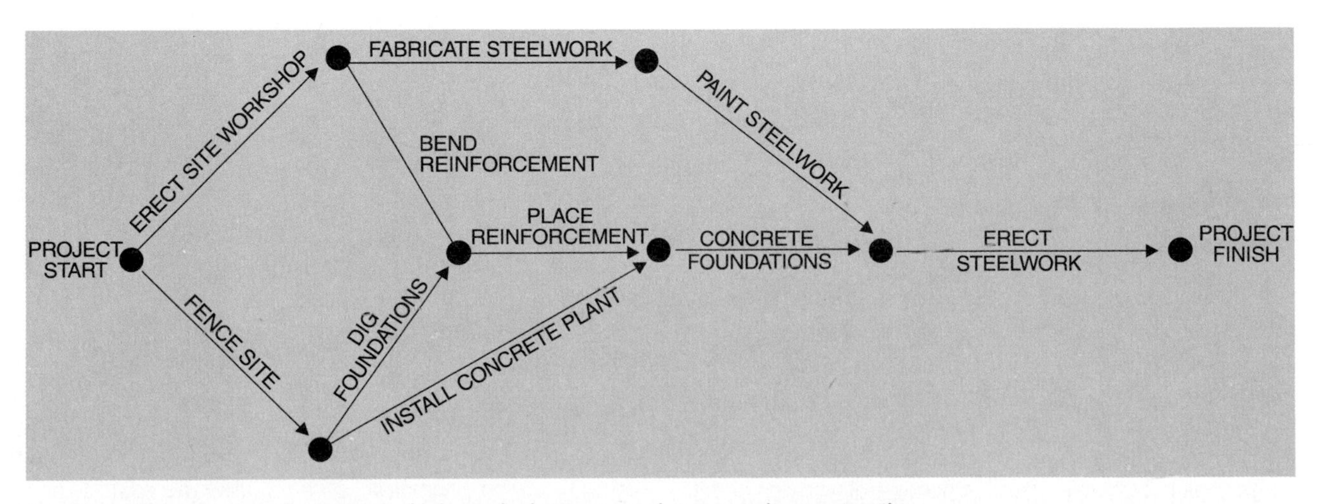

Fig. 10-4. The critical path method of job analysis traces major events in construction.

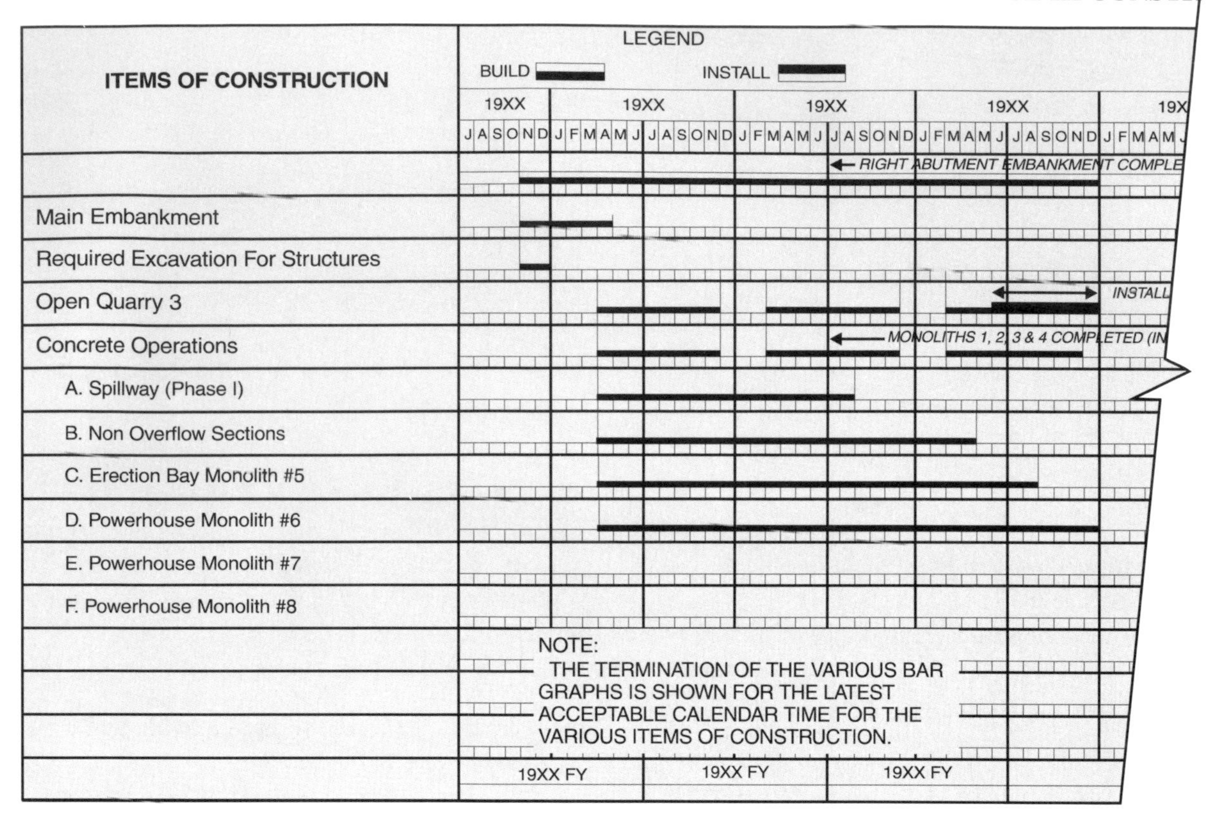

Fig. 10-5. Black bars are time modules for building of parts of a power house. Note from the legend at the top, the placement of a heavy inked section of bar indicates either building or installing.

INSTALLATION OF TAINTER GATES

ITEM	COST	PERCENTAGE OF JOB	TIME (IN DAYS)
Delivery from subcontractor	$50,000	50	216
Store at site	1,000	1	10
Weld fitting	10,000	10	14
Move to erection site	2,000	2	4
Hoist to position	5,000	5	4
Place and fit	12,000	12	14
Prime coat	3,000	3	6
Three coats, paint	17,000	17	36

Fig. 10-6. A builder may prepare a list of the work to be done, the cost, and the time required to build or install. This list breaks down the tasks for installing tainter gates for a dam and power plant.

PROGRESS CHART

Rather than showing time elapsed or the amount of work done at the site, the **progress chart** shows the cost of the job, Fig. 10-7. The heavy line represents progress the contractor planned to make before work started. The thin, broken line is the progress actually made.

To plot the graph on the progress chart, place a fairly heavy dot directly above some amount of time elapsed and directly across from the percentage of dollars spent at that time. Place similar dots representing the percent of cost spent at other periods of time. Then draw a series of straight lines connecting each dot to the next. This line is called the graph line.

The numbers on the chart are called coordinates. **Coordinates** are simply a set of numbers that lead you to a point. Look at Fig. 10-7. The progress chart represents a subcontractor who is assembling tainter (radial) gates for the construction of the Truman Dam. Looking at the graph, you can see that the brown vertical line from number 7.2 intersects with the brown horizontal line coming from 50. Numbers 7.2 and 50 are coordinates. They determine where a point will be placed on the chart. In this case, the contractor was to deliver to the construction site in a little over seven months. In reality, it took eight months. This is shown by the second, broken line. The contractor had to adjust progress to bring the graph lines back together. The adjustment was an increase in the labor force, thereby decreasing the time necessary to do the next task.

Progress charts may need to be revised and updated several times during construction. It is often better to set up a new chart after the job is under way to keep the schedule accurate.

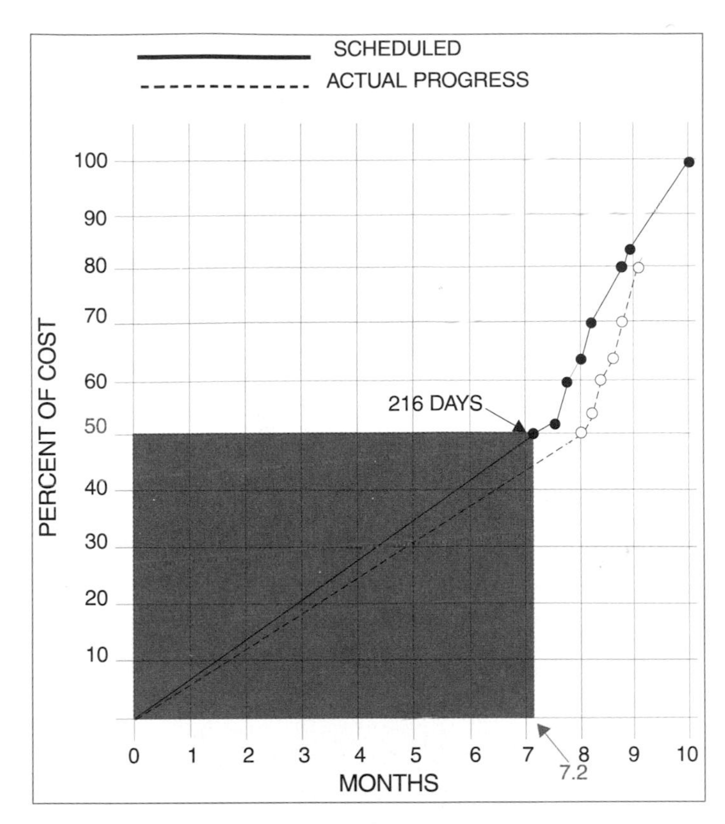

Fig. 10-7. The progress chart above was set up by a contractor to keep track of money and time spent on each part of the job. The heavy black line was prepared to estimate cost and time. The broken line plots actual progress. The colored lines show how location of points is determined by the intersection of vertical and horizontal lines extending from coordinate numbers.

SUMMARY

Time is a very critical resource to all contractors. Lost time on a construction project can never be made up. It is the job of the construction scheduler to accurately estimate how much time it will take to complete a particular task or job.

The scheduler has several tools to use to help in estimating how much time it will take to complete a job. The critical path method, the bar chart, and the progress chart are common methods to estimate and keep track of time in the construction industry.

KEY WORDS AND TERMS

All of the following words and terms have been used in this chapter. Do you know their meaning?

Bar chart
Construction scheduler
Coordinates
Critical path method (CPM)
Estimate
Progress chart
Scheduling

TEST YOUR KNOWLEDGE

Do not write in this book. Please write your answers on a separate sheet of paper.

1. List the three methods commonly used in estimating the time to complete a job.
2. Define scheduling.
3. What does CPM mean?
4. What information does the line represent on the CPM method of scheduling?
5. What do the three numbers that are under each line on a CPM chart represent?
6. Which scheduling method is often used on very large projects?
7. Why is it important for all workers to show up to work on time?

APPLYING YOUR KNOWLEDGE

1. Set up a CPM chart for a simple construction project such as making a new sidewalk.
2. Visit with a contractor and ask to see actual progress charts for a major construction project.
3. Write out a detailed explanation of how a contractor can get a construction project back on schedule once it has fallen behind due to a delay in materials.
4. What does your answer to question three do to the total profit for the contractor? Why?

DOUGLAS STEAD

At the start of a new day, this desk will be staffed by information personnel. They can direct visitors quickly and thoroughly.

11

CHAPTER

MATERIALS IN CONSTRUCTION

> *The information provided in this chapter will enable you to:*
> - *Explain the differences between primary processing and secondary processing.*
> - *Classify construction materials into the groups: polymers, metals, ceramics, and composites.*
> - *Describe the physical properties of polymers, metals, and ceramics.*
> - *Recognize the difference between ferrous and nonferrous metals.*

Every construction project is intended to perform certain functions. Whether or not the completed project can perform these functions depends upon the materials selected and how they were used. Knowledge of design and of the materials available allows the builder to select those materials that best fit the needs of the structure being built. Selection of materials must be based on a good understanding of the material. The builder must know how the material will perform over a long period of time and how much maintenance it will require during the life of the structure.

Improvement of construction materials is moving at a rapid rate. Consequently, it is important to have some understanding of how a variety of materials are classified in addition to being able to work with them. Knowing how materials are classified can give you a better understanding of how to work with materials as they are used on the construction site.

Materials can be classified in a variety of ways. One of these ways is simply determining whether the material is a solid, a liquid, or a gas. In the construction world, we are most interested in materials that are solid at the common range of temperatures that the structure would endure. Although, gasses and liquids also play a very vital part in construction. Some construction materials, such as solvents, fuels, and paints, are liquid and must be properly used in order to complete a construction project. Gasses, such as oxygen and acetylene, used in cutting torches, are also vital to construction work. Although we will deal mostly with solid construction materials in this resource section, you should be aware that construction companies rely on many forms of materials in order to complete the job.

Solid construction materials may be classified by their makeup, Fig 11-1. These materials may be:

- Polymers–wood, plastic, or asphalt.
- Metals–aluminum, steel, or copper.
- Ceramics–stone, glass, brick, or tile.
- Composites–materials made up of two or more materials bonded together, such as fiberboard, particleboard, or fiberglass.

Composite will be dealt with in the following chapter.

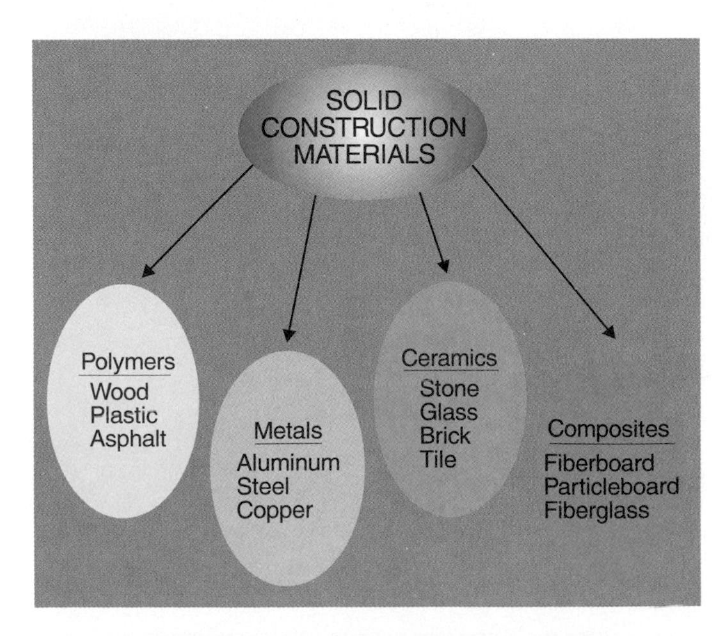

Fig. 11-1. Classifications of solid construction materials.

MATERIAL PROCESSING

Construction materials must be processed into standard shapes and sizes. They must also have some consistency from one piece to the next before arriving at the job site. This form of processing is called **primary processing,** Fig. 11-2. In primary processing, raw materials are gathered by using one of three methods. The materials are harvested (timber), mined (iron or copper ore), or obtained through drilling (oil and gas).

Once the raw materials are collected, they must be processed into usable industrial materials such as lumber, pipe, nails, and fuel oil. This form of processing is called **secondary processing.** Secondary processing provides us with the construction materials needed to complete the construction project, Fig. 11-3.

POLYMERS

Polymers are created by two or more molecules combining through a chemical reaction. Polymers are extremely common and useful construction materials. Synthetic polymers, such as plastic, and natural polymers, such as wood, can be found almost anywhere in modern construction projects.

A

B

Fig. 11-2. A–Portable drilling rigs are used for exploratory searches for oil and gas. B–Minerals are mined from the ground in an impure form. Future processing transforms the minerals into a desirable state. (Montana Power)

Fig. 11-3. Secondary processing turns logs into lumber. This lumber is air drying, one of the final stages in its processing.

PLASTICS

Plastics are synthetic polymers. They are widely used in construction today. Plastics are used in some insulations, vapor barriers, carpeting, electrical insulation, glues and adhesives, and weather and damp proofing. Pipes for both supply and waste plumbing are often made of plastic.

Synthetic polymers are manufactured by using chemicals from oil, coal, or wood to produce long chain molecules. These molecules cling together to form any variety of items used in construction. These long chain molecules can be varied by a chemist to produce plastic with a variety of properties desired by a manufacturer. The use of plastics in the future will be greatly increased as new materials are found and recycling becomes more important and economical, Fig. 11-4.

WOOD

Wood has always been one of our most important construction materials. Lumber production is one of the United States largest industries. Sawmills have been in production since 1610. Forest products are used for fuel, furniture, ships, and buildings. Americans have come to depend upon wood products from the great forests, Fig. 11-5.

Millions of board feet of lumber are produced annually. One board meter is equal to 10.763 board ft. Lumber is widely used in modern homes. Wood for pulp and paper makes life more comfortable and interesting. Wood helps insulate walls. Used on outside walls it can make them more pleasing to the eye, Fig. 11-6. Wood is the base material for many products that make life more pleasant.

Fig. 11-4. New thermoplastic roof panels are being installed. These panels look like hand-split cedar shakes, yet can be installed more quickly, resist mildew, and will not rot. They also help preserve trees. (GE Plastics)

Fig. 11-6. Do you like the feel that wood siding has given this home? (Red Cedar Shingle and Handsplit Shake Bureau)

Fig. 11-5. Trees are used in a myriad of products. They are one of our most important resources.

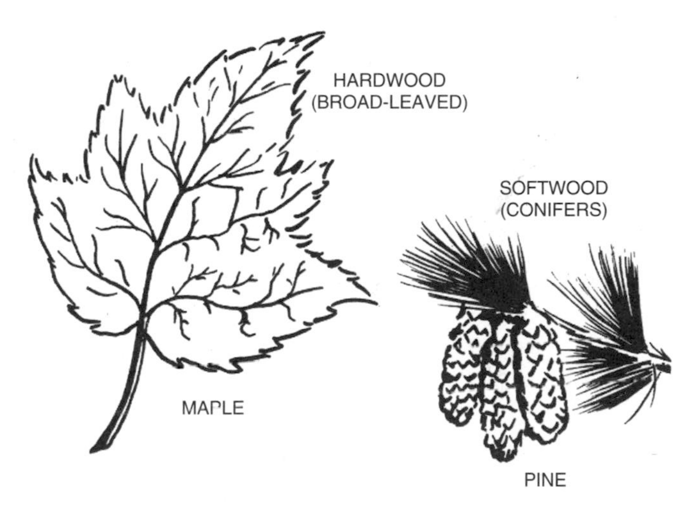

Fig. 11-7. Hardwoods are produced by broad-leaved trees. Softwoods are produced by conifers. (Frank Paxton Lumber Co.)

Classification of wood

Woods are generally classified for construction purposes as either hardwood or softwood, Fig 11-7. These terms may be a little misleading. There is no direct relationship to the relative hardness of the wood. Balsa wood is classified as a hardwood, but it is softer than most woods in the softwood classification. **Softwoods** are woods produced by evergreens or conifers (cone bearing trees). **Hardwoods** are woods produced from the broadleaf or deciduous trees. Deciduous trees are trees that lose their leaves.

Hardwoods. Hardwoods are generally used for wood furniture, decorative interior paneling, and as interior trim. See Fig. 11-8. Some hardwoods are extremely hard and strong, but they are seldom used as structural materials. Their scarcity, high cost, and beauty place these woods at such a premium they are used for interior purposes only.

Softwoods. Softwoods are the evergreen, cone-bearing, and needle-leaved trees. Wood or lumber from these trees is generally used for construction. The wood provides framing for homes and forms for concrete. Softwood lumber is often laminated to serve as structural members and as plywood covering. See Fig. 11-9 for more information on the classifcation of woods.

Fig. 11-8. Fine doors, trimmings, and tables are made from hardwoods.

Turning logs into lumber

To transform logs into usable lumber, the logs are first cut in the forest and hauled to the sawmill, Fig. 11-10. There the bark is removed. Bark removal has two important functions. First, it removes most of the dirt and rocks from the log so the saws are not damaged when cutting the logs into lumber. Second, it allows the waste from the sawmill operation, such as edgings and slabs, to be used for paper pulp.

Inside the mill shed, logs are positioned on a riderless carriage. Air-operated, this carriage can hold logs up to 20 ft. (6 m) long. The carriage carries the log past the headrig (headsaw) or a band mill (band saw), which cuts the log into cants. A **cant** is a log that has been slabbed (cut) to square it up on two or four sides. See Fig. 11-11.

The cants are then moved to the gang saw. The gang saw has many separate saw blades, cutting several boards at the same time. A vertical resaw machine cuts an individual plank into separate boards. See Figs. 11-12 to 11-15.

From the gang saws and vertical resaws, the lumber travels on chain conveyors to trimmer saws. The usable waste is sent to chippers. Waste utilization is a part of any modern mill. Waste wood is reduced to chips. The chips are, in turn, converted into paper, paper boxes, hardboard, particleboard, and many other products. Waste that cannot be used in any other product is burned to produce energy for the mill.

SPECIES	Comparative Weights	Color	Hand Tool Working	Nail Ability	Relative Density	General Strength	Resistance to Decay	Wood Finishing	Cost
HARDWOODS									
APITONG	Heavy	Reddish Brown	Hard	Poor	Medium	Good	High	Poor	Medium High
ASH, brown	Medium	Light Brown	Medium	Medium	Hard	Medium	Low	Medium	Medium
ASH, tough white	Heavy	Off-White	Hard	Poor	Hard	Good	Low	Medium	Medium
ASH, soft white	Medium	Off-White	Medium	Medium	Medium	Low	Low	Medium	Medium Low
AVODIRE	Medium	Golden Blond	Medium	Medium	Medium	Low	Low	Medium	High
BALSAWOOD	Light	Cream White	Easy	Good	Soft	Low	Low	Poor	Medium
BASSWOOD	Light	Cream White	Easy	Good	Soft	Low	Low	Medium	Medium
BEECH	Heavy	Light Brown	Hard	Poor	Hard	Good	Low	Easy	Medium
BIRCH	Heavy	Light Brown	Hard	Poor	Hard	Good	Low	Easy	High
BUTTERNUT	Light	Light Brown	Easy	Good	Soft	Low	Medium	Medium	Medium
CHERRY, black	Medium	Medium Reddish Brown	Hard	Poor	Hard	Good	Medium	Easy	High
CHESTNUT	Light	Light Brown	Medium	Medium	Medium	Medium	High	Poor	Medium
COTTONWOOD	Light	Greyish White	Medium	Good	Soft	Low	Low	Poor	Low
ELM, soft grey	Medium	Cream Tan	Hard	Good	Medium	Medium	Medium	Medium	Medium Low
GUM, red	Medium	Reddish Brown	Medium	Medium	Medium	Medium	Medium	Medium	Medium High
HICKORY, true	Heavy	Reddish Tan	Hard	Poor	Hard	Good	Low	Medium	Low
HOLLY	Medium	White to Grey	Medium	Medium	Hard	Medium	Low	Easy	Medium
KORINA	Medium	Pale Golden	Medium	Good	Medium	Medium	Low	Medium	High
MAGNOLIA	Medium	Yellowish Brown	Medium	Medium	Medium	Medium	Low	Easy	Medium
MAHOGANY, Honduras	Medium	Golden Brown	Easy	Good	Medium	Medium	High	Medium	High
MAHOGANY, Philippine	Medium	Medium Red	Easy	Good	Medium	Medium	High	Medium	Medium High
MAPLE, hard	Heavy	Reddish Cream	Hard	Poor	Hard	Good	Low	Easy	Medium High
MAPLE, soft	Medium	Reddish Brown	Hard	Poor	Hard	Good	Low	Easy	Medium Low
OAK, red *(average)*	Heavy	Flesh Brown	Hard	Medium	Hard	Good	Low	Medium	Medium
OAK, white *(average)*	Heavy	Greyish Brown	Hard	Medium	Hard	Good	High	Medium	Medium High
POPLAR, yellow	Medium	Light to Dark Yellow	Easy	Good	Soft	Low	Low	Easy	Medium
PRIMA VERA	Medium	Straw Tan	Medium	Medium	Medium	Medium	Medium	Medium	High
SYCAMORE	Medium	Flesh Brown	Hard	Good	Medium	Medium	Low	Easy	Medium Low
WALNUT, black	Heavy	Dark Brown	Medium	Medium	Hard	Good	High	Medium	High
WILLOW, black	Light	Medium Brown	Easy	Good	Soft	Low	Low	Medium	Medium Low

Fig. 11-9. Wood selection chart.

SPECIES	Comparative Weights	Color	Hand Tool Working	Nail Ability	Relative Density	General Strength	Resistance to Decay	Wood Finishing	Cost
SOFTWOODS									
CEDAR, Tennessee Red	Medium	Red	Medium	Poor	Medium	Medium	High	Easy	Medium
CYPRESS	Medium	Yellow to Reddish Brown	Medium	Good	Soft	Medium	High	Poor	Medium High
FIR, Douglas	Medium	Orange-Brown	Medium	Poor	Soft	Medium	Medium	Poor	Medium
FIR, white	Light	Nearly White	Medium	Poor	Soft	Low	Low	Poor	Low
PINE, yellow longleaf	Medium	Orange to Reddish Brown	Hard	Poor	Medium	Good	Medium	Medium	Medium
PINE, northern white (*Pinus Strobus*)	Light	Cream to Reddish Brown	Easy	Good	Soft	Low	Medium	Medium	Medium High
PINE, ponderosa	Light	Orange to Reddish Brown	Easy	Good	Soft	Low	Low	Medium	Medium
PINE, sugar	Light	Creamy Brown	Easy	Good	Soft	Low	Medium	Poor	Medium High
REDWOOD	Light	Deep Reddish Brown	Easy	Good	Soft	Medium	High	Poor	Medium
SPRUCES *(average)*	Light	Nearly White	Medium	Medium	Soft	Low	Low	Medium	Medium

Fig. 11-9. Wood selection chart. (Continued)

Fig. 11-10. Logs are piled up to wait their turn on the lot of a sawmill.

Fig. 11-11. Logs being cut into cants.

Methods of cutting

The manner in which a piece of wood is cut from the log affects its appearance and strength. Lumber sawed in slices parallel to one side of the log is said to be **plain sawed** or **flat sawed,** Fig. 11-16. Flat-sawed lumber is produced with less waste. It is therefore cheaper. Growth ring patterns are more distinct, while the grain and decorative patterns of many hardwoods are brought out when plain sawed.

Fig. 11-12. Cants are sliced into boards.

Fig. 11-13. Safety is important in the sawmill. Hardhats, jumpsuits, and gloves protect this worker.

Fig. 11-14. Boards of equal length and width gather at the end of the line.

Fig. 11-15. The boards are stacked outside in the open air.

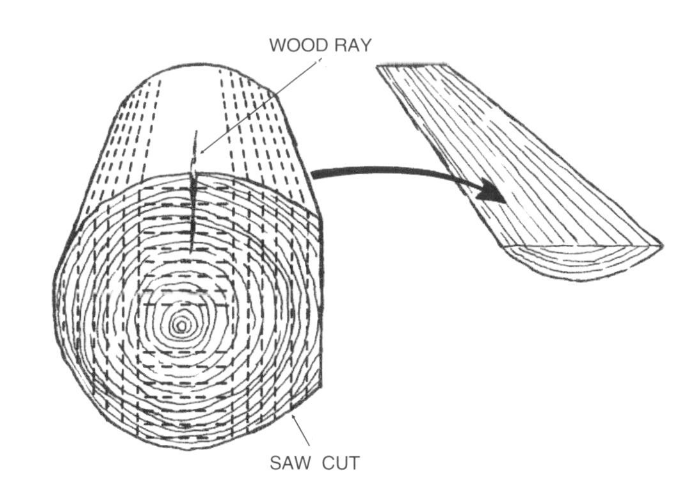

Fig. 11-16. Plain- or flat-sawed lumber is cut tangent to the annual growth ring. A *tangent* cut runs at right angles to a line running through the center of the log. (Frank Paxton Lumber Co.)

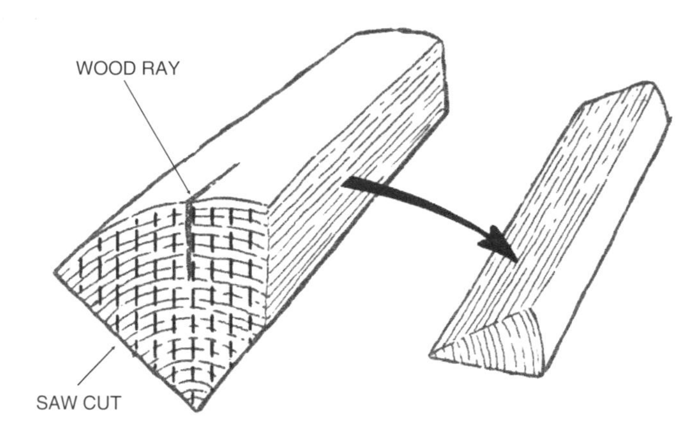

Fig. 11-17. Quarter-sawed lumber is cut more perpendicular to the annual growth rings.

But, flat- or plain-sawed lumber has certain disadvantages. The lumber is more likely to warp, check, split, and cup.

If the log is first quartered and boards are cut from each quarter perpendicular (at right angles) to the exterior of the log, it is said to be **quarter sawed,** Fig. 11-17. Quarter-sawed lumber is more expensive because of the added labor and waste in cutting. However, quarter-sawed lumber swells and shrinks less than plain-sawed lumber. During seasoning it develops fewer cracks and checks.

Seasoning lumber

The process of drying lumber to the point where it is ready to be used is called **seasoning.** Freshly sawed (green) lumber contains as much as one hundred percent of its dry weight in moisture.

Air drying. To dry properly, lumber must be stacked in covered piles. Each layer is separated by one-inch strips to allow airflow between layers. See Fig. 11-18. Drying time to bring lumber to the proper moisture content varies with the area and the weather conditions. In the United States, softwood is seldom stored longer than three or four months.

The recommended moisture content in wood varies from seven percent in dry areas to 18 percent in damp areas. It is helpful to store lumber on the job site for a period of time. This practice allows the lumber to reach a point at which moisture content is the same as the surrounding air. This is called **equilibrium.** At this point the shrinking and swelling of the wood will be at a minimum.

Kiln drying. Lumber may be kiln dried to speed up reduction of moisture. It is placed in ovens, or **kilns,** and exposed to an elevated temperature and controlled humidity.

Fig. 11-18. One-inch strips between the boards allow the wood to dry quickly and evenly.

Drying temperatures of 70° to 120°F (20° to 50°C) for periods of time ranging from four to ten days are needed for softwoods. More drying is needed for hardwoods. Kiln drying reduces cracking and checking in certain types of lumber.

Lumber grades

Lumber is **graded** according to strength, appearance, or usability. This has been the subject of extensive testing. The National Bureau of Standards, the Department of Agriculture, and various lumber associations have spent many years in establishing lumber grades. Grades and sizes are provided to protect the user and to establish a degree of uniformity.

When lumber is cut from the log, the pieces are very different in appearance and strength. Much depends on the number and size of imperfections. Grades are based on the number, size, and location of these defects. Defects include knots, checks, splits, and pitch pockets. The highest grades are practically free from these imperfections. Lower grades allow more and more imperfections.

ASPHALT

Asphalts are strong, durable, highly waterproof, and easily available. These bituminous materials are also highly resistant to the action of most acids, alkalies, and salts. Asphalts are used in the construction industry to construct roads, to build pads that contain chemicals used in manufacturing processes, and to build roofing systems. They are also used as sealants and coatings.

Asphalt is produced by the distillation of crude oil and is a dark brown to black material. It is the residue left from refining petroleum. In paving projects, asphalt is used to cement crushed rock together to make a smooth, hard surface that is nearly waterproof and can carry the expected traffic loads. Asphalt materials are also used in the construction of roofs. In this case, they are used to cement

Fig. 11-19. Asphalt produces a strong, waterproof, and durable top for highways and other roadways.

together various roofing papers. This process makes a waterproof mat to withstand the elements, Fig. 11-19.

METALS

Traditionally, metals are classified as ferrous and nonferrous. **Ferrous metals** contain a large percentage of iron. **Nonferrous metals** contain little or no iron.

When two or more metals are combined while in the molten state, they form a metallic substance called an **alloy.** Alloys, likewise, are classed as either ferrous or nonferrous, depending on the percentage of iron they contain. Ferrous metals are used in construction because of strength characteristics. Steel, for example, is an alloy of iron, Fig. 11-20.

Iron probably has had more influence on civilization than any other material. Iron knife blades and swords have been found in ancient tombs dating back to 3000 BC. These

Fig. 11-20. Steel is the most common metal used in building construction. Large commercial structures are often framed with steel.

weapons indicate that the civilizations of Assyria, Babylon, and Greece had learned to separate iron from ore and to forge it into useful tools.

It is believed that iron was first forged in open limestone fire pits. Ore mixed with charcoal was added to the fire, and limestone was added to act as *flux* to promote separation of impurities. The ore mixed with impurities melted and flowed to the bottom of the pit, where it formed a spongy mass. This iron, for the most part, was made into weapons, armor, and small tools.

STEEL

The making of steel involves the oxidation and removal of unwanted elements from **pig iron,** the product of the blast furnace. This second step in steel production adds certain other elements to produce a desirable composition. The molten metal is worked until the sulfur and phosphorus contents are acceptably low. These elements cause undesirable characteristics in steel. Generally, carbon and other elements are added after the steel has been poured from the furnace into the ladle.

Most steel is made by adding scrap steel to the pig iron melt. The average charge in the United States is about 50 percent scrap, which makes steel a recyclable material.

Structural mill

Large structural steel members are made by rolling the red hot steel. The steel is passed through a succession of rolls that gradually form each member into the desired shape, Fig. 11-21.

Structural shapes are formed by passing the steel back and forth through one set of rolls in the structural mill. The space between the rolls decreases with each pass of the steel, until the member has reached the desired thickness and shape, Fig. 11-22.

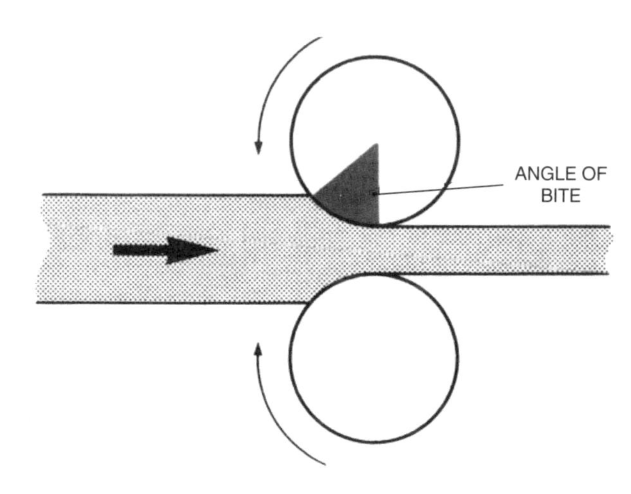

Fig. 11-22. Mill rolls reduce the thickness of the steel slab. The maximum amount of reduction is determined by the *angle of bite.* Smooth rolls can take a 30 degree bite.

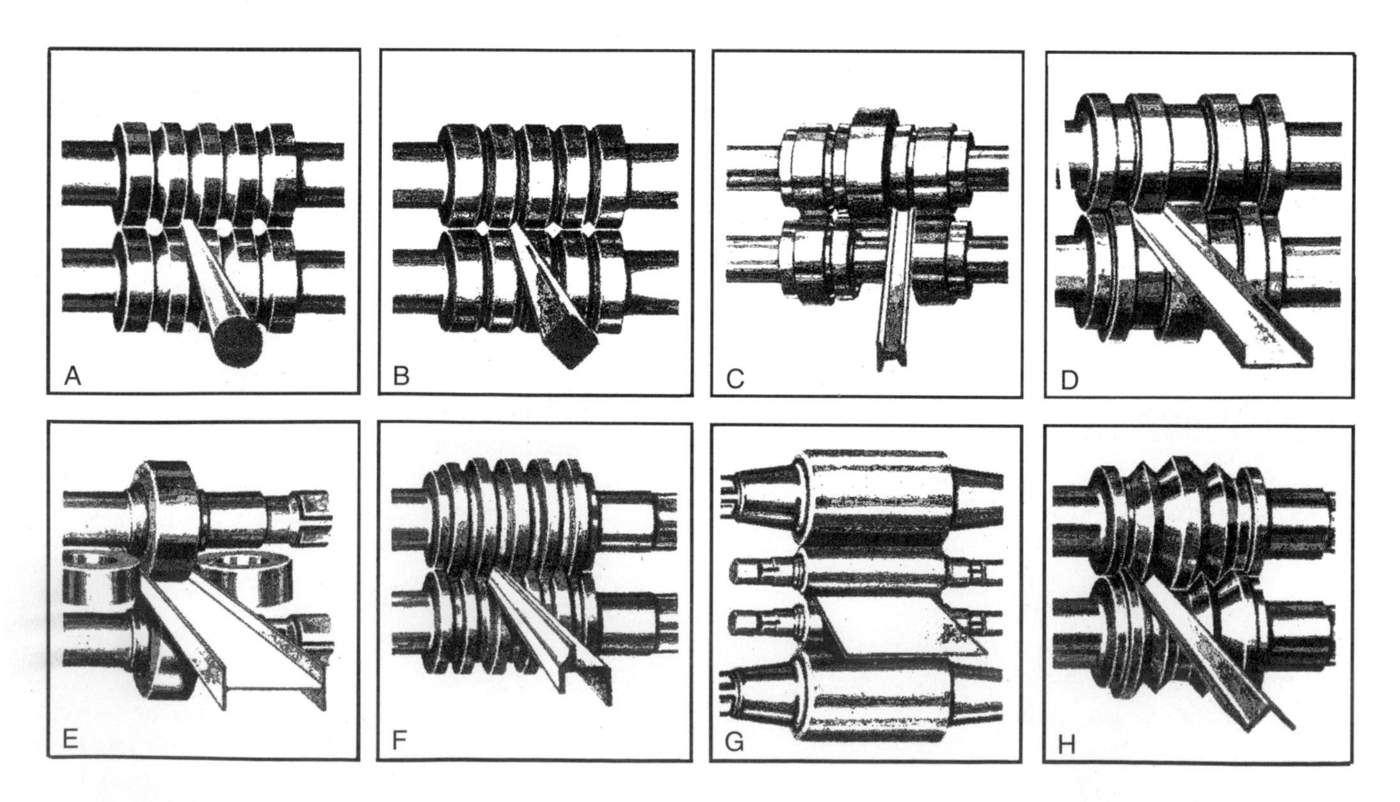

Fig. 11-21. Roll arrangement and passes cut in the rolls to produce: A–Rounds. B–Squares. C–Rails. D–Channels. E–Columns. F–Special sections. G–Sheets. H–Zee-bars.

Pipe

Pipe is an important structural material for plumbing and support columns. It is classified according to the process by which it is manufactured.

The two principal forms of pipe are seamless pipe and welded pipe. **Seamless pipe** is produced by forcing a steel rod over the pointed nose of a piercing mandrel. The mandrel punches a hole in the center of the rod, while rolls on the outside control the diameter.

Welded pipe is produced by forming a flat strip into a U-shape, then closing it into a tube. The seam is welded by passing an electric current between the edges.

Wire

Wire for construction is produced by drawing a steel rod through a series of dies, Fig. 11-23, each reducing the diameter. Wire is usually thought of as round. However, it also comes in square, rectangular, polygonal, and many other shapes. A variety of strengths, hardnesses, and finishes are available to the construction industry.

Galvanizing

Steel sheets may be protected against rust and corrosion by a galvanized zinc coating. **Galvanizing** serves as a double protection. It forms a mechanical barrier against moisture and prevents oxidation, or rusting, of the steel base.

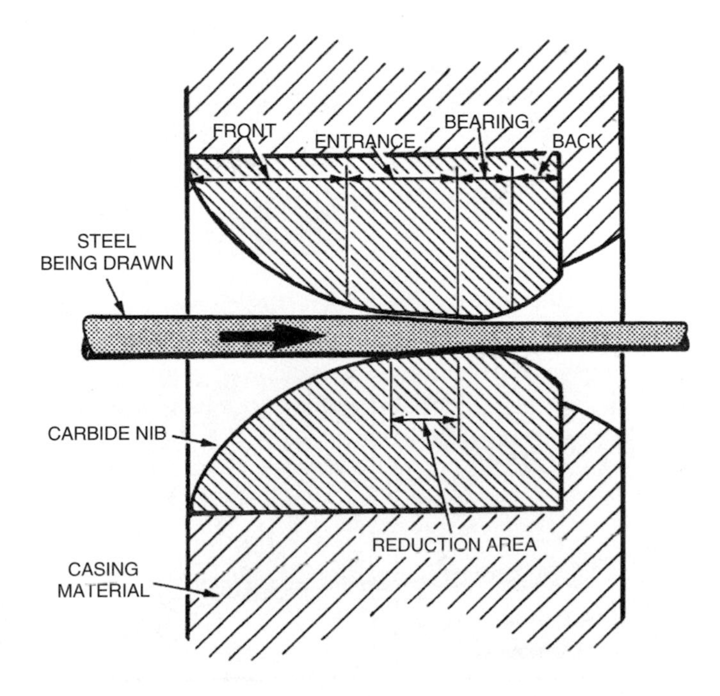

Fig. 11-23. Steel enters a wire drawing die and is pulled against a carbide nib. The emerging wire is smaller than the steel fed into the die.

ALUMINUM AND COPPER

Aluminum and copper are the two most widely used nonferrous metals. The ores of these two metals are processed to produce them in their pure form.

Copper is processed into construction materials such as copper pipe, electrical wire, and flashing. It is alloyed with other metals to produce brass faucets and plumbing fittings. Aluminum is generally available in sheets for flashing. In addition, it is processed into extrusions for window and door frames, specialty hardware, roofing, and reflective insulation coverings.

These two metals have the distinct advantage that they will not rust like iron or steel. However, they must not be placed in contact with each other. Variations in temperature will create a small electrical current through the metal. This current causes an electrochemical reaction that can destroy the metals.

HARDWARE

Hardware is a term that applies to a large number of metal items used in the construction industry. In general, **hardware** consists of utility items that are used in putting together the finished building. These are items such as nails, bolts, framing anchors, and many other items such as ash doors and fireplace dampers. Rough hardware goes into the construction of the basic building.

There is a very large variety of specialty hardware designed for applications where a particular need has developed in the construction industry. Companies have been formed that specialize in the manufacture of these items. Fig. 11-24 shows some of these items.

Nails

Everyone has driven a nail into some construction material. Nails have been used as fastening devices in construction since the ancient Egyptians began using hand-wrought nails. In the late nineteenth century, a wire-nail machine was developed in France. With this breakthrough, and other advancements in technology, these fasteners were mass-produced. Hundreds of different sizes and types were developed, Fig. 10-25.

Sizes of nails

Nails of various sizes are used in almost all construction projects. Nail size refers to the nail's length and to the diameter of its shank. Its length "size" is indicated by the term **penny.** In any given penny size, nail length will always be the same, regardless of head or shank style. See

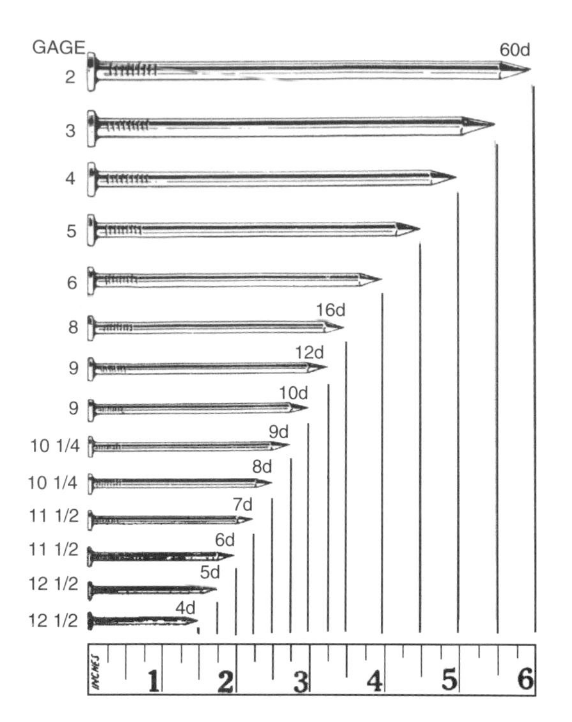

Fig. 11-24. Nails are a typical piece of hardware used in the construction industry. Shown are the sizes of common wire nails. (National Forest Products)

The shank diameter of nails varies with the length and style of each particular nail. See Fig. 11-27. The shank size is specified by a wire gauge size. Nails are made from wire cut to length, and the head is fashioned on the cut length. The nail retains the wire gauge size from which it was cut.

Types of nails

Nails are grouped into four general classes: common, box, casing, and finishing. Several subclasses of nails are to be found and used in construction. Nails have been developed for special purposes for use in special places under certain conditions.

Common and box nails. Common and box nails are for normal construction, particularly wood framing. Smooth box nails will have smaller diameters and thinner heads than common nails of the same penny size. Since their smaller diameter has less tendency to split the lumber, box nails are recommended for most uses.

Usually, 16d box nails are used in general framing. The 6d and 8d sizes are most frequently used for applications such as wall sheathing and roof sheathing. The size of the nail used depends on the thickness of the sheathing. Likewise, the length of the nail is determined by the thickness of lumber or plywood used. Two-thirds of the length of the nail should penetrate the second piece of wood.

WIRE NAIL CHART						
			COMMON NAIL	BOX NAIL	CASING NAIL	FINISHING NAIL
SIZE	mm	LENGTH INCHES	GAUGE NO.	GAUGE NO.	GAUGE NO.	GAUGE NO.
2d	25	1	15	15 1/2	15 1/2	16 1/2
4d	38	1 1/2	12 1/2	14	14	15
6d	50	2	11 1/2	12 1/2	12 1/2	13
8d	63	2 1/2	10 1/4	11 1/2	11 1/2	12 1/2
10d	76	3	9	10 1/2	10 1/2	11 1/2
12d	82	3 1/4	9	10 1/2	10 1/2	11 1/2
16d	88	3 1/2	8	10	10	11
20d	102	4	6	9	9	10
30d	114	4 1/2	5	9	9	---
40d	126	5	4	8	8	---

Fig. 11-25. This chart gives the size and gauge details on common, box, casing, and finishing nails.

Fig. 11-26. However, the diameter "size" of the nail shank *will* vary.

The "penny" designation is an old English term that may have come from the cost of 1000 nails of a particular size. Today, the term penny indicates only the length of the nail. In its abbreviated form, penny is written as "d" (8d, for example).

Scaffold nails. Scaffold, double-headed, or duplex nails can be used to save time and trouble in many operations where the fastener must be removed, Fig. 11-28. Use in the

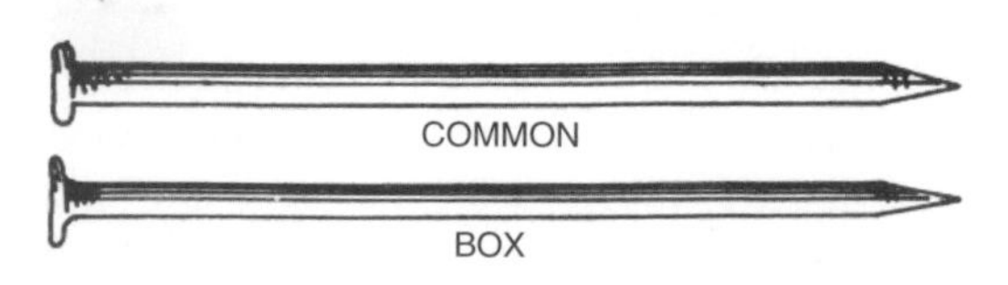

Fig. 11-26. Common and box nails of the same penny size have the same length. (American Plywood Assoc.)

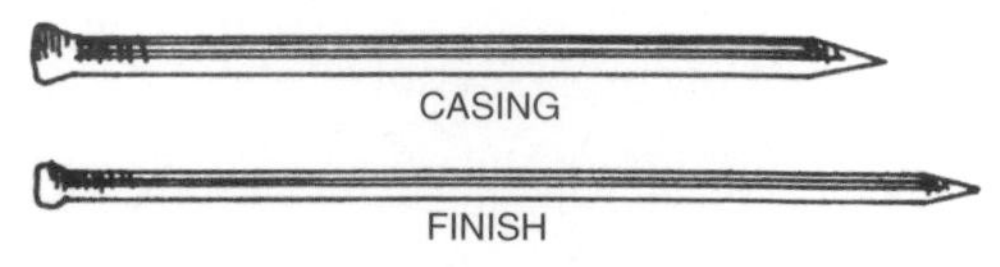

Fig. 11-27. The shank diameter varies with the style of the nail. This casing nail has a greater shank size than the longer finish nail beneath it.

scaffolding, bracing, concrete forms, and temporary layout work makes the double-headed nail indispensable for both strength and ease of removal. The 8d and 16d sizes are most commonly used for scaffolds and bracing and any other temporary fastening.

Nonstaining nails. Nails, because of their iron content, will cause a discoloration of the wood when exposed to the weather. For long service and freedom from staining, nonstaining nails are used. They are necessary when exterior exposure is combined with the need for good appearance in siding, fascias, soffits, exterior trim, and wood decks, Fig. 11-29. Galvanizing is the most common nail coating, and it offers good protection against staining. Nails are also made of aluminum, bronze, and stainless steel, but they are more expensive.

Bolts and screws

Bolts are used to fasten wood to wood in heavy timber construction. They are also used to secure wood to metal and wood to concrete. Bolts are manufactured in sizes from 1/4 in. to 1 1/4 in. (6 mm to 32 mm) in diameter and from 3/4 in. to 30 in. long (19 mm to 800 mm).

Bolts and anchor straps are used to secure wood sills to a concrete or masonry foundation, Fig. 11-30. Square and hexagonal shaped headed bolts, and square and hexagonal shaped nuts are the most common, Fig. 11-31. However, special bolts, like the anchor bolt with an L-shaped end are manufactured for specific jobs. The bent end forms a better bond with concrete and increases holding power.

Fig. 11-28. Scaffold nail. Note the double head for ease of removal. (American Plywood Assoc.)

Fig. 11-29. Construction featuring natural exposed wood, requires the use of nonstaining nails. (California Redwood Assoc.)

Wood screws are made to secure wood to wood and metal to wood. These fasteners are available with various shapes of heads: flat, round, oval, and pan. See Fig. 11-32. Usually screw heads are slotted or Phillips to fit the driver, Fig. 11-33. Most screws are made of steel. However, screws also are made of aluminum, brass, and steel.

The length of a wood screw is measured in inches. The diameter of the shank is designated by a wire gauge size. The larger the gauge number, the larger the shank diameter. A No. 0 wood screw is 0.06 in. (2 mm) in diameter. A No. 24 wood screw is 0.372 in. (9 mm) in diameter.

When purchasing screws, the following information is needed:

- Length.
- Diameter size.
- Head style and type.
- Material.

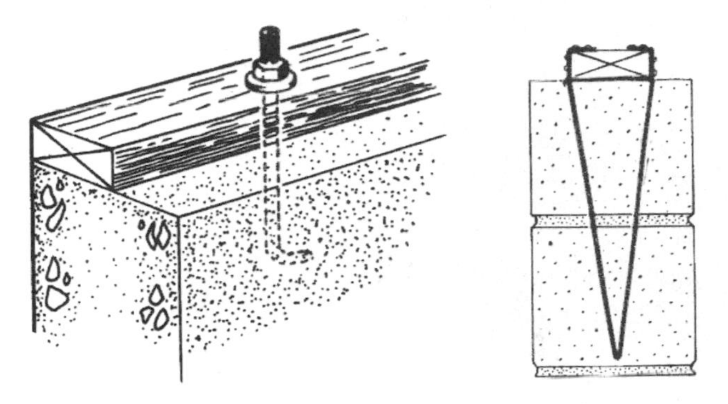

Fig. 11-30. Anchor devices hold a wood sill to a foundation. On the left is an L-shaped bolt. On the right is an anchor strap. (TECO)

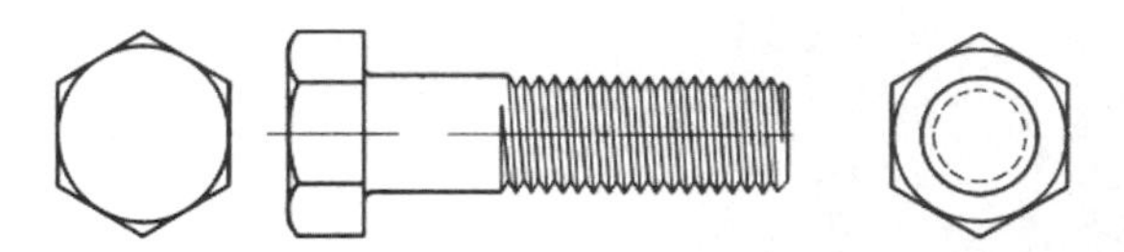

Fig. 11-31. Hexagonal head bolt and nut.

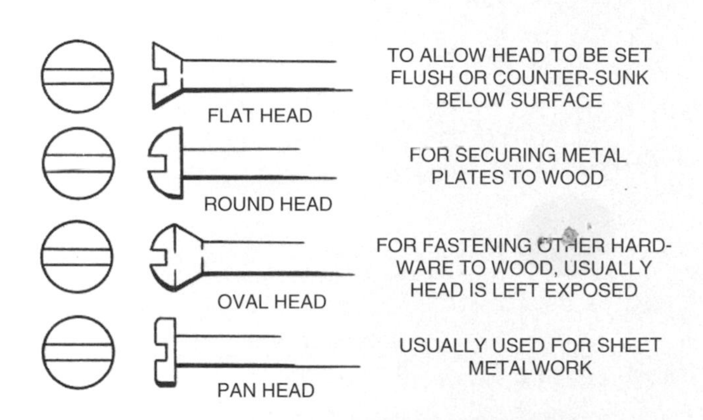

Fig. 11-32. Four screw head shapes and their usage.

CERAMICS

The term *ceramic* is a Greek word for potters clay. Today, **ceramic** is used to describe a wide range of materials that are inorganic and have a crystalline structure. Ceramics are excellent construction materials because they are not greatly affected by heat, weather, or chemicals. They have high melting points and can be easily formed into desired shapes. They are also extremely hard. They comprise construction materials such as cement, ceramic tile, stone, porcelain, glass, and bricks, Fig. 11-34.

CEMENT

Cement is the powdery substance used in conjunction with water, sand, and gravel to make concrete. Concrete is discussed along with other composites in the following chapter.

Knowledge of how to make hydraulic cement was lost in the middle ages. It was rediscovered in the 1750s by an Englishman, John Smeaton. In 1824 in England, Joseph Aspdin developed and patented a hydraulic cement that was superior to any other of the time. He called it **portland cement** because it looked like a grayish limestone mined on the isle of Portland. Aspdin found that if a carefully controlled mixture of limestone and clay was burned at a higher temperature than had been used before, the resulting cement had better hydraulic qualities. At this higher temperature, the limestone and clay fused together forming a new material.

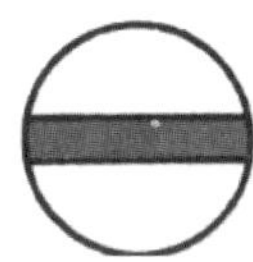

Fig. 11-33. Slotted and Phillips are the two most common head types for screws.

Portland cement today has the following basic composition:

Lime	60 - 65 percent
Silica	10 - 25 percent
Iron oxide	2 - 4 percent
Aluminum	5 - 10 percent

Most of the ingredients in portland cement are found freely in nature. Though, these ingredients cannot always be used in their natural state.

Manufacture of portland cement

Portland cement is manufactured by mixing and pulverizing the materials and then feeding them into a kiln. These kilns are steel cylinders 20 ft. (6 m) in diameter and 300 to 700 ft. (91 to 213 m) long, Fig. 11-35. Fuel is fed into one end of the kiln and ignited. As the raw material rolls in the rotating and slightly sloping cylinder, the heat fuses the raw material into green/black clinkers. The **clinkers** have specific chemical and physical properties which give cement its hydraulic characteristics. The clinkers, direct from kiln, are ground into fine powder. Some of the powder is shipped in bulk to ready-mix concrete plants. The remainder is packed into paper bags containing one cubic ft., 94 lb. (44 kg), of cement. These are shipped to building supply dealers.

STONE

The natural materials most readily available for construction have generally determined the character of the architecture produced by any culture. Stone has been used as a structural material, as a finish material, and as roofing throughout the centuries.

With the development of new materials and new structural techniques, stone is now used primarily as a decorative material. See Fig. 11-36. The range of color, texture, and finish is almost inexhaustible. Some variety of stone is available in nearly all areas of the world.

Fig. 11-34. Ceramic materials are used for an assortment of purposes. They can satisfy structural and decorative needs. Brick, glass, and tile are all ceramics. (Lowden, Lowden and Co., Velux-America Inc.)

Fig. 11-35. Enormous kilns, such as the one shown, mass-produce cement.

Classification of stone

Geologically, stone is classified as igneous, sedimentary, and metamorphic. The characteristics of each type have a definite bearing on their durability and use.

Igneous rock is produced by heat and pressure. Such rock is produced naturally through volcanic activity and the pressure exerted by shifting of the earth's surface.

Sedimentary rock is made up of silt (bits of eroded igneous rock) or the skeletal remains of marine life that have been deposited by ancient seas. Such rock always lie in layers called strata.

Metamorphic rock is formed by the gradual change in the character and structure of igneous and sedimentary rock. Three forces produce metamorphic rock:

- Pressure and heat.
- Water action, which dissolves and redeposits minerals.
- Action of hot magma on old rocks. (Magma is molten material from within the earth.)

All types of stone are used in buildings. The most important ones used in the construction industry are granite, sandstone, slate, limestone, and marble.

Fig. 11-36. Stone materials are used for decoration. Stone comes in an assortment of colors. (General Electric)

BRICK

Building **bricks** are small masonry blocks of inorganic nonmetallic material hardened by heat or chemical action, Fig. 11-37. Building bricks may be solid or may have core openings to reduce the weight. Bricks are produced in a wide variety of colors, shapes, and textures.

Classes of brick

Bricks are usually classified as:

- Adobe: made of natural, sun dried clays and a binder.
- Kiln burned: made of clay or shale and burned to harden them.
- Sand-lime: mixtures of sand and lime hardened under pressure and heat.
- Concrete: made of cement and aggregates.

Brick types and sizes

Bricks are available in many different sizes and types. The most commonly used bricks are made of solid clay. They are divided into subclasses such as common building brick, face brick, special brick, and custom brick.

CLAY BUILDING TILE

Structural clay tiles are made of burned clay. Larger than brick, they have large, cored openings and thin walls. Building clay tile has many uses in the construction industry. The hollow-celled units are produced in loadbearing and nonloadbearing forms. Structural clay tile may be used

Fig. 11-37 Bricks can be used as a structural material, or they may be used as an attractive siding like the one being applied to this house.

to fireproof structural steel or as a furring or filling material inside curtain walls.

Clay tile may be glazed with a transparent or colored glaze. This produces a glass-like, moisture-proof surface.

Terra-cotta

Terra-cotta, meaning "baked earth," is a nonloadbearing burned clay building unit. Similar to brick, it has been used for centuries as an ornamental substitute for stone. It is attractive as floor tile and as a veneer for inside and outside walls.

GYPSUM BLOCK

Gypsum blocks are hollow units made of gypsum and a binder of vegetable fiber, mineral fiber, or wood chips. Never used where they would be subjected to moisture, they are suitable for nonloadbearing, fire resistant interior partitions. Another use is as a fire proofing material around structural steel beams and columns. They are usually finished with a plaster coating. Easily cut and light weight, gypsum blocks are relatively inexpensive to install.

GLASS

Today, windows are made of a transparent material called glass. Actually, glass was used as a clear, protective glazing material for windows as early as the first century AD.

The first important architectural use of glass came in the Middle Ages in the form of jeweled stained glass windows installed in the Gothic cathedrals of Europe. Then, centuries passed before new concepts in building design offered a vastly expanded role for glass.

Only in the Twentieth Century, with new and improved methods for making glass, has the potential for flat glass been tapped for the architecture/construction field, Fig. 11-38. Today, glass is a multipurpose design material and a functional building material used in numerous places to create better and brighter living conditions.

Fig. 11-38 Glass decorates the inside of a home with outdoor landscaping. (Four Seasons Greenhouses)

Special glass

Special glass is available for special purposes. It is designed with a given property for a specific reason. Some possibilities are to increase the impact resistance, to reduce heat transmission, color may be added, or its reflection qualities could be increased. These are some of the many properties that can be controlled during the manufacture of glass.

Tempered glass. Heating glass almost to its melting point, then chilling it rapidly produces **tempered glass.** This process creates a glass that is three to five times as strong as ordinary glass.

Insulating glass. The increased demand for weather-conditioned buildings has led to the development of double-glazed and even triple-glazed glass. Often referred to as **insulating glass,** these units consist of two or more sheets of glass separated by an air space, Fig. 11-39. The air between the sheets of glass is free of moisture and greatly slows down the transfer of heat. The insulating air also helps in overcoming the problem of moisture condensation that often forms on the inside of glass in cold weather.

Low heat transmission glass. One specialized glass, **low heat transmission glass,** is treated on one side with a very thin metallic coating. This does not allow as much heat to be transferred through it. In this way it helps reduce heat loss in winter and heat gain in the summer. The metallic film is so thin that you can still see through the glass.

Glass block. Two glass shells can be fused together to form **glass blocks.** The shell is airtight to hold a partial vacuum inside. This partial vacuum provides good insula-

Fig. 11-39. Fiberglass insulates the walls of homes effectively, but windows can be a source of tremendous heat loss. Insulated glass windows protect against these losses and save energy. (Lowden, Lowden and Co.)

tion. Additional advantages are low maintenance and controlled daylight. Glass block may be used as windows, interior partitions, screens, or entire walls.

SUMMARY

Today's construction activities require a great variety of materials. These materials can be classified as gases, liquids, or solids. Because gases and liquids must be contained in pipes or containers, they are often overlooked as construction materials. The most common construction materials are solids such as wood, concrete, steel, and asphalt. Solid materials can be further divided into polymers, metals, ceramics, and composites.

Most construction materials have to be processed into a standard shape, size, and quality product before they can be used in the construction industry. This processing provides the architect and engineer with a large variety of materials from which to choose. This selection provides an almost infinite number of possibilities to be used to meet the needs of the construction industry.

KEY WORDS AND TERMS

All of the following words and terms have been used in the chapter. Do you know their meaning?

Adobe
Alloy
Asphalt
Brick
Cant
Ceramic
Clinker
Concrete
Equilibrium
Ferrous metal
Flat sawed
Galvanizing
Glass block
Graded
Gypsum block
Hardware
Hardwood
Igneous rock
Insulating glass
Kiln
Kiln burned
Low heat transmission glass
Metamorphic rock
Nonferrous metal
Penny
Pig iron
Plain sawed
Plastic
Polymers
Portland cement
Primary processing
Quarter sawed
Sand-lime
Seamless pipe
Seasoning
Secondary processing
Sedimentary rock
Softwood
Tempered glass
Terra-cotta
Welded pipe

TEST YOUR KNOWLEDGE

Do not write in this book. Please write your answers on a separate sheet of paper.

1. List the three basic classifications for all construction materials.

2. List several examples for each of the above material classifications.
3. Trace the production of lumber from its raw material to its final form. Identify the primary and secondary processing steps.
4. List three materials that are classified as polymers.
5. List three materials that are classified as metals.
6. List five materials that are classified as ceramics.
7. What type of rock is made up of silt and the skeletal remains of marine life?
8. Give three special properties that can be added to glass.

APPLYING YOUR KNOWLEDGE

1. Design a display that shows the different construction materials used in building construction.
2. Make a display of wood materials showing the raw material and then examples of processed wood used in a variety of wood products.
3. Set up a demonstration for your class to show the difference between ferrous and nonferrous metals.
4. With the help of your teacher, design and build a material testing device to determine the breaking strength of different samples of wood.

This stockpile of materials is waiting to be wholesaled to contractors.

12

CHAPTER

MATERIALS II - COMPOSITES

The information provided in this chapter will enable you to:
- *Explain the difference between a composite material and other construction materials.*
- *Describe the makeup of plywood.*
- *Explain why plywood is stronger than solid wood of the same thickness.*
- *List and describe four types of pressed board materials.*
- *List the ingredients in concrete.*
- *Explain the importance of the proportions of the mix in concrete.*

Composites are made up of two or more materials that are bonded together with some adhesive. By making one material from a combination of other materials, the resulting composite can combine the good qualities of all the materials used. Plywood is an example of this. Wood splits easily with the grain. In the composite, plywood, the wood grain of the plywood components are run perpendicular, at 90 degree angles, to the components on either side of them. This makes a material that is very strong in all directions.

Composites can be made up of natural materials, such as wood, or can be made up of synthetics, such as plastic. Some composites are a combination of natural and synthetic materials.

PLYWOOD

Ancient Egyptians cut thin strips of wood and formed plywood sheets as long ago as 3000 BC. Sheets discovered in excavations have survived for centuries.

Today the use of plywood is very much dependent on the improvements of adhesives. **Plywood** is construction material made of thin sheets of wood (veneers) bonded together with glue. Each sheet of veneer is assembled so that its grain is at a right angle to the layer next to it, Fig. 12-1. Plywood is stronger than solid wood of the same thickness.

Plywood resists splitting, checking, and does not swell or shrink as much as solid wood. Plywood is also less likely to warp and twist than solid wood of the same size. It is easy to work with ordinary tools. This helps speed up construction.

PLYWOOD SIZES

Plywood is generally produced in 4 ft. by 8 ft. (122 cm by 244 cm) sheets. Standard thicknesses are 1/4, 5/16, 1/2, 5/8, 3/4, 1, and 1 1/8 in. (6, 8, 12, 16, 18, 25, and 28 mm). Plywood sheets have three, five, or seven plies, or veneers. Having an odd number of sheets leaves the grain of both outside layers running the same direction.

MANUFACTURE OF PLYWOOD

Cut veneers are dried and trimmed into standard size sheets as shown in Fig. 12-2. Defects like small knots are cut out and patched with thin wooden plugs.

The inside of the plywood sheet, called the **core,** is one of three types of core buildup.
- Several layers of veneer.
- Solid wood.
- A composition made up of wood particles.

Fig. 12-3 illustrates two types of core.

Plywood that has a core of solid wood is called **lumber core plywood.** It is used where low warpage is needed, such as in cabinet doors.

The composite or particleboard cores are exceptionally warp free. Their dimensional stability makes them valuable for sliding and hinged cabinet doors. Wall paneling may also have particleboard cores.

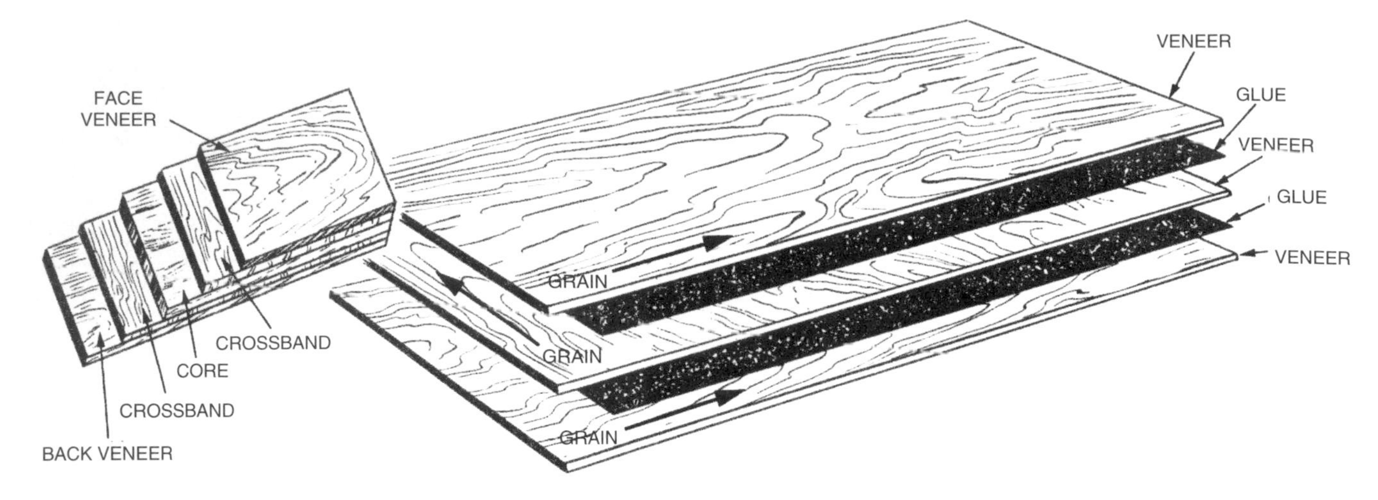

Fig. 12-1. Each plywood layer has the grain running at 90 degrees to the next layer. Such sheets are stronger than solid wood of the same thickness. (Frank Paxton Lumber Co.)

Fig. 12-2. The operator grades and cuts dried veneer after it comes out of the in-line drying machine. (Weyerhaeuser Co.)

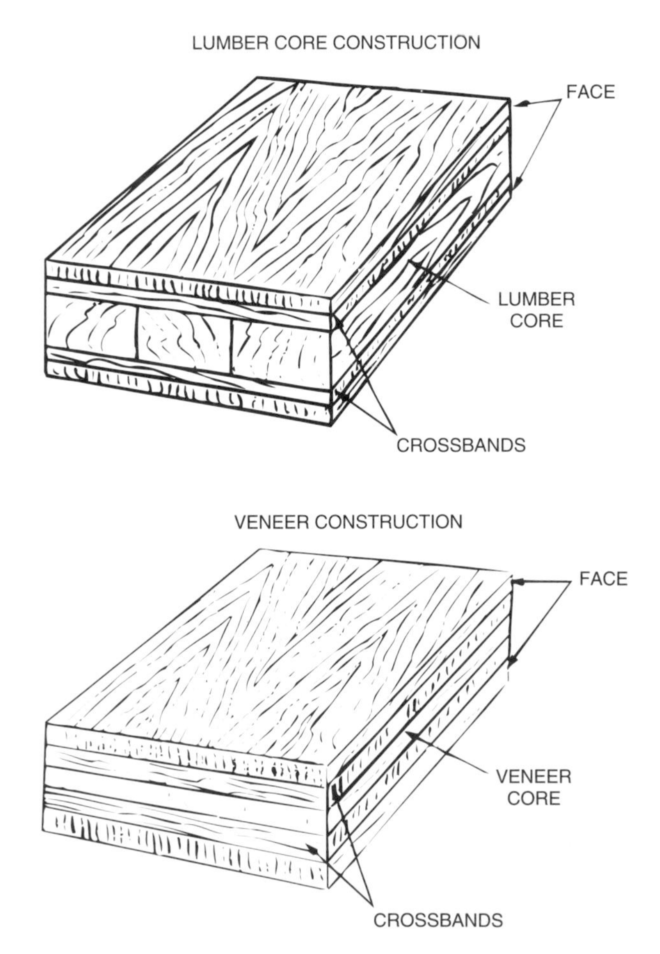

Fig. 12-3. Two methods of core preparation for plywood.

TYPES OF PLYWOOD

Plywood is either exterior or interior type. It depends on the type of glue used to bond the plies together.

Exterior plywood will hold its original form, shape, and strength with repeated wetting and drying. The plywood is made with a hot-pressed adhesive that is insoluble, so it is not affected by water.

Interior plywood is bonded together with glue that is not waterproof. However, such plywood will maintain strength when subjected to occasional moisture.

HARDWOOD PLYWOOD

The term hardwood plywood covers a wide range of products. Hardwood plywood is manufactured with a core of softwood and outer layers of hardwood veneer.

Many expensive and rare hardwoods have color, figure, or grain characteristics that make them highly prized for paneling, cabinetwork, and furniture.

PLYWOOD GRADES

Each layer of plywood veneer is graded for allowable defects, Fig. 12-4. Grades are based on the number of defects present (knots, splits, checks, stains, and open sections.) Repairs are permitted in some grades of veneer.

Softwood plywood is graded for appearance. Sanded plywood is graded primarily by the appearance of its front and back faces. The grade is designated by a letter.

- N–Special order cabinet grade veneer 100 percent heartwood, free of any knots.
- A–Highest standard grade. No open defects.
- B–Presents a solid face, tight knots, repairs allowable.
- C–Plugged, splits limited to 1/8 in. (3 mm) wide, holes limited to 1/4 in. by 1/2 in.
- D–Poorest grade, open space, no repairs.

The grade of the panel is indicated by two letters. An A-C panel is one having an A-grade on one face and a C-grade on the second face.

In exterior plywood, the core plies must be grade C or better. Interior plywood panels may have D grade core plies.

Construction plywood grades

Unsanded plywood panels used for sheathing, subfloors, underlayment under resilient floor coverings, and structural applications are graded according to strength. These grades have been established to meet the needs of construction. The grade is stamped on every panel, Fig. 12-5.

Most softwood is Douglas Fir, although Southern Pine, Western Larch, Western Hemlock, White Fir, Cedar, and other species are used. The American Plywood Association has assembled several species into groups according to their strength and stiffness. See Fig. 12-6.

PRESSED BOARDS

Pressed boards may be composed of any vegetable, mineral, or synthetic fiber mixed with a binder and pressed into a flat sheet. Pressed boards are widely used in the construction industry as insulation, sheathing, and finished panels.

The pressed boards may be soft-textured panels of loosely held fibers. These have little strength but good insulating properties. Other panels may be pressed into dense sheets that are water resistant or flame retarding.

FIBERBOARD

Cane fiberboard is made from the stalks of sugar cane. Caneboard, known as Celotex, is used for insulation in walls and roofs. Given a finish, it is used for bulletin boards.

In order to increase its water resistance, the board is impregnated with asphalt. This asphalt impregnated board is also used for roof insulation or in locations where moisture is present.

PARTICLEBOARD

Some panels are made of wood fibers that are bonded with glue and pressed into sheets under high pressure and temperatures, Fig. 12-7. They are called **particleboards.** The small particles or chips are positioned in a random crisscross arrangement to form boards of outstanding uniformity in thickness and dimensional stability. Particleboard may be used for cabinetwork and underlayment. It can be engineered for strength and durability. It is ideal backing material for plastic laminates and wood veneers.

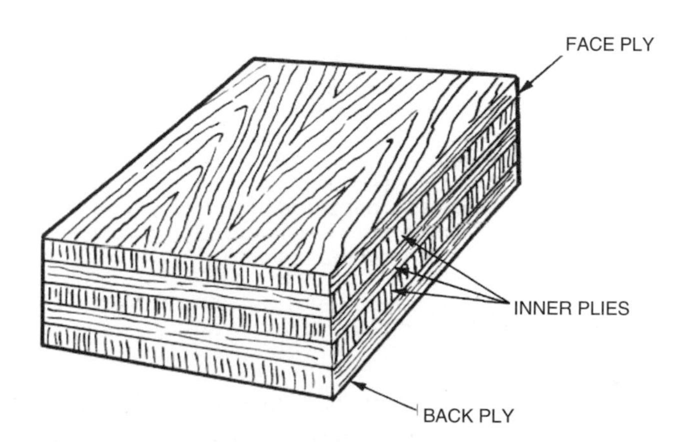

INTERIOR GRADES			
PANEL GRADE DESIGNATION	FACE	MINIMUM BACK	INNER PLY
A–A	A	AA	D
A–B	A	B	D
A–C	A	C	D
A–D	A	D	D
B–B	B	B	D
B–C	B	C	D
EXTERIOR GRADES			
A–A	A	A	C
A–B	A	B	C
A–C	A	C	C

Fig. 12-4. Softwood plywood surface grades in relation to inner ply grades.

Typical APA Performance-Rated Panel Registered Trademarks

APA RATED SHEATHING

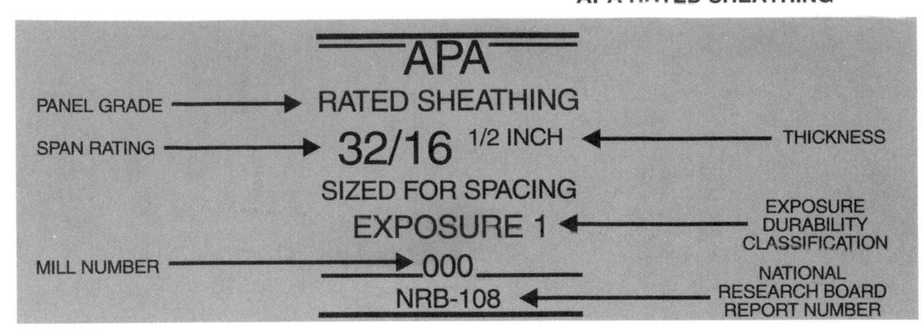

Specially designed for subflooring and wall and roof sheathing. Also good for broad range of other construction and industrial applications. Can be manufactured as conventional veneered plywood, as a composite, or as a nonveneered panel. For special engineered applications, veneered panels conforming to PS 1 may be required. EXPOSURE DURABILITY CLASSIFICATIONS: Exterior, Exposure 1, Exposure 2. COMMON THICKNESSES: 5/16, 3/8, 7/16, 1/2, 5/8, 3/4.

APA RATED STURD-I-FLOOR

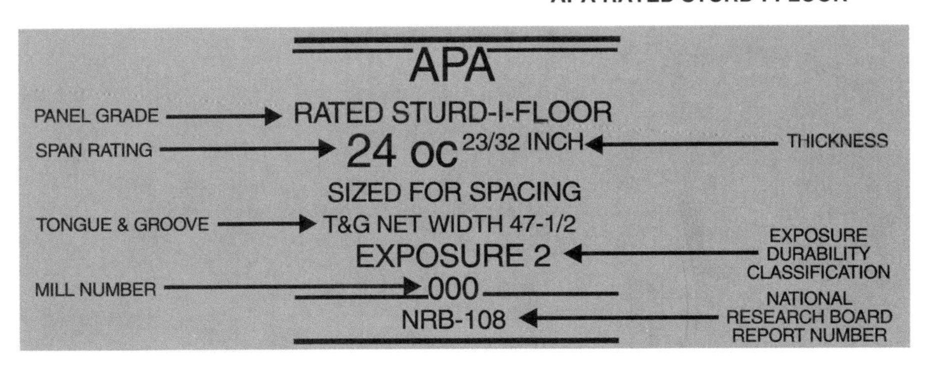

Specially designed as combination subfloor-underlayment. Provides smooth surface for application of resilient floor coverings and possesses high concentrated and impact load resistance. Can be manufactured as conventional veneered plywood, as a composite, or as a nonveneered panel. Available square edge or tongue-and-groove. EXPOSURE DURABILITY CLASSIFICATIONS: Exterior, Exposure 1, Exposure 2. COMMON THICKNESSES: 19/32, 5/8, 23/32, 3/4.

Fig. 12-5. The American Plywood Association, working with the U.S. Department of Commerce, has developed specifications for structural plywood. Companies meeting the standards are permitted to carry the trademark, several of which are shown. It is an assurance of quality for the builder and owner. (American Plywood Assoc.)

Group 1	Group 2		Group 3	Group 4	Group 5
Birch	Cedar, Port Orford	Maple, Black	Alder, Red	Aspen	Fir, Balsam
Yellow	Douglas Fir 2	Meranti	Cedar	Bigtooth	Poplar, Balsam
Sweet	Fir	Mengkulang	Alaska	Quaking	
Douglas Fir 1	California Red	Pine	Pine	Birch, Paper	
Larch, Western	Grand	Pond	Jack	Cedar	
Maple, Sugar	Noble	Red	Lodgepole	Incense	
Pine, Caribbean	Pacific Silver	Western White	Ponderosa	Western Red	
Pine, Southern	White	Spruce, Sitka	Spruce	Fir, Subalpine	
Loblolly	Hemlock, Western	Sweet Gum	Redwood	Hemlock, Eastern	
Longleaf	Lauan	Tamarack	Spruce	Pine	
Shortleaf	Red		Black	Sugar	
Slash	Tangile		Red	Eastern White	
Tanoak	White		White	*Poplar, Western	
	Almon			Spruce, Englemann	
	Bagtikan				*Black Cottonwood

Fig. 12-6. Classification of softwood species by groups.

HARDBOARD

Hardboard panels are made of wood chips that have been exploded, leaving the fibers and lignin (nature's basic building block of wood). These are fused under heat and pressure into a hard, long-lasting board. Hardboard is manufactured in thicknesses of 1/10 to 5/16 in. (2.5 to 8 mm) with the faces smooth or in a screen pattern. Various additives may be used to give it special characteristics. Hardboard may be used for interior purposes. Its smooth surface is ideal for painting or veneering.

Embossed vinyl film is sometimes attached to a hardboard back producing a material with a wood-like appearance. A wood grain resemblance is printed on while the texture of wood pores is pressed into the surface of the film. The result is a durable surface resistant to wear and stain.

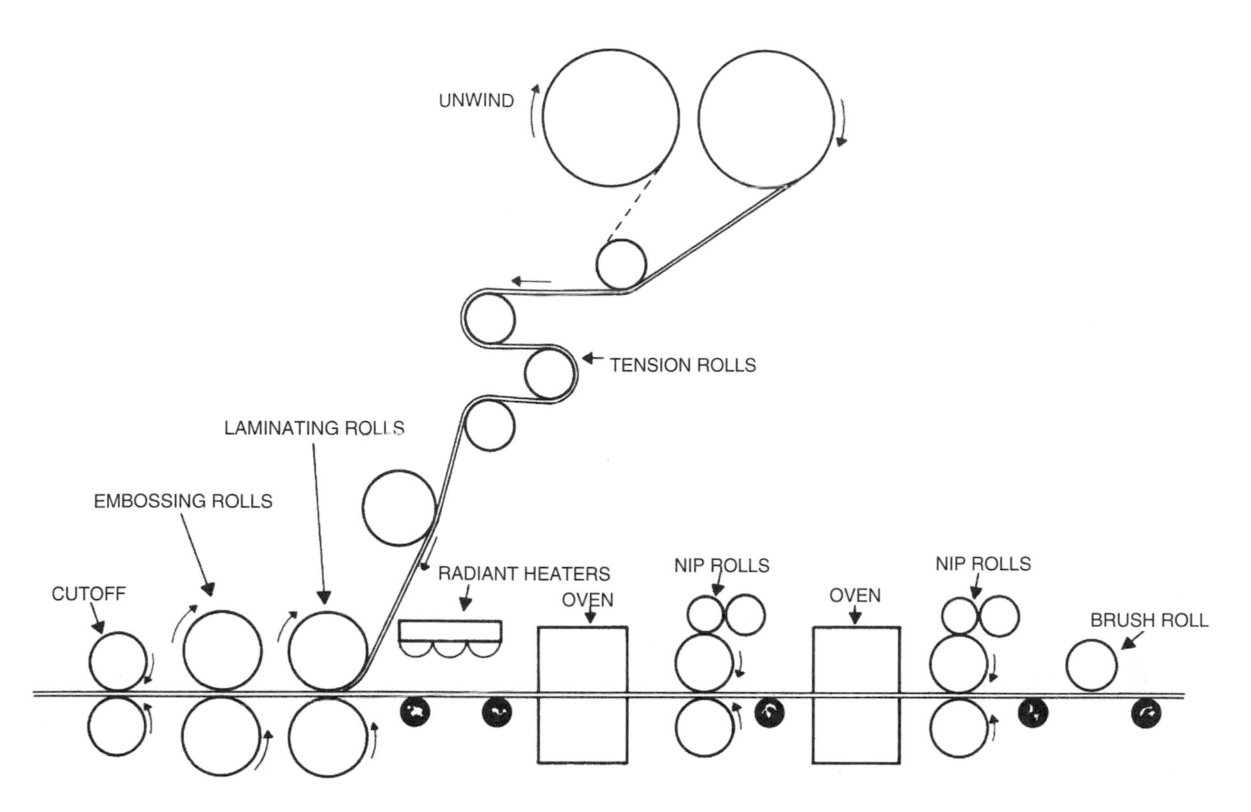

Fig. 12-7. Production steps for making particleboard. (National Particleboard Assoc.)

WAFERBOARD

Waferwood, or **waferboard,** is produced from flakes of wood that are about 1 1/2 in. square. They are bonded into sheets of varying thicknesses with a waterproof adhesive. The 4 x 8 sheets are available in seven thicknesses from 1/4 in. through 3/4 in. Waferboard is suitable for roof and floor sheathing and side walls and soffit coverings, Fig. 12-8. In some instances it is also used as a finishing material for walls and ceilings.

Fig. 12-8. The large wood flakes are very identifiable in waferboard sheathing. (Waferboard Assoc.)

CONCRETE

Concrete has long been used as a construction material. It first became popular with the Romans and is widely used today. See Fig. 12-9.

Around 100 BC, Roman builders developed an excellent concrete that allowed them to erect vast structures. Some of these buildings endure even to this day, though in various states of ruin.

The main ingredient of their concrete was a special volcanic ash. It was mined on the slopes of Mt. Vesuvius near the village of Pozzuoli.

When this ash was mixed with limestone, burned, and combined with water, it became a very strong cement. It was then mixed with small pieces of rock to form a strong and lasting concrete. Roman builders also found that this cement would harden under water. We call this **hydraulic cement.**

Today, the characteristics of concrete vary widely. Qualities of the mix can be changed depending on the composition of the aggregates (sand and stones), chemical composition, and physical properties of the cement. But in every instance, concrete is a mixture of cement, water, sand, gravel, and stones of various sizes. See Fig. 12-10.

The term cement is given to any material that will bond two or more nonadhesive (nonsticking) substances together. In concrete construction, the term means portland cement.

HARDENING AND SETTING

When portland cement is mixed with enough water and left undisturbed, the mixture loses its fluid state and becomes a solid. Concrete does not harden by drying. It hardens because of a chemical reaction. Water and cement combine chemically to form a new compound. This process is called **setting.**

Initial setting may take place after only a few minutes, or it may take several hours. When the cement has been combined with aggregates and has set, it will continue to harden (gain strength) for several months or even years.

Fig. 12-9. Concrete has been used in the making of roads for thousands of years.

WATER RATIO

If too little water is used in the mix, some particles of cement will not be chemically changed. If too much water is used, the excess water may be trapped in the mix. In either case, the concrete will be weakened. The temperature of the materials and moisture in the air (humidity) affect the exact amount of water the mixture needs.

A 94 lb. (44 kg) bag (one cubic ft.) of cement requires 2.5 to 3 gal. (9 to 11 liters) of water for a complete chemical combination of materials. The use of exactly this amount is not practical for field conditions. Usually, 4 to 8 gal. (15 to 30 liters) must be used for each bag of cement. The extra water serves as a lubricant to carry the cement into small pores of the aggregate.

DESIGN OF MIX

The water/cement ratio of a given concrete determines its strength to a great extent. However, to produce concrete that is economical as well as strong, the proper aggregate type, grade, size, and proportions must be used.

Proper portions of the various sizes of aggregates can be determined by trial mixes, either in the field or in the laboratory. A concrete with too much coarse aggregate may contain too many voids. A mixture with too much sand may be smooth but is not economical. It will not attain its greatest strength. See Fig. 12-11.

Proportions of the mix

In the past, the proportion of cement, sand, and aggregate was thought to control the various strengths of con-

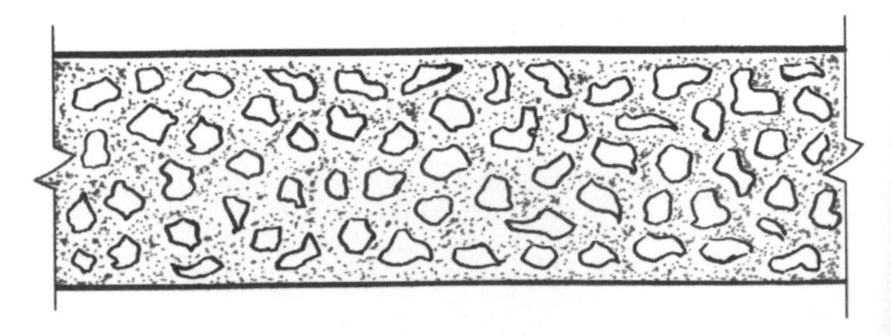

Fig. 12-10. Sand, gravel, and stones of various sizes are dispersed through a section of concrete. Cement bonds the aggregate together.

Fig. 12-11. Too much coarse aggregate can leave weakening voids in concrete. Too little aggregate is not cost effective. Quantity and size of aggregate must be carefully chosen. (Lowden, Lowden and Co.)

crete. Little or no importance was placed on the water/cement ratio. The water/cement ratio has been found to govern much of the strength of finished concrete. A 1:3:4 mix contains, by volume, one part cement, three parts sand, and four parts rock. Enough water is added to this dry mix to make it flow into the forms.

Fig. 12-12. When large quantities of concrete are required, the concrete will be mixed in a central batching plant. The concrete is then shipped to the site ready to pour.

Mixing of concrete

Another factor in the workability and strength of concrete is the method used to mix the ingredients. It is essential to mix thoroughly. Sometimes concrete is mixed on the site. In other cases, the concrete is mixed at a central batching plant and is hauled to the site in a ready-mix truck, Fig. 12-12.

The time required to mix the ingredients varies. It depends upon the size and efficiency of the mixer. Mixing time should not be less than one minute for concrete of medium consistency. A concrete mix will not remain usable as long during hot weather.

Mixing is calculated from the time all solid materials are mixed together. Prolonged mixing will not greatly add to the strength of concrete as long as additional water is not added to increase the slump.

Slump test

Consistency of fresh concrete is measured in terms of its slump. **Slump** is the distance the wet concrete will settle under its own weight. See Fig. 12-13.

To test the slump of concrete, samples are taken directly from the mixer and placed in a cone-shaped container. The cone is turned upside down into a pan to release the concrete. No longer supported by the cone, the wet concrete settles or slumps. If the concrete sample does not have the specified slump, the mix must be redesigned. Water should never be added to the slump. It will reduce the resulting strength.

The minimum slump recommended for various types of work are:

- Mass concrete for slabs on soil, 1 to 3 in. (2.5 to 7.5 cm).
- Reinforced foundation walls, 1 to 3 in. (2.5 to 7.5 cm).
- Columns for building, 1 to 4 in. (2.5 to 10 cm).
- Beams, 1 to 4 in. (2.5 to 10 cm).

AGGREGATES

Aggregates are sand and larger particles of stone, rock, or other material that make up about 65 to 80 percent of the total volume of concrete. They do not react chemically with cement and water. Sizes range from fine sand to pieces 1 1/2 in. (3.5 cm) in diameter and larger. The quality of the concrete is affected in several ways by the aggregate:

- The strength of concrete is no greater than the strength of the aggregate.
- The size and shape of aggregate affects the flowability of the concrete.

Fig. 12-13. After the wet concrete is dumped from the cone and given time to settle, the distance the concrete has settled is measured. This is the slump of the sample.

REINFORCEMENT

Concrete has great compression strength (strength from a squeezing action). A square inch of concrete can be designed to support loads of 10,000 lb. (4500 kg) or more.

However, concrete has little tensile strength (resistance to pulling action). In **reinforced concrete,** steel and concrete are combined to take advantage of the high compressive strength of concrete and the high tensile strength of steel. Concrete is cast around reinforcement steel. As it hardens, it grips the steel to form a bond with it. This bond becomes stronger as the concrete hardens.

PRECAST CONCRETE

If concrete members for construction can be cast on the ground and then lifted or tilted up into position, the cost of concrete work is greatly reduced. Work done at ground level is simpler and less expensive.

Intricate shapes with various textures and colors may be cast in wood, metal, plastic, or plaster forms. The best quality concrete is achieved in a precast concrete manufacturing yard. See Fig. 12-14. These precast units are transported to the construction site and hoisted into place, as in Fig. 12-15.

Slabs can also be precast for wall sections, then tilted up into position as shown in Fig. 12-16. Floors and roof slabs are cast at ground level and then lifted into position.

Fig. 12-14. Concrete can be cast into an infinite number of interestingly shaped and sized pieces.

Fig. 12-15. These precast beams are being hoisted into position on bridge piers. (Northwest Engineering)

PRESTRESSED CONCRETE

Prestressed concrete products are usually precast. A prestressed concrete beam, for instance, is a structure of

Fig. 12-16. Tilt up construction from start to finish. A–Forms for the wall sections are put up. B–Concrete is placed into the forms. C–The concrete sections are removed from the forms. D–The precast sections are stacked and stored. E–Wall sections are lifted into place. F–Several sections are joined together to form a structure. (Morrison Knudsen Corp.)

concrete and reinforced steel. The steel is placed in tension by applying hydraulic pressure before concrete is poured. After the concrete is cured, tension is released causing the concrete to be squeezed together.

A prestressed structural unit may be designed to use only half the concrete and a quarter of the steel used in the conventional structural unit.

CONCRETE BLOCK

Hollow masonry units of portland cement, sand, and fine gravel aggregates are called **concrete blocks,** Fig. 12-17. Concrete blocks are used for interior and exterior load-bearing and nonload-bearing walls, partitions, and backing for veneer. The weight, color, and texture of a concrete block depends largely on the type of aggregate used in its manufacture. Blocks made with sand and gravel weigh from 40 to 45 lb. (18 to 20 kg) for each 8 in. by 8 in. by 16 in. unit. Blocks are strong and durable. Standard units have the typical light gray color of concrete.

DETAILED BLOCKS

Detailed blocks have a patterned face. Some have vertical and horizontal grooves to simulate brick. Others have triangular, or rectangular indented areas. Such block may be laid in different directions to create interesting patterns.

Fig. 12-17. Concrete blocks have a multitude of load-bearing uses in home construction.

FUTURE COMPOSITES

New and innovative composites are always on the horizon. People are looking to create cheaper, lighter, and stronger materials for different jobs. Many new ceramics and plastics are being produced for all sorts of construction tasks.

A new factor that is also influencing composites is the need to recycle our waste materials. One material that has been explored is called chunkrete. **Chunkrete** is a concrete that uses large chips of wood in the aggregate. There are many sources of waste wood. Material from demolished houses, old pallets, and tree waste often head into landfills.

A massive quantity of new concrete is poured every year. Some of that material could be replaced with a lighter concrete such as chunkrete. In addition to the recycling advantages, chunkrete is lighter in weight than conventional concrete. This reduces the transportation cost of the material. Chunkrete is also better at sound and vibration reduction than common concrete.

Chunkrete is not as strong as a concrete with a stone aggregate and cannot support as heavy a load. Yet when concrete fails structurally, it can bear no more weight. The wood in chunkrete allows the substance to retain some load-bearing ability. See Fig. 12-18. Various uses for chunkrete have been researched. Some suggested uses for chunkrete are in highways, airport pavements, sidewalks, and many landscaping projects.

There are many other composites that are being researched at this time. Some research will not produce results for many years. Other projects will be producing new products just around the corner. What new materials can you see arising in your future?

Fig. 12-18. The two samples shown were compressed until they suffered failure. The more conventional concrete (bottom) has greater compressive strength. But when it failed, it failed completely. The chunkrete (top) still retained significant compressive strength after compressive failure.

SUMMARY

Many important construction materials are composites. A composite material is one that is made up of two or more materials bonded together. Plywood and concrete are two of the most important composites.

Plywood has been around for thousands of years. Each layer in plywood has the grain running at a right angle to the layers next to it. This makes plywood very strong. Plywood is one of several composites made of wood glued together. Fiberboard, particleboard, hardboard, and waferboard are other useful wood composites.

Concrete is made up of an aggregate mixed with cement. The water ratio is very important to produce a strong material. Concrete can be made to support even greater stresses through reinforcement and prestressing techniques. Chunkrete is a new composite that replaces the stone aggregate of concrete with wood chunks.

KEY WORDS AND TERMS

All of the following words and terms have been used in this chapter. Do you know their meaning?

Aggregate
Chunkrete
Composite
Concrete blocks
Core
Detailed blocks
Exterior plywood
Hardboard
Hydraulic cement
Interior plywood
Lumber core plywood
Particleboard
Plywood
Pressed boards
Prestressed concrete
Reinforced concrete

Setting
Slump
Waferboard

TEST YOUR KNOWLEDGE

Do not write in this book. Please write your answers on a separate sheet of paper.

1. Veneers of plywood are held together with glue. True or False.
2. Interior plywood is designed to be waterproof. True or False.
3. List each of the grades for plywood in the order of worst to best.
4. Describe briefly the materials that form hardboard.
5. What problems can occur if to much water is used in the mix for concrete?
6. Define slump.
7. What is the difference between precast concrete and prestressed concrete?

APPLYING YOUR KNOWLEDGE

1. With the help of your teacher, design and build a testing device to test the breaking strength of plywood versus a solid piece of wood of the same thickness.
2. Take a trip to a lumberyard. Examine the variety of pressed boards. Note the prices and sizes of the boards available. Ask a store worker where each construction material would be used.
3. Mix up several small samples of concrete. Vary the water ratio and the size of the aggregate used. Perform slump tests on each sample.

Concrete block foundations are common in residential construction.

13

CHAPTER

CAPITAL IN CONSTRUCTION

The information provided in this chapter will enable you to:
- *Describe two different types of contractors.*
- *List four different kinds of mortgages.*
- *Calculate simple interest on a loan.*
- *List the four provisions common to most contracts.*
- *Explain the process of bidding.*
- *Define the terms: bonds, stocks, appropriations, liens, foreclosure, insurance, and workmen's compensation.*

Fig. 13-1. A new project may look wonderful in idea form. Without financial backing, it will remain just an idea.

Capital is another of the seven major resources in construction that needs to be carefully considered. Costly resources are needed to construct a project of any size. A considerable amount of money is needed to pay for the labor force, the materials, and the energy required to complete a construction project. Because of this great need for capital, most individuals and companies secure a loan from a lending institution to help pay for the construction.

A company may choose to use their capital many ways. The company may purchase materials or equipment rather than invest a large amount in a building. The financial world is very complicated and requires much study to completely understand the system. In this chapter on capital, we will discuss financial resources and how important capital is to modern construction.

FINANCING

When a building project is initiated, Fig. 13-1, the first questions to arise are: "How much will it cost?" and "Where will the money come from?" Money is obtained through a number of means that are collectively known as **financing.**

LOANS

Seldom is a corporation or family able to pay *cash* for a new building or home. Therefore, a source of money other than the regular operation budget is required to finance the structure. In most cases, a loan is secured to pay for the cost of construction. With a **loan,** the occupant pays a periodic fee, called interest, for the privilege of using these funds until the loan is repaid.

When a loan is required for a construction project, a lending institution is contacted during the early planning stages, Fig. 13-2. If the loan is approved, most lending institutions issue a **letter of commitment.** This is a legal confirmation of a loan. This written commitment guarantees the owner that the money will be available at the time and phase of construction it is needed.

The letter of commitment usually states the loan amount, term (length of time), interest rate, and the time limit for accepting the loan. In addition, the letter contains any other conditions that are agreeable to the lending institution and borrower.

Fig. 13-2. Financing for a construction project can be arranged through a bank.

Mortgages

One method of securing real estate loans is with a mortgage. A **mortgage** is a contract that pledges the property as loan security, Fig. 13-3. It is a way of guaranteeing repayment of the loan while the buyer uses the property, Fig. 13-4. Possession is

Fig. 13-4. A mortgage allows families to enjoy a home while buying it over a period of time.

retained as long as the buyer makes the monthly payments promised in the real estate mortgage note, Fig. 13-5.

For years there were three basic types of home mortgages:

- Conventional loans from banks or savings and loans.
- Loans insured and guaranteed by the Federal Housing Administration.

ILL. S. & L. LEAGUE, Form No. 1 (Short)

MORTGAGE

THIS INDENTURE WITESSETH: That the undersigned ______________________

____

____

of the ______________ County of ______________, State of Illinois, hereinafter referred to as the Mortgagor, does hereby Mortgage and Warrant to

____

a corporation organized and existing under the laws of the ______________, hereinafter

referred to as the Mortgagee, the following real estate, situated in the County of ______________ in the State of Illinois, to wit:

...ereby secured, or (b) preparations for the ... foreclosure ... after the accrual of the right to foreclose, whether or not actually com...ced; or (c) preparaxions for the defense of or intervention in any suit or proceeding or any threatened or contemplated suit or proceeding which might affect the premises or the security hereof, whether or not actually commenced. In the event of a foreclosure sale of said premises there shall first be paid out of the proceeds thereof all of the aforesaid items.

IN WITNESS WHEREOF, the undersigned have hereunto set their hands and seals this ______________

day of ______________, A. D. 19___.

______________ (SEAL) ______________ (SEAL)

______________ (SEAL) ______________ (SEAL)

STATE OF ILLINOIS
COUNTY OF ______________ } ss.

I, ______________, a Notary Public in and for said county, in the State aforesaid,

DO HEREBY CERTIFY that ______________________

____

personally known to me to be the same person(s) whose name(s) (is) (are) subscribed to the foregoing instrument, appeared before me this day in person and acknowledged that ___________ signed, sealed and delivered the said instrument as ______________ free and voluntary act, for the uses and purposes therein set forth, including the release and waiver of the right of homestead.

GIVEN under my hand and Notarial Seal, this ___________ day of ______________, A. D. 19____

____
Notary Public

My Commission Expires ______________

Fig. 13-3. This printed form is one type of mortgage agreement.

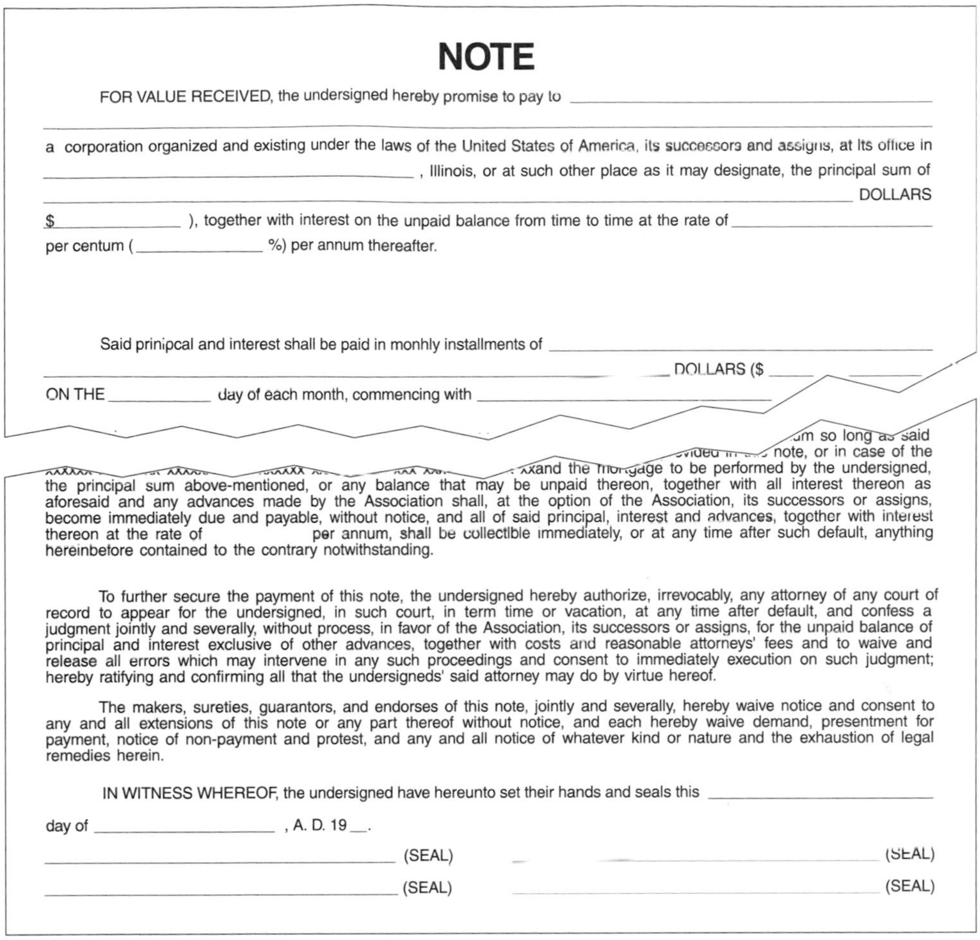

NOTE

FOR VALUE RECEIVED, the undersigned hereby promise to pay to ____________________

a corporation organized and existing under the laws of the United States of America, its successors and assigns, at its office in ____________________, Illinois, or at such other place as it may designate, the principal sum of ____________________ DOLLARS

$__________), together with interest on the unpaid balance from time to time at the rate of ____________________ per centum (__________%) per annum thereafter.

Said prinipcal and interest shall be paid in monhly installments of ____________________ DOLLARS ($__________

ON THE __________ day of each month, commencing with ____________________

... so long as said ... note, or in case of the ... and the mortgage to be performed by the undersigned, the principal sum above-mentioned, or any balance that may be unpaid thereon, together with all interest thereon as aforesaid and any advances made by the Association shall, at the option of the Association, its successors or assigns, become immediately due and payable, without notice, and all of said principal, interest and advances, together with interest thereon at the rate of per annum, shall be collectible immediately, or at any time after such default, anything hereinbefore contained to the contrary notwithstanding.

To further secure the payment of this note, the undersigned hereby authorize, irrevocably, any attorney of any court of record to appear for the undersigned, in such court, in term time or vacation, at any time after default, and confess a judgment jointly and severally, without process, in favor of the Association, its successors or assigns, for the unpaid balance of principal and interest exclusive of other advances, together with costs and reasonable attorneys' fees and to waive and release all errors which may intervene in any such proceedings and consent to immediately execution on such judgment; hereby ratifying and confirming all that the undersigneds' said attorney may do by virtue hereof.

The makers, sureties, guarantors, and endorses of this note, jointly and severally, hereby waive notice and consent to any and all extensions of this note or any part thereof without notice, and each hereby waive demand, presentment for payment, notice of non-payment and protest, and any and all notice of whatever kind or nature and the exhaustion of legal remedies herein.

IN WITNESS WHEREOF, the undersigned have hereunto set their hands and seals this ____________________ day of ____________________, A. D. 19__.

____________________ (SEAL) ____________________ (SEAL)

____________________ (SEAL) ____________________ (SEAL)

Fig. 13-5. A mortgage note spells out the terms of a real estate loan, how the money will be repaid, and the rate of interest.

- Loans insured and guaranteed by the Veterans Administration.

All of the mortgages listed had a fixed rate of interest. The rate extended over a period of 20 or 30 years.

Rising housing costs and an unstable interest rate have changed the way mortgages are written. Lenders are unwilling to finance long-term loans at fixed rates. New types of mortgages are being written:

- **Graduated payment mortgage.** The monthly payments are low at first but become higher at a future date. This concept is based upon the idea that the buyer's income will increase as monthly payments increase.
- **Adjustable rate mortgage.** The interest rate is allowed to move up or down according to a national interest rate index.
- **Renegotiable rate mortgage.** This type is automatically renewed every three to five years. Payments could go up or down, depending on whether interest rates are higher or lower.
- **Wrap-around mortgage.** This is a mortgage made up partly by assuming an old mortgage on the property that is at a lower rate of interest. Then, a new mortgage that is at a higher interest rate is added. The interest rate is adjusted so that it lies somewhere between the two rates.

Interest

Interest is the price you pay for the use of money. The agency that grants the loan will determine the monthly repayment, Fig. 13-6.

Interest payments on a loan are calculated by multiplying the **principal** (amount of loan) by the interest rates. The interest is given by the annual percentage rate (APR). For example, suppose you are borrowing $10,000 for a

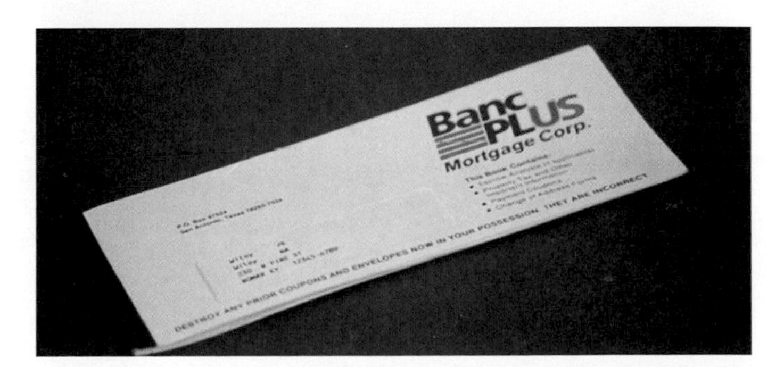

Fig. 13-6. Loan repayment books provide a monthly schedule for paying off large loans. Banks that provide loans on new cars and homes often use loan repayment books.

year at 15 percent interest. The interest would be $10,000 x .15 = $1500. The total to be repayed at the end of the year would be $11,500. This represents the original $10,000 plus $1500 in interest.

A lending institution might also add a special fee or fees. One of these is an **origination fee,** which is equal to one or two percent of the mortgage. Another fee, used frequently when interest rates are unstable, is a two or three percent charge for any type of long-term mortgage. It compensates the lender for the difference in the agreed rate and the escalating (rising) interest rates as the mortgage matures. These extra charges are called **points.** The buyer or seller may pay these fees depending on the market.

Repayment of loan

Most private real estate loans are made for a period of 10 to 30 years, Fig. 13-7. The loan arrangement should provide for a systematic reduction of the principal for the duration of the mortgage term. Each installment payment is recorded, Fig. 13-8. Remember, however, that the faster the loan is repaid, the less the total cost of the project will be.

In addition, a good loan should have a specific maturity date that permits complete repayment of the principal and interest through a regular planned **amortization.** This is the gradual removal of the financial commitment by periodic payments.

INSTALLMENTS PAID

PAID		PAID TO	INTEREST	PRINCIPAL	BALANCE
1					
2					
3					
4					
5					
6					
53					
54					
55					
56					
57					
58					
59					
60					

Fig. 13-8. Amortization is a planned reduction of the loan by regular installments.

BONDS

Bonds are used by some municipalities, school districts, and other public institutions to secure money for construction. **Bonds** are sold to individuals or lending institutions with the understanding that the seller will repurchase each at a later date for an additional amount of money. They become a guarantee for payment, Fig. 13-9. In effect, a bond is another type of loan with interest.

Fig. 13-7. Real estate loans make the expansion of industry possible. This Boeing plant is being enlarged by 50 percent for the production of a new line of planes.
(Boeing Commercial Airplane Group)

STOCKS

Industrial and commercial structures may be financed by the issuance of **stock** certificates, shares of the company, Fig. 13-10. Repayment, therefore, is made through dividends paid by the company to the shareholder. The shareholder, in return, becomes owner of a certain percentage of the corporation.

APPROPRIATIONS

Federal, state, and local governments finance much of the larger construction projects of a public nature. These projects require huge sums of money, Fig. 13-11. Dams, which require large amounts of money and time to construct, usually are financed by government **appropriation.** Buildings to house state-supported institutions of higher education are constructed by appropriations made by state legislators, Fig. 13-12. Many construction projects are financed through joint appropriation by the various levels of government, Fig. 13-13.

Money for appropriations comes from the working budgets of the governmental levels. The working budgets

Fig. 13-9. Often when large public projects are started, bonds will be sold to provide the financing. Shown is Terminal 1 at O'Hare International Airport. (Chicago Dept. of Aviation)

Fig. 13-10 Sample stock certificate.

Fig. 13-11. Large sums of money are needed to build launch sites and keep them running. The money to finance these operations is appropriated by the government. (NASA)

Fig. 13-12. State legislators can allocate money to build and expand institutions of higher learning.

are made up of money collected from the taxpayers. Money for appropriations also may come from the sale of bonds.

CONTRACTING

A great deal of money has to exchange hands to ensure that a project will be built to specifications. Labor must be paid, materials and supplies purchased, and equipment rented or bought, Fig. 13-14. This has to be within a budget that is fair to the contractor and the client. Furthermore, the client should be able to make the payments on the project so the loan can be reduced.

A system of contracting has been worked out through the years that serves these needs very well. A **contract** is simply an agreement between two or more parties that

Fig. 13-13. Construction for transportation requires public money.

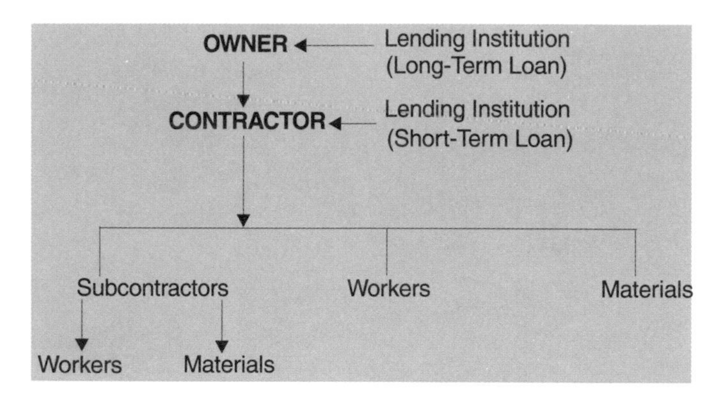

Fig. 13-14. The flow of money for a construction project.

describes in detail the responsibilities of each party. It covers the quality of construction, the completion dates, and the payments.

KINDS OF CONTRACTORS

There are basically two types of contractors, the general contractor and the subcontractor. A **general contractor** obtains an overall contract for building a residence, a road, a dam, or a commercial building, Fig 13-15. The **subcontractor** specializes in a particular type of construction enterprise, such as plumbing, electrical, concrete, or framing.

Often, a general contractor will secure a bid and then subcontract the entire job out to subcontractors. In this case, the general contractor manages the entire job but relies on the subcontractors to do the actual construction work. There are many general contractors who started out working for a subcontractor. They then became a subcontractor themselves and eventually a general contractor.

LICENSING CONTRACTORS

Most states require the general contractors and subcontractors to have a license in order to participate in bidding

Fig. 13-15. On large projects, the general contractor will set up a temporary office on the site. Mobile homes make convenient temporary offices.

for projects. These licenses are carefully controlled by state officials and unions to ensure that the contractors have the knowledge required for their profession. Most states have extensive tests that must be passed before an individual can become a contractor.

The courts state that the purpose of a **contractor's (builder's) license** is to protect the public against the consequences of incompetent workmanship and deception.

Contractors are licensed as a means by which the general public can determine the competence of a builder. A legislative body may rule that building permits should be issued only to duly licensed contractors. Therefore, it may be unlawful for any unlicensed person to act in the capacity of contractor.

The state legislature has the power to regulate the business of construction. However, the legislature may delegate this authority to local city or county governmental agencies. Then, the local governmental agency has the power to enact legislation that regulates construction in their area. This regulation must, however, be reasonable and not discriminatory.

Contractors failing to abide by these regulations may have their licenses revoked. They may also be prevented from filing a special claim against the property under construction for work performed. A license may be revoked for violation of a particular provision of a statute or law, such as abandonment of a construction project, or for failure to complete the building project for the agreed price.

CONTRACTS

A contract is a binding agreement between two or more parties. A contract between parties may be as simple as a handshake, or it may be very complicated with the signing of many papers. In general, all contracts have the following provisions as shown in Fig. 13-16.

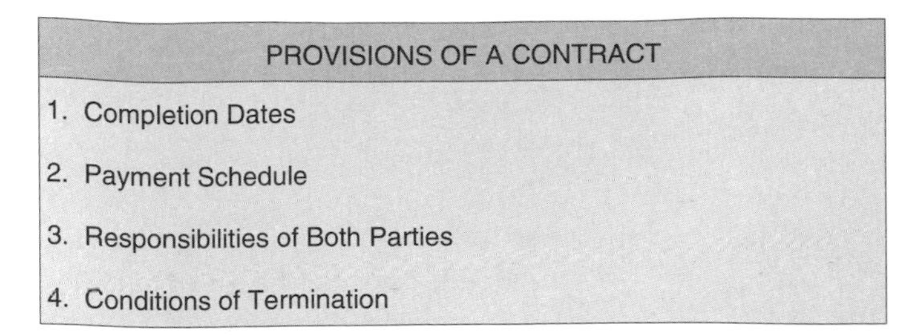

PROVISIONS OF A CONTRACT
1. Completion Dates
2. Payment Schedule
3. Responsibilities of Both Parties
4. Conditions of Termination

Fig. 13-16. Building contracts have four provisions in common.

- Completion schedule. The **completion schedule** details when construction can begin and when it is to be completed. In some cases, stiff financial penalties are assessed if the project is not completed on time.
- Payment schedule. On jobs that will extend over a long period of time, contractors will have their money invested in materials and labor. To allow the contractors to redeem their investment as early as possible, a **payment schedule** is arranged. A percentage of the total amount of the contract is paid to the contractor at specified times. For example, 20 percent would be paid when the foundation is complete and another 20 percent when the structure is enclosed. A third 20 percent would be paid when all utilities ares installed, and the remaining 40 percent when the project is completed.
- Responsibilities of the parties. The specific activities and services that owners and contractors are liable for need to be clearly spelled out in the contract. These are the **responsibilities of the parties.** Having this information in writing helps eliminate problems between the parties before they arise. Sometimes lawyers must be called in to handle disputes between the owner and the contractor.
- Conditions of termination. There are a large number of conditions under which either party may end the contract without completion of the project. These are the **conditions of termination.** These must be carefully detailed before the contract is signed.

Types of contracts

There are two major types of contracts between an owner and a contractor. They are called the fixed contract and the cost-plus contract, Fig. 13-17. There are a large number of variations of these two basic types of contracts.

The most common type is the fixed contract, Fig. 13-18. The **fixed contract** states that the builder agrees to furnish all materials and labor to complete the project for a *set* sum of money. The advantage of the fixed contract is the owner knows exactly how much the construction project will cost. Labor or material prices may have major changes. This could cause the contractor to lose money or possibly make a greater profit than expected. With this type of contract, the contractor will often give an estimate that is higher than what the project should cost. This is to cover unforeseen construction problems.

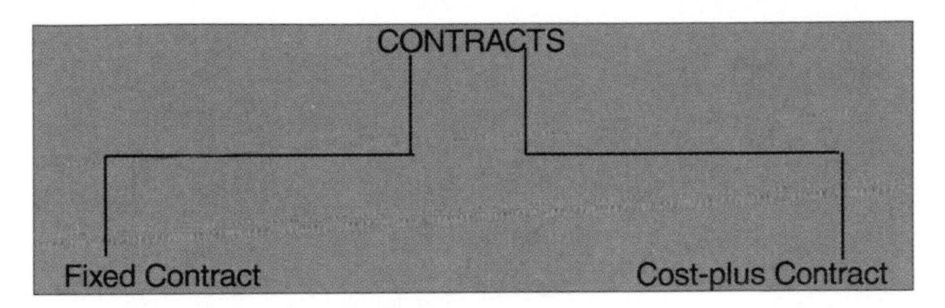

Fig. 13-17. There are two main types of contracts.

The **cost-plus contract** states that the contractor will receive a set percentage or a fixed amount over and above the total costs of completing the project. This type of contract protects the contractor from any unforeseen changes that take place during the construction of the project. However, with this type of contract the owner is not always protected. Contractors do not have the incentive to keep costs down that they would have with a fixed contract.

ESTIMATING

Building construction **estimating** is the careful determination of probable construction cost of a given project. Many items influence and contribute to the cost of a project, Fig. 13-19. Each must be compiled and analyzed. Since the estimate is prepared before construction is started, the estimator spends a great deal of time studying the working drawings and specifications.

Many consider estimating the most important function in contracting. It is from this estimate that the contractor enters into competitive bidding to win the contract. Competition in construction bidding is keen. Sometimes many firms bid for the same project. In order to compete and stay in business, a contractor must be a low bidder. The contractor must offer to build the structure for less money than anyone else. However, the estimate must include a profit. Otherwise, it becomes impossible to stay in business.

In a small company, the contractor will do the estimating. In larger companies a specialist is hired to do it. This individual works from drawings supplied by the architect or designer. The ability of the estimator to visualize all the different phases of construction becomes a critical element in the success of the company.

Estimating the cost of a construction project must be done very carefully because there are so many variables. Each variable can affect the actual cost of constructing the project. Fig. 13-20 illustrates a large project requiring considerable care in estimating costs.

Many of the variable costs in construction are unpredictable. Weather, material availability, subcontractors charges, and transportation are such variables.

However, many other variables are predictable to a high degree. Material cost, labor wages, equipment cost, and administration costs are in this group. Regardless of the variables involved, the construction estimator must prepare a cost analysis as accurately as possible. The cost estimate can spell the difference between profit or loss on a construction contract.

The successful estimator organizes the estimate into specialty divisions of:

- Earthwork costs.
- Materials costs.
- Labor costs.
- Mechanical equipment cost.
- Equipment depreciation.
- Overhead and contingencies.
- Profit.

Fig. 13-19. Every feature of this home and the surrounding landscaping was taken into account when the contractor submitted a bid. Some extra cost was added in so the contractor could make a profit. (Velux-America)

EARTHWORK

One of the first items to consider in estimating construction is the type of soil found at the construction site. The

CONSTRUCTION CONTRACT

THIS AGREEMENT, made and entered into this ______________ day of

______________________, 19 ____ , by and between ______________________

______________________________, hereinafter called the OWNER, and

____

hereinafter called the CONTRACTOR, WITNESSETH:

That the Contractor and the Owner in consideration of the agreements, covenants and payments hereinafter set forth, agree as follows:

1. The Contractor shall furnish all the materials and perform all the work including any and all sub-contracting

set forth and herein contemplated, including the running of lines for service facilities.

21. This contract is subject to the ability of Owner to obtain a construction loan on the afore described property in the sum of

____.

____

OWNER

____

CONTRACTOR

____

Fig. 13-18. A fixed construction contract is shown in abbreviated form.

Fig. 13-20. In large construction projects, estimation must be done very carefully. There are many areas for error to creep into the estimation.

Fig. 13-21. The cost of moving excavated earth from a construction site is part of the building cost. Cost is based upon volume, not area. (John Deere)

estimator may begin by studying the soil borings provided in the drawings. This would help determine the difficulty in removing the soil. If a great deal of rock were found, for example, more time would be allowed for removal.

Next, the quantity of rock or soil to be removed, Fig. 13-21, would be estimated. You will recall from your math courses that volume is measured in cubic yards or cubic meters. It is calculated by multiplying the length times the height times the width. Fig. 13-22 shows how to find volume and area for various shapes of excavations.

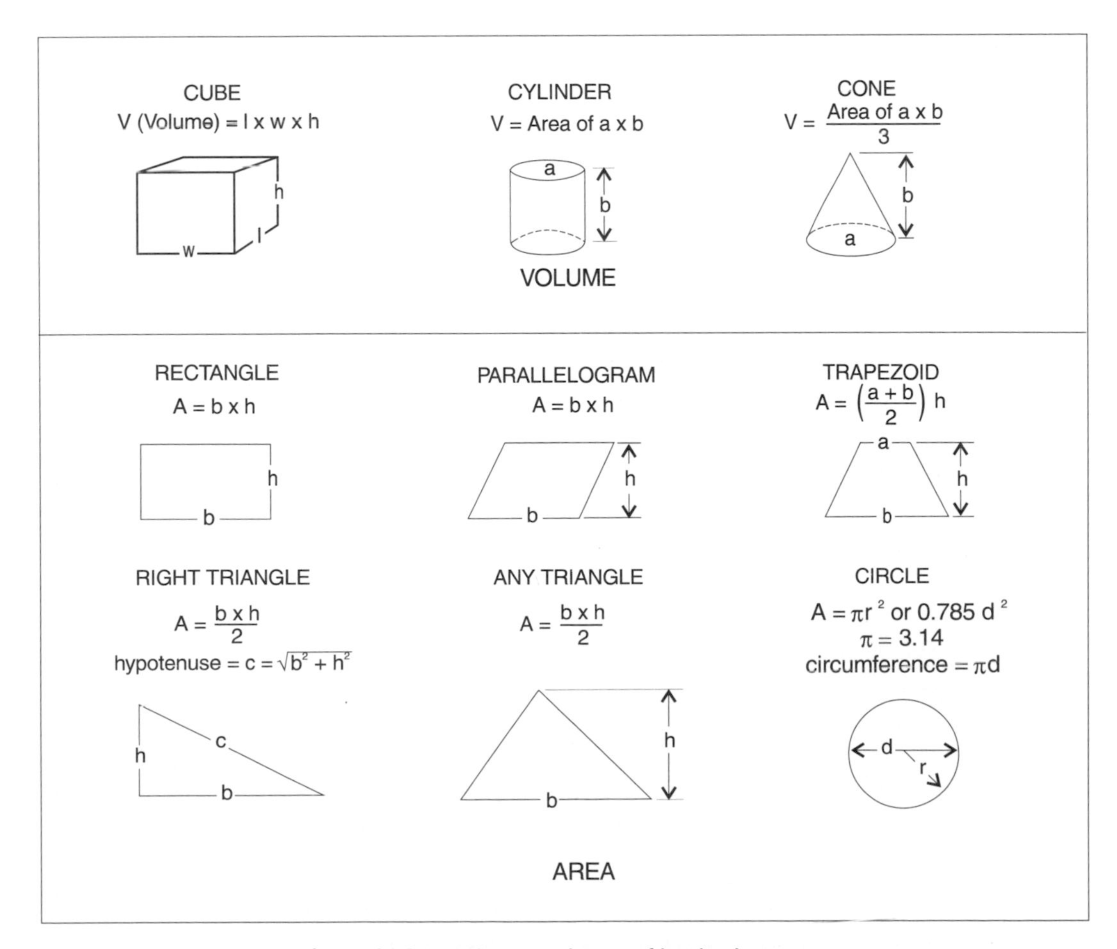

Fig. 13-22. Methods used to calculate volume and area of basic shapes.

The estimator must bear in mind that excavated earth takes up more space than settled (compacted) earth. Thus, it will take up more room in the dump truck, and the storage space for excavated earth must be larger than excavated space.

MATERIAL

Material is the single largest cost item of most construction projects. Consequently, it is important for the estimator to have a strong understanding of the materials that will be used and their price.

Most often, quantities must be computed after studying the working drawings. The estimator studies the plans and specifications counting quantities of each kind of material. After the quantities of each kind of material are determined, the cost is computed. Prices of materials are found in catalogs or by calling suppliers.

LABOR

Labor cost, Fig. 13-23, is calculated by multiplying the time by the hourly wage of the workers. This eliminates any effects of workdays and workweeks that are longer or shorter than normal.

Estimators first identify the job or operation to be performed. They estimate the worker-hours required to do the unit of work. They include all trades involved in the job–masons, carpenters, machine operators, and laborers.

Fig. 13-23. Labor is a large percentage of any contractor's cost. An estimator calculates how many hours it will take to survey a site and multiplies the figure by the salaries of the surveying crew. This sum is added to the total cost of the project.

All workers will probably not spend exactly the same number of hours on the job. They may only be doing one part of the job. Once the hours worked by each trade have been estimated, they are multiplied by the wages per hour and totaled.

To complete the estimator's task, many variables must be considered. The skills and mental attitude of the worker may affect the length of time required to complete a certain piece of work. Allowances should be made for variation in wages, working conditions, availability of skilled and unskilled workers, climatic conditions, and supervision during the construction of the project.

In addition, a worker seldom works continuously for 60 minutes of each hour. The actual time will range from 30 to 50 minutes per hour. Time must be taken to start up the operation, for clean-up after the operation, coffee breaks, trips to the restroom, and stops for a drink of water. These are some of the many variables which interrupt work. Waiting for materials and the scheduling of work also contribute to the difficulty of preparing the construction estimate of labor costs.

The estimator is responsible for deciding the number of hours required to perform a given unit of work. This can be done only through experience. If labor costs are uncertain, past project records are the most accurate source for preparing an estimate.

MECHANICAL EQUIPMENT

Estimates of mechanical equipment such as electrical, plumbing, heating/cooling, and other special fixtures are usually submitted by a single subcontractor. To work up the estimate, they go through the plans and specifications systematically, item-by-item counting the quantity of each, Fig. 13-24.

The subcontractor must identify each piece of equipment and obtain the manufacturer's price. Experience is the best guide to the hours of labor required to install each piece.

EQUIPMENT DEPRECIATION

As soon as a piece of construction equipment has been purchased, it begins to lose some of its value, Fig. 13-25. When this equipment is used on the project, it begins to wear out. One day, the equipment will become completely worn out or obsolete. If an allowance for **depreciation** is not included in the estimate, when the piece is worn out, there will be no money set aside to replace it.

Depreciation can be figured on a yearly basis. For estimating the depreciation costs, you should assign the equipment a useful life. This is given in years or units of production.

Fig. 13-24. All mechanical equipment built into the home must be counted in the estimate. This stove was installed with the kitchen. A subcontractor priced the equipment for the bid. (Whirlpool)

Fig. 13-25. These trucks, conveyors, and tractors will all wear out in time. New equipment requires a large outlay of money. A contractor must start planning now for their replacement in the future. (Morrison Knudsen Corp.)

If a unit of equipment had an original cost of $20,000 and an estimated life of five years, the depreciation value would be $4000 per year. Depreciation of each piece of equipment must be estimated as part of the cost of completing the project.

OVERHEAD AND CONTINGENCIES

Overhead constitutes a large percentage of costs on the job. Failure to allow enough money for overhead has caused many firms to lose money and even go out of business.

Home office overhead

Home office overhead costs are not readily chargeable to any one project. They represent the cost of doing business and the fixed expenses that must be paid by the contractor. Fixed expenses include rent on office space, electricity, heat, office supplies, postage, insurance, taxes, and telephone services. Salaries for office employees are also a fixed expense.

Field office overhead

Field office overhead expenses are all the costs that can be readily charged to a project. Expenses incurred at the spot where construction is taking place are such costs. Construction workers' wages, building permits, photographs, surveys, clean-up, and winter construction expenses of temporary enclosures are good examples of job overhead.

Contingencies

Contingencies are those items of expense that are left out, not foreseen, or forgotten on the original estimate. Almost every construction project has these items, Fig. 13-26. In some cases, the items left out could not have been anticipated at the time the estimate was prepared.

Fig. 13-26. On any large and complex project, it is possible to overlook certain unforeseen costs. Contractors allow for such additional costs in their bids. (Morrison Knudsen Corp.)

Contingencies should not be the excuse for a poor estimate. The proper approach is to be as careful and thorough as possible in listing all items from the working drawings and specifications.

PROFIT

Profit is the amount of money added to the total estimated cost of the project. This amount of money should be clear profit. It compensates the company and its owners for the use of their money and for the risk they have taken in financing the company and managing the construction job.

Profit can be figured in several ways. Following are two acceptable methods:

- A percentage is added to each item as it is estimated. Varying amounts are allowed for the different items. Amounts such as 8 to 15 percent might be added for laying concrete, 3 to 5 percent for subcontracted work.
- A percentage is added to the total estimated price of materials, labor, overhead, and equipment. Twenty to 25 percent is normal on small projects, 5 to 10 percent on a larger project.

BIDDING

For individuals or companies to obtain the most for their money when building a project, they will ask a number of contractors to submit a **bid** for the desired work. Bid forms, like the one shown in Fig. 13-27, may be furnished by the owner. The bids are generally sealed so that each contractor does not know how much another contractor will bid to complete the job. This prompts all contractors to submit their lowest bid for the job.

This competitive atmosphere allows a project to be built with the highest quality and at the lowest price, Fig. 13-28. The owner does not have to accept the lowest bid if it is felt that the work done will not be of satisfactory quality. Every contractor must maintain a quality reputation, or no one will accept their bid even if it is the lowest.

BID LETTING

Before bidding can begin, word must be spread to contractors that a project is ready for bidding. There are several ways to let this be known. Public advertising is required for many public contracts. The advertisement is generally placed in newspapers and trade magazines. Often, notice is also posted in public places.

Private owners often advertise in the same way to attract a larger cross section of bidders. Included in the advertisement is a description of the nature, size, and location of the project. The owner, availability of bidding documents, and bond requirements are also included. In addition, the time, manner, and place that the bids will be received are given.

On large projects a set procedure is followed by both the owner and the contractor. This bid-letting process is set by law in most states. The first step in this process is to advertise for bids, Fig. 13-29. This is usually done with advertisements in the classified section of papers or trade journals.

A company wishing to bid on a project needs to ask for a set of plans. They study the blueprints and specifications in order to make an accurate bid. They must turn in their bid by a specified date.

A public meeting is held, and all bidders are invited to witness the bid-opening process. The job is usually awarded to the contractor who is the low bidder. At that time, the contractor will be asked to post bond. Look again at Fig. 13-29. Posting bond ensures that if something happens to the contractor the job will still move forward to completion. It protects the owner from financial loss. The contract is then signed and the notice to begin construction is issued.

OTHER FINANCIAL CONSIDERATIONS

There are a number of other important financial considerations that affect the progress of a construction project. Knowledge of liens, foreclosure proceedings, and the many types of insurance can prove helpful.

LIENS

A **lien** is a legal claim against the property of another for the satisfaction of a debt. Most liens arise when one party fails to do something that they are obliged to do under the provisions of the construction contract. Most liens are filed for lack of payment.

If the property owner pays the construction bills promptly when due, there is no need for the contractor to file a lien with the courts. Liens are only filed against the property under construction when the contractor, subcontractor, workers, or material supply firms are not paid. Legal measures may then be needed to enforce payment.

Mechanic's lien

A **mechanic's lien** is a special claim that the law permits to be filed against the real estate involved in the construction when persons furnishing materials or labor are not paid. See Fig. 13-30. Mechanic's liens involve the contractor, subcontractor, laborers or workers, and the material supply firms.

STANDARD FORM 21
DECEMBER 1965 edition
GENERAL SERVICES ADMINISTRATION
FED. PROC. REG. (41 CFR) 1-16401

REFERENCE
Invitation No.
DACW41-71-B-0010

BID FORM
(CONSTRUCTION CONTRACT)

Read the Instructions to Bidders (Standard Form 22)
This form to be submitted in

DATE OF INVITATION
13 August 19

NAME AND LOCATION OF PROJECT	NAME OF BIDDER *(Type or print)*
Construction of Harry S. Truman Dam, Stage III Harry S. Truman Reservoir, Missouri	

____
(Date)

TO: DEPARTMENT OF THE ARMY
Kansas City District, Corps of Engineers
700 Federal Building
Kansas City, Missouri 64106

In compliance with the above-dated invitation for bids, the undersigned hereby proposes to perform all work for construction of Harry S. Truman Dam, Stage III, Harry S. Truman Reservoir, Missouri,

in strict accordance with the General Provisions (Standard Form 23-A), Labor Standards Provisions Applicable to Contracts in Excess of $2,000 (Standard Form 19-A), specifications, schedules, drawings, and conditions, for lump sum and unit prices set forth in the attached Bidding Schedule.

The undersigned agrees that, upon written acceptance of this bid, mailed or otherwise furnished within 60 calendar days after the date of opening of bids, he will within 10 calendar days after receipt of the prescribed forms, execute Standard Form 23, Construction Contract, and give performance and payment bonds on Government standard forms with good and sufficient surety.

The undersigned agrees, if awarded the contract, to commence the work within 10 calendar days after the date of receipt of notice to proceed, and to complete the work within the completion dates after the date of receipt of notice to proceed specified in the SPECIAL PROVISIONS.

RECEIPT OF AMENDMENT: The undersigned ackknowledges receipt of the following amendments of the invitation for bids, drawings, and/or specifications, etc. (Give number and date of each):

The representations and certifications on the accompanying STANDARD FORM 19-B are made a part of this bid.

ENCLOSED IS BID GUARANTEE, CONSISTING OF	IN THE AMOUNT OF

NAME OF BIDDER *(Type or print)*	FULL NAME OF ALL PARTNERS *(Type or print)*
BUSINESS ADDRESS *(Type or print) (Include "ZIP Code")* County of ______________	
BY *(Signature in ink. Type or print name under signature)*	
TITLE *(Type or print)*	

DIRECTIONS FOR SUBMITTING BIDS: *Envelopes containing bids, guarantee, etc., must be sealed, marked, and addressed as follows:*

MARK:
Bid under DACW 41-71-B-0010
To be opened 24 September 19
Receipt of Amendments No. ____
__________acknowledged.

ADDRESS: DEPARTMENT OF THE ARMY
Kansas City District,
Corps of Engineers
700 Federal Building
Kansas City, Missouri 64106

Fig. 13-27. Both sides of a bid form used in bidding on a public project.

Fig. 13-28. This contractor submitted the winning bid for this residence. The winning bid is not always the lowest bid. A contractor's reputation is important as well.

The mechanic's lien is filed with the county clerk where the construction land is located. This lien constitutes a powerful factor in collecting past due bills for labor and material.

FORECLOSURE

Foreclosure results when a mortgage is in default (the payments are not being paid). A property owner secures money by pledging the property as payment. If the owner becomes delinquent in this debt, the lenders will take the legal action of foreclosure. The property is used to satisfy the lender as payment of the debt.

INSURANCE

Insurance is another important consideration in construction. Contractor's buy several different forms of **insurance,** which protect them against great financial loss in a variety of areas. Builder's risk, public liability, and workmen's compensation are three common forms of insurance.

Builder's risk insurance

Standard **builder's risk insurance** protects all parties against physical damage to the insured property during the construction period. This policy covers damage resulting from any of the perils (fire, wind, vandalism) named in the policy. It provides reimbursement based upon actual loss or damage rather than any legal liability that may be incurred.

Most builder's risk insurances are based upon the **completed value concept.** This method assumes that the value of a project increases at a constant rate during the course of construction, Fig. 13-31. While the policy is written for the value of the completed project, the premium is based upon a reduced or average value. The coverage provided for the project is over the actual work completed with the standard materials at any given time.

Legals

INVITATION TO BID
EAST KAGY BOULEVARD
ROAD IMPROVEMENTS
IN
GALLATIN County, MT

Notice is hereby given that sealed bids will be received, publicly opened and read aloud at 1:30 p.m., Local Time, on March 3, at the Gallatin County Commissioners Office, Gallatin County Courthouse, Bozeman, MT 59715 by the Gallatin County Commissioners, hereinafter referred to as "Owner".

In general, the improvements of which bids are requested will consist of the following: Road Improvements to consist of constructing approximately 3100 L.F. and paving approximately 2300 L.F. of County road along East Kagy Boulevard, with placement of 60,000 C.Y. of borrow. Project located at Bozeman, Montana.

Said improvements shall include all tools, equipment, materials and labor to complete said work. Each proposal or bid shall be submitted on bid forms furnished by the Engineer, and must be enclosed in a sealed envelope and addressed to **Gallatin County Commissioners**. The envelope shall be plainly marked on the outside **East Kagy Boulevard Road Improvements**. Each proposal or bid shall be delivered to the Gallatin County Commissioners, Gallatin County Courthouse, P.O. Box 1905, Bozeman, Montana 59715, by the time and date listed above.

The Contract Documents, including drawings and specifications are available at the office of Morrison-Maierle/CSSA, 601 Haggerty Lane, P.O Box 1113, Bozeman, Montana 59715, on the payment of $25.00 nonrefundable for each complete set.

All bids must be accompanied by lawful monies of the United States or a Cashier's Check, Certified Check, Bid Bond, Bank Money Order or Bank Draft, drawn and issued by a National Banking Association located in the State of Montana, or by any Banking Corporation incorporated under the Laws of the State of Montana, in an amount equal to, but not less than ten percent (10%) of the total bid amount, payable to the order of Gallatin County as liquidated damages in the event said successful Bidder shall fail or refuse to execute the contract in accordance with the terms of his bid. After a contract is awarded, the successful Bidder will be required to furnish a separate 'Performance Bond' and a 'Payment Bond', each in the amount of one hundred percent (100%) of the contract amount.

Gallatin County reserves the right to reject any or all proposals or to waive any formality or technicality in any proposal in the interest of Gallatin County. No Bidder may withdraw his proposal for a period of sixty (60) days after the date of opening thereof.

Each Bidder must agree to conform to all federal, state and local regulations relative to employment of labor.

JANE JELENSKI,
Gallatin County
Commissioners

First Publication: 2-12
Second Publication: 2-18

Fig. 13-29. The first step in the bidding process is to advertise the project. The classified section of the newspaper is often used.

Fig. 13-30. Many people and several firms (contractor and subcontractors) have money invested in materials, labor, and equipment services during the course of a building project. If construction bills are not paid, mechanics liens may be filed against the property under construction

Within the framework of the builder's risk policy are the following perils: fire and lightning; vandalism and malicious mischief; extended coverage for windstorms, hailstorms, smoke, explosion; and riot or civil commotion. Additional perils might include collapse, landslide, water damage, breakage, or theft. It is also possible to obtain separate coverage for flood and earthquakes.

Public liability insurance

Public liability insurance is protection against liability for the injury or death of a person. A contractor is liable in money damages for wrongful or negligent conduct that causes injury or death to persons rightfully on the building site or adjacent property. Therefore, the contractor must be protected with public liability insurance. Negligence invariably involves some type of wrongdoing. A claimant, other than an employee, must prove that the contractor was negligent and that this negligence was the direct cause of the injuries.

Where injury to children is involved, there is a doctrine of law called the **attractive nuisance.** See Fig. 13-32. Small children, who are unaware of the dangers, are attracted to the construction site by the mounds of dirt, tractors, and equipment. Precautionary steps must be taken to prevent the children from being injured.

Workmen's compensation insurance

Workmen's compensation insurance protects the contractor against claims resulting from injury or death of his workmen through an industrial accident. In most states, laws prescribe that the employer must provide protection for the employee who is injured while on the job, Fig. 13-33.

The amount of compensation paid to an injured employee under workmen's compensation never equals his or her normal earnings. Most states fix the amount paid by a regular schedule that is part of its workmen's compensation laws. The compensation paid to the injured employee is not affected by the insurance company paying all medical, surgical, hospital, or even burial expenses.

In some states, compensation laws are elective. The employer and employee are privileged to jointly accept the provisions of the law or ignore them. In other states, no choice is granted. Both the employer and employee are bound to the provisions of the law.

Fig. 13-31. Using the completed value concept, when a project is half finished, it is insured for half of its final value.

Fig. 13-32. Construction sites are not playgrounds, yet mounds of dirt, equipment, and unfinished construction attract children. For this reason, public liability insurance must be purchased by the contractor.

Fig. 13-33. Many safety precautions are used on construction projects. Yet, injuries are still possible. All contractors must have workmen's compensation insurance. (Morrison Knudsen Corp.)

SUMMARY

Capital is needed to pay for construction. Contractors may use their own capital, or money may be borrowed from lending institutions. The exchange of money for work and materials provides jobs for construction workers and contractors.

Large projects need a considerable amount of money. This often requires money to be loaned to individuals. Large projects also require contractors to make competitive bids on the clients' projects. Proposed jobs are advertised to the construction industry and sealed bids are often requested at a specified date. The bid opening is a public meeting where the contractor is selected by the client.

A mortgage is one method with which an individual can secure a loan to build a home or other project. The mortgage is set up for a certain percentage of the final cost of the project with a predetermined repayment schedule. The mortgage allows an individual to borrow capital in order build a home.

There are two major types of construction contractors. There are general contractors, who oversee the entire project and subcontractors, who specialize in a particular trade or construction activity.

Most contracts have a completion date, a payment schedule, the responsibilities of the parties, and conditions for the termination of the contract. Contracts can be of a fixed type where the contractor agrees to complete a particular job for a set price. Contracts can also be of a cost-plus type where the contractor is paid a specified percentage of the total cost of materials and labor to complete the job.

KEY WORDS AND TERMS

All of the following words and terms have been used in this chapter. Do you know their meaning?

Adjustable rate mortgage
Amortization
Appropriation
Attractive nuisance
Bid
Builder's risk insurance
Bond
Capital
Completed value concept
Completion schedule
Conditions of termination
Contingencies
Contract
Contractor's license
Cost-plus contract
Depreciation
Estimating
Field office overhead
Financing
Fixed contract
Foreclosure
General contractor
Graduated payment mortgage
Home office overhead
Insurance
Interest
Letter of commitment
Lien
Loan
Mechanic's lien
Mortgage
Origination fee
Payment schedule
Points
Principal
Profit
Public liability insurance
Renegotiable rate mortgage
Responsibilities of the parties
Stocks
Subcontractor
Workmen's compensation insurance
Wrap-around mortgage

TEST YOUR KNOWLEDGE

Do not write in this book. Please write your answers on a separate sheet of paper.

1. What are the two major types of construction contractors?
2. List the four provisions that are most common in any construction contract.
3. State three sources for an individual to obtain the capital required to build a new home.

4. Explain the difference between stocks and bonds.
5. Give the three separate areas that are covered by builder's risk, public liability, and workmen's compensation insurance.
6. Calculate the interest on a $100,000 loan at 9.75% per annum for one year. How much interest would you pay on this loan for one year? How much would you pay over 20 years?
7. Estimate the number of cubic yards of concrete needed for a sidewalk 4 ft. wide, 50 ft. long, 4 in. thick.
8. Define depreciation.

APPLYING YOUR KNOWLEDGE

1. Search the classified section in your local paper and locate advertisements for bids. Carefully read the bid and determine what is to be built and its location. Follow up after the bid is let and find out who was the successful contractor.
2. Invite a loan officer to tell your class about the many different ways to obtain capital.
3. Form a group and select a small construction project, such as a short section of sidewalk or park benches. Make preliminary designs of the project and then advertise for bids within the class. The class should divide into several construction companies. Have a bid opening and determine what company provided the best and lowest bid for the project.
4. Locate a subcontractor. Invite that individual to your class to tell you about the advantages and disadvantages of contracting.
5. Write an essay explaining why you think insurance is important to the contractor and to the worker.
6. Prepare an estimate for a small construction project such as a storage shed.

Lending institutions, with their capital, allow blueprints and ideas to become construction projects.

14

CHAPTER

TOOLS, MACHINES, AND EQUIPMENT IN CONSTRUCTION

The information provided in this chapter will enable you to:

- *List and discuss the four major processes we can perform on materials.*
- *List several tools or pieces of equipment that can be used to separate construction materials.*
- *Describe four methods used to combine materials.*
- *Explain how construction materials can be formed into desired shapes.*
- *Define the conditioning processes used on some materials.*
- *Demonstrate five layout tools and tell where they might be used on the construction site.*
- *Explain the difference between chip removing tools and shearing tools.*
- *Recognize lifting and holding tools and state where they are used.*

Tools and equipment comprise another important resource for construction. Even if you have all of the other resources present, little construction activity could be accomplished without the use of tools. A tool does not have to be a sophisticated piece of machinery. It only needs to extend human power and be under human control. In this chapter, we will discuss major classes of tools and how they are used to make our work easier.

PROCESSING WITH TOOLS AND EQUIPMENT

All construction processes can be placed into categories that depend on what they accomplish. There are four major processes you can perform on material. These processes are: separating, combining, forming, and conditioning, Fig. 14-1. No matter what you may be constructing, these four processes allow you to transform materials into a structure that you desire.

SEPARATING

If you divide a piece of material into two pieces, you are performing a **separating** process. To be performing a separating process, it makes little difference if you use a jackknife, a hand saw, a power saw, a laser beam, a high pressure water jet, or any other piece of equipment. The result in all cases, making two pieces out of one, is separating, Fig. 14-2. It is true that some of the tools listed above would cut much faster than others. Some of the tools would cost a great deal more than others. But, the end result is separating. Therefore, we call the above tools separating tools.

Managers of the separating activity must decide if they have enough capital, time, and labor to use the equipment that separates the material at the fastest rate. The money to purchase the latest equipment is not always available.

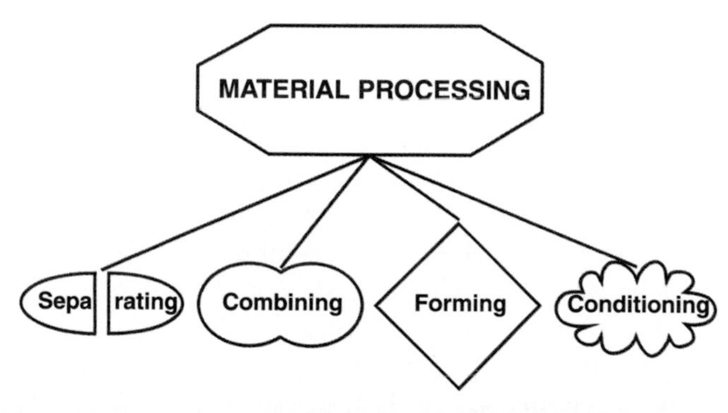

Fig. 14-1. There are four categories of material processing.

Fig. 14-2. The pipe is undergoing a separating process. An abrasive cutoff saw is one type of separating tool.

Fig. 14-3. The wood framing of a home is held together with mechanical fasteners. Nails and screws are most common. (Lowden, Lowden and Co.)

Therefore, people get along with what they have until they can afford better equipment.

New equipment and tools are being introduced into the marketplace every day. If you are working for a contractor, you would be expected to work with this new equipment after a very short training period. You, your background, and your ability to understand the operators manual enable you to do this. Understanding the basics of the separating process will help you learn to use new tools and equipment as they are available.

COMBINING

A second major process necessary in the construction field is combining. **Combining** consists of attaching two or more pieces of material together to form a single piece. Types of combining processes are:

- Combining with mechanical fasteners.
- Adhesion.
- Cohesion.
- Jointery.

There are many different kinds of mechanical fasteners. Nails, screws, bolts, and rivets name just a few that are available. In all cases, **mechanical fasteners** attach the materials together using some type of clamping action that makes the pieces behave as one unit, Fig. 14-3.

Adhesion is the process of combining pieces using their ability to stick to a different material. The materials that bond the pieces together are called adhesives. Examples of some common adhesives are: solder, asphalt tars, cement, and glues. Notice that a dissimilar material is used to hold other materials together, Fig. 14-4.

The process of **cohesion** is the combining of materials where the adhesive is made of the same substance as the materials being bonded together, Fig. 14-5. The most common example is a welding action where the welding rod is of the same material as the parts being welded.

Jointery is the process of holding materials together with a specialized joint, Fig. 14-6. No mechanical fasteners or adhesives are used. Some examples are the joining of sheet metal with a locked joint or a Pittsburgh seam. Threading pipe together is also jointery.

In the construction field, there are numerous examples of joining materials together. However, all of these joining methods will fall into one of the categories discussed: mechanical fasteners, adhesion, cohesion, or specialized joints.

FORMING

A third process that can be used in processing materials for construction is forming, Fig. 14-7. In the **forming** process, material is molded into a particular shape and size. It is often done with materials such as concrete, asphalt, or clay tile.

The construction of a hydroelectric dam is an excellent example of forming. In this case workers spend a great amount of their time in the construction of forms used to contain concrete in a particular shape while the concrete is setting. After the concrete is set, the forms are removed and then reset for the next "pour." This process continues until the top of the dam is reached. The forms must be made very strong to withstand the tremendous pressure exerted by fresh concrete.

A

B

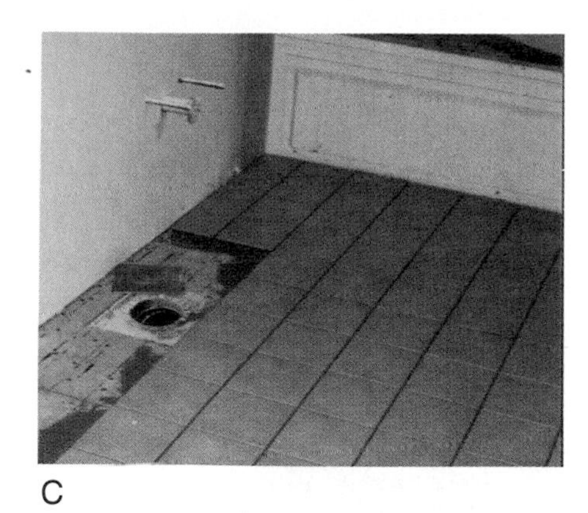

C

Fig. 14-4. Adhesives are used in many areas of construction. A–Mortar locks bricks to build walls. B–Wood flooring is glued in place. C–Tile is adhered to wood floorboards. (Lowden, Lowden and Co.)

Fig. 14-5. Welding is the most common form of cohesion. (Santa Fe Pacific)

Construction of buildings, both residential and commercial, often requires extensive use of concrete in the footings and foundations of the structure. Specialized forming panels are often used that allow the work to proceed quickly. Also, special holding devices are used to properly position and hold the panels in place while the concrete is being placed and while it cures.

Solid materials, such as wood or metal, may also be formed for use in construction. The architectural design may require wood beams to be of a particular shape or curvature. The most common way to obtain these beams is to form a number of thinner boards in the desired shape

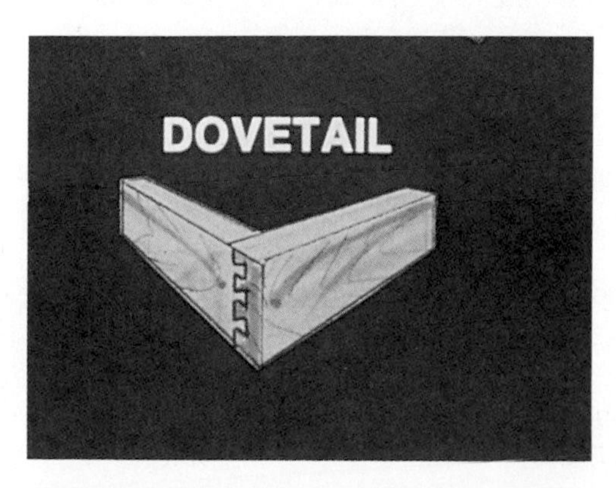

Fig. 14-6. Dovetail joints lock two pieces of wood together with the use of a special cut. (AGS&R Communications)

and hold them together with glue. This process is called laminating and is extensively used to make beams in a variety of shapes.

Steel I beams, sheet metal duct work, and extruded aluminum window frames are other examples of formed materials commonly used in the construction industry, Fig. 14-8.

CONDITIONING

Conditioning materials involves changing the characteristics of the materials but not changing the physical dimensions of the material. Examples of conditioning include the hardening of a metal, the curing of concrete, or magnetizing steel. In these cases, the internal structure of the material is being changed, but the size or shape of the item is affected little if at all.

Asphalt used for paving a highway is an example of a material that requires some conditioning before it is ready for use. In this case, the asphalt is first laid down on the road bed. Then it is conditioned when it is compacted into a solid mass, Fig. 14-9. Without this conditioning process the asphalt would soon break up. It would not be a suitable road surface. In this conditioning process the only change in the asphalt is pressing it together with the use of a heavy steel wheeled roller.

COMBINATION OF PROCESSES

Seldom is only one process used when constructing a project. You must learn to use all of the processes individually and in combination with one another.

For example, in the construction of a concrete sidewalk you would first have to excavate the top soil, a separating process. Then, you would have to build and set the concrete forms to contain the concrete, a forming process. Next, the concrete would have to be mixed by combining the proper amount of water, sand, gravel, and cement. Lastly, you would have to condition the concrete by troweling to compact the surface and allow it to cure before it can be used.

Nearly all construction requires the use of a combination of the four basic methods of processing materials. You will discover this as you observe various construction sites.

PROCESSING TOOLS AND EQUIPMENT

There are a number of ways tools and equipment can be classified. A widely accepted set of classifications are:

- Layout and measuring tools.
- Separating tools.

Fig. 14-8. I beams are a common support in industrial buildings. They are one of many formed materials.

Fig. 14-7.These cement blocks were produced by a forming process.

Fig. 14-9. Asphalt paving is compacted into a solid mass. This conditioning process creates a durable road surface.

- Fastening or combining tools.
- Finishing tools.
- Lifting and holding tools.
- Mixing tools.

LAYOUT AND MEASURING TOOLS

In order for us to accurately construct dams, bridges, roads, and buildings, we need a system to transfer distances and angles from the specifications to the construction site. The tools that do this are called **layout and measuring tools,** Fig. 14-10. They include tools as simple as a pencil and as complicated as a laser controlled earthmover.

Common layout and measuring tools are tape measures, squares, chalk lines, levels, and transits. To use many of these tools you need some basic math skills. With these skills the layout tools are capable of producing excellent results.

SEPARATING TOOLS

Separating tools or separating equipment have two purposes. They are used to remove unwanted material, and they are used to make two or more pieces out of one. These tools can be further classified as chip removing tools and shearing tools.

With **chip removing tools,** some of the material is lost by the formation of chips, Fig. 14-11. Saws, cutting torches, and laser cutting equipment all fall under this category. Some amount of material, even when cutting with lasers, is lost. The combination of the two pieces formed by this process will not equal the size of the original material.

A separating process that does not produce any chips is **shearing.** Cutting with scissors is an example of this separating process where no material is lost, Fig. 14-12. Many construction processes use this type of separating. Glass is cut with a shearing process. When glass is etched and then shaped, no material is lost.

FASTENING AND COMBINING TOOLS

Fastening and combining tools include those tools used to fasten materials together. Hammers, screwdrivers, wrenches, and welding equipment name just a few of the

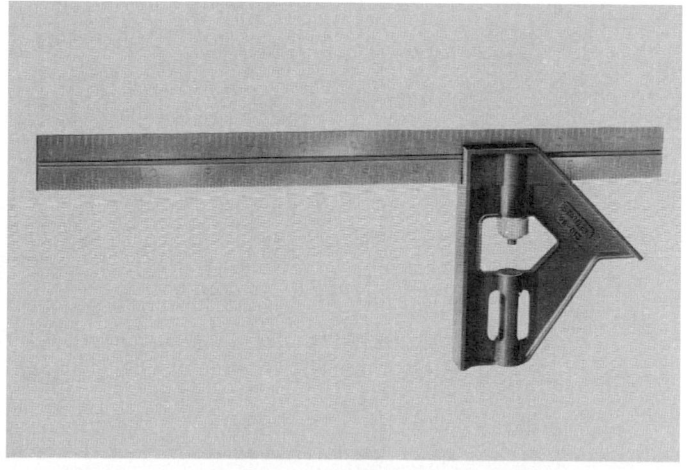

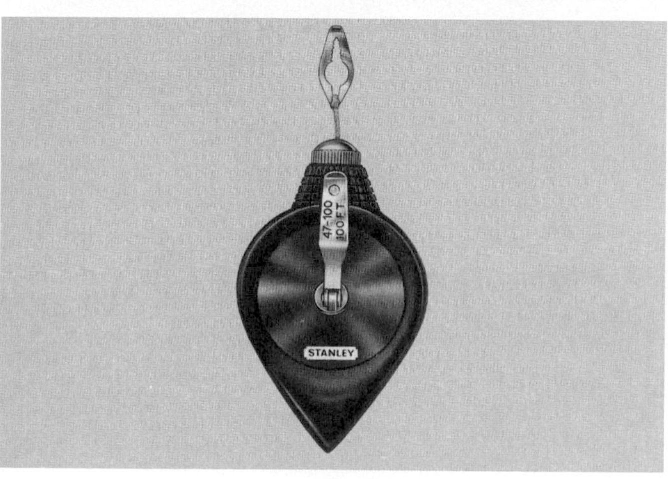

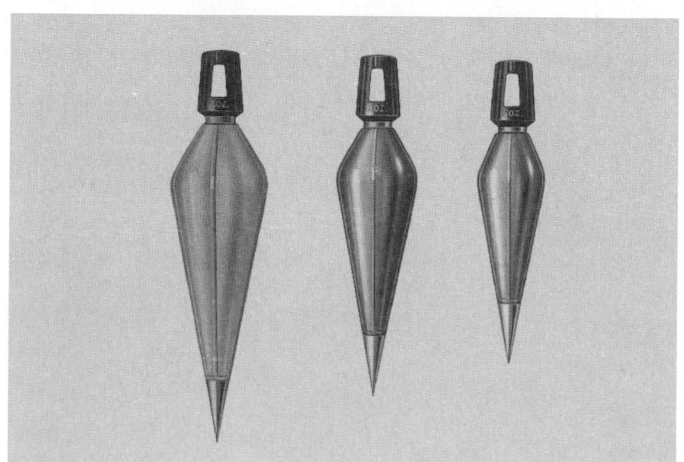

Fig. 14-10. Before construction begins, a site undergoes extensive planning and examination. Layout and measurement tools are the first to see use. Shown are a combination square, a framing square, a chalk line reel, and several plumb bobs. (Stanley)

many tools in this classification, Fig. 14-13. These tools are used to assemble the construction project. In skilled hands, these tools are very efficient.

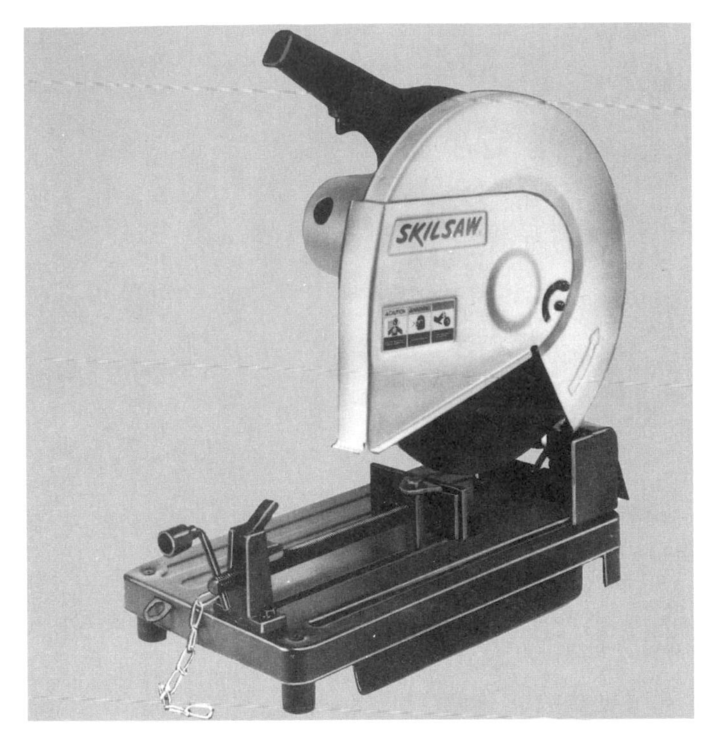

Fig. 14-11. All of these separating tools leave chips during the cutting process. (Skil Corp., Stanley, Dremel)

FINISHING TOOLS AND EQUIPMENT

On any construction job there are a great many finishing details to attend to. An assortment of **finishing tools** have arisen to complete these jobs. Concrete is finished with trowels to make it smooth and easy to care for. Asphalt is compacted with the use of a steel wheeled roller. This makes the asphalt more dense and water resistant.

LIFTING AND HOLDING TOOLS

On large and small construction projects there are times when materials must be lifted some distance. On small buildings, such as homes, this lifting is often done by human muscle. On large projects, the lifting must be done with the use of heavy cranes, Fig. 14-14. This equipment can be classified as **lifting and holding tools.** There are times when we must temporarily hold large pieces of material in place while permanent fastening methods are being applied.

MIXING EQUIPMENT

Mixing is an important process that makes sure various ingredients are conditioned to provide a strong material. Concrete, asphalt, paint, mortar, and drywall compound are examples of materials that must be thoroughly mixed in order to produce a quality finished product. To be certain that the mixing of these materials is done correctly, **mixing tools** were developed. Fig. 14-15 illustrates this activity.

SUMMARY

All material processing can be divided into four major categories: separating, combining, forming, and conditioning. These four basic processing procedures are performed

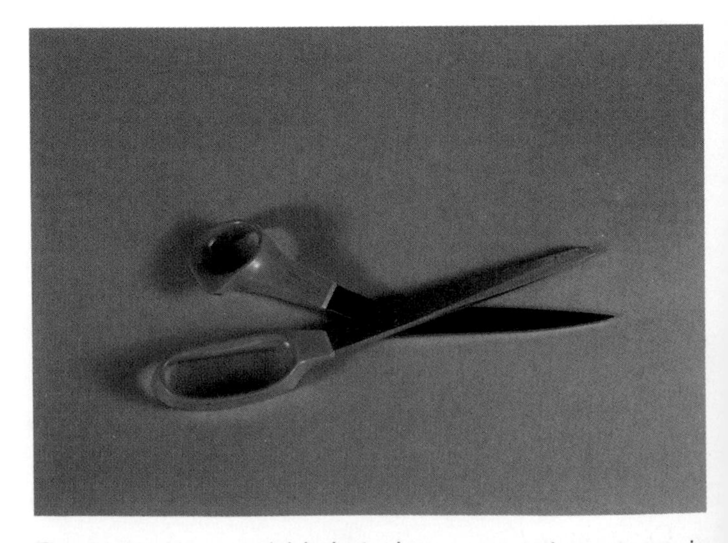

Fig. 14-12. No material is lost when a separating process is performed with a shearing tool.

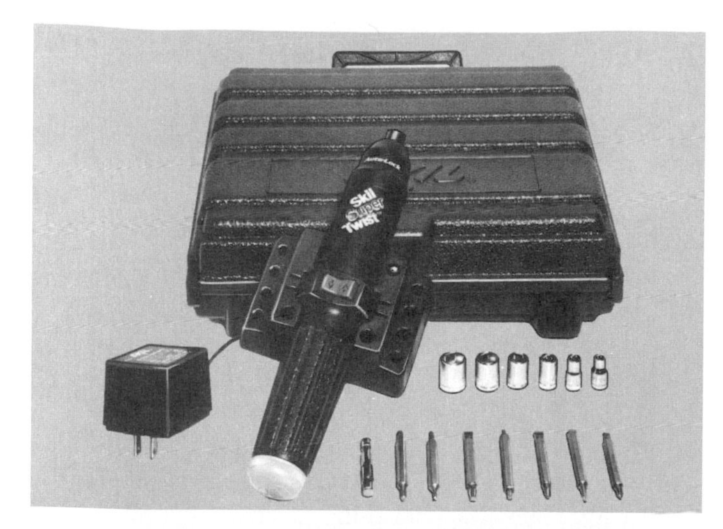

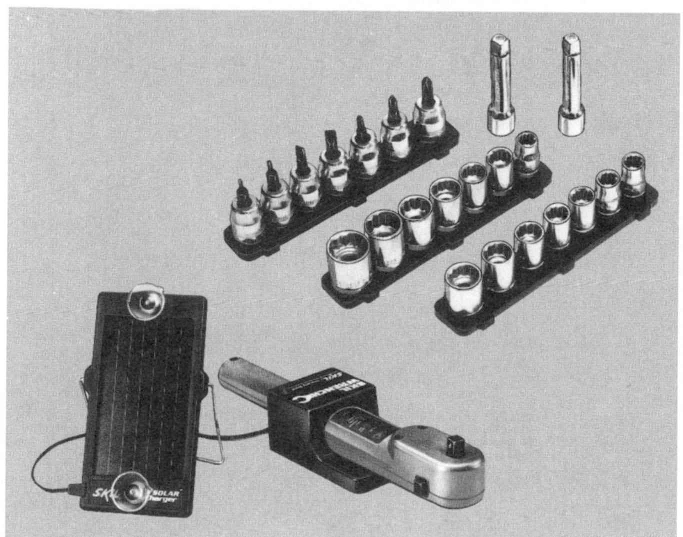

Fig. 14-13. Many fastening tools now come in a motorized form. (Skil Corp.)

Fig. 14-14. Many items are too bulky or too heavy to be held in place by workers. Lifting and holding tools, such as the large crane shown, are used for these tasks.

Fig. 14-15. This cement mixer supplies small quantities of cement on-site.

on the construction site by skilled workers. These workers have developed specific skills through education and experience with the tools of the trade.

Tools and machines that perform these processes can be further classified into the categories of: layout and measuring tools, separating tools, fastening and combining tools, finishing tools, lifting and holding tools, and mixing tools. Tools from all of these areas are required to complete most construction projects.

KEY WORDS AND TERMS

All of the following words and terms have been used in this chapter. Do you know their meaning?

Adhesion
Chip removing tool
Cohesion
Combining
Conditioning
Fastening and combining tool
Finishing tool
Forming
Jointery
Layout and measuring tool
Lifting and holding tool
Mechanical fastener
Mixing tool
Separating
Separating tool
Shearing

TEST YOUR KNOWLEDGE

Do not write in this book. Please write your answers on a separate sheet of paper.

1. List the four major classifications of material processing.
2. What kind of a tool is a chain saw?
3. Soldering is an example of which combining process?
4. List two methods used in construction to form materials into the desired shape.
5. Give two examples of material conditioning.
6. Most construction projects will require tools from only one or two classifications. True or False.

APPLYING YOUR KNOWLEDGE

1. Using a tool catalog as a resource, make a bulletin board by cutting out pictures of the various types of tools used in the construction industry. Place the tools into the proper classifications.
2. Visit a hardware store and look for new types of tools that are not available to you in your school.
3. Examine the interior and exterior of your school. Make a listing of the various items that have been produced by forming.
4. Contrast the fastening devices of ancient construction with those used in today's construction industry.
5. Explain to someone out of your class the four major classifications of material processing and then provide them with examples.

What tasks do you think these machines are performing?

15

CHAPTER

ENERGY IN CONSTRUCTION

The information provided in this chapter will enable you to:

- *Discuss the important role energy plays in the construction world.*
- *Explain why wood needs less energy to be converted into a useful construction material than does steel.*
- *Provide two examples why solar orientation is important to energy conservation.*
- *List and explain the three methods of heat transfer.*
- *Recognize the importance of air infiltration and the efficiency of heating.*
- *Define R, K, U, and Btu.*

The last of the seven resources that plays a very important part in the construction industry is energy. All energy comes from the sun or from the earth in one form or another. The sun is our major source of energy, supplying well over 90 percent of all the energy we use on earth. This **solar** activity accounts for wind, water, and all the sources of energy that have been derived from plants or animals. This includes all of our fossil fuels.

FORMS OF ENERGY

Some forms of energy are renewable and some are nonrenewable. A **renewable** energy source is a source that can be replenished easily. Trees, corn, and other plants that are used for fuel can be planted and harvested. As long as we take care of our soil, these sources of energy are renewable. Other commonly used renewable energy resources are: wind, hydroelectric, solar, geothermal, biomass, and some types of nuclear energy. These sources can supply us with energy far into the future.

Nonrenewable energy resources are those resources that cannot be easily replenished. Once they are used up, they are gone. These resources include oil, coal, and natural gas. It took many thousands of years for the earth to produce its current supply of oil and coal. Our current technology cannot replace the oil and coal that are being used up.

ENERGY AND THE CONSTRUCTION INDUSTRY

Large amounts of energy are needed to construct various projects that humankind wants or needs. The homes we live in are one example. Also, it takes a good deal of energy to heat and cool these projects. Thus, it is important for us to have an understanding of this important resource.

Construction, as well as the other technologies, requires a plentiful and easily controlled energy source. Often, on the construction site this energy is supplied to us through electricity, petroleum, or pneumatic (air) sources, see Fig. 15-1.

ENERGY USAGE PRIOR TO CONSTRUCTION

Little thought is given to the amount of energy used to produce industrial materials supplied to the construction site. Yet, a tremendous amount of energy goes into the processes of producing and transporting materials such as concrete and steel. This large consumption will have to be strongly addressed in the future. As our nonrenewable energy sources are depleted, more energy efficient methods must be used in material production. A smaller and more expensive supply of energy can drive up the price of materials. This would dramatically increase the cost of construction projects.

Generally speaking, wood materials need less energy to be processed into usable forms than do such materials as steel or masonry. This result is due to the great amount of heat required to refine the raw materials of steel and masonry before they are in a usable form. Wood on the

Fig. 15-1. Shown is a temporary power pole set up for a construction site. When the work is finished, the electric company will come and remove the box and meter.

Fig. 15-2. Little energy is needed to transform wood into a usable structural material. Most houses built today are of wood frame construction. (Lowden, Lowden and Co.)

other hand is already in a usable form. It need only be cut to standard dimensions before it can be used. This is one of the reasons why over 90 percent of the homes built today use wood frame structures as opposed to other materials, Fig. 15-2. Although energy savings are not critical at this time, they well may be in the future.

Fig. 15-3. Large construction projects require a great deal of energy. Energy is needed to produce the cement for this project. More energy is necessary to transport the cement and other materials to the site. Finally, additional energy is used in assembling the materials. (Tom Wood)

ENERGY USED DURING CONSTRUCTION

Today there are many types of power equipment used during the construction of a project. Most heavy equipment on a construction site uses diesel fuel for energy. This equipment includes bulldozers, backhoes, and large wrecking equipment.

Lights and power tools are often powered by electricity. Electrical generators are set up, or power lines are run in, for these purposes. In some cases, an auxiliary heating source is required to keep materials from freezing. These sources may be powered by propane or natural gas.

It goes without saying that without modern energy resources we would not be constructing projects at nearly as fast a rate as we currently are. Equipment such as bulldozers and backhoes allow unfriendly terrain to be made usable in very short order. In fact, many projects would be *impossible* to construct without large amounts of energy being available, Fig. 15-3.

ENERGY USAGE IN CONSTRUCTION PROJECTS

In addition to the quantity of energy necessary to build a structure, more energy is necessary to make the structures comfortable for human use. Any home, store, or office complex will have heating/cooling and electrical systems installed.

Keeping a living or working area at a comfortable temperature is one of the largest consistent power draws of any structure. Understanding **heat transfer** is important when designing and constructing homes and buildings.

HEAT TRANSFER

Transferring energy in the form of heat from one place to another is necessary in order for us to maintain temperatures that are within our comfort range. Depending on where you live, there is a need to supply extra heat or to provide some means of cooling. Heat energy can be transferred in three ways: conduction, convection, and radiation, Fig. 15-4.

- **Conduction** is the movement of heat in a solid material from one molecule to the next. No matter is transferred, just energy. Conduction is the method by which the handle of a metal frying pan gets hot when you heat it on a stove.
- **Convection** is the transfer of heat in a gas or liquid caused by the movement of the fluid. Currents are created between warmer and colder areas. These convection currents work towards balancing the temperature in a room.
- **Radiation** is the movement of heat by radiant energy (electromagnetic waves). One common example is the heating of the earth by the sun. Radiant energy from the sun heats the earth and all objects that it touches. A radiant heater will heat people and other objects in a room without directly heating the air.

In building construction we must be aware of how to use all three of these methods of heat transfer. All three methods may be used to heat a structure. The method or methods chosen will affect the shape of the structure as well as the materials chosen with which to build it.

MEASURING HEAT TRANSFER

The most common way engineers use to measure the heating or cooling capacity of equipment is with the use of the **British thermal unit (Btu).** One Btu is equal to the amount of heat needed to raise the temperature of one pound of water one degree Fahrenheit. The output for a standard heating unit of a home with 1200 square ft. would be about 100,000 to 150,000 Btu. Of course, the location and climate of the home would have a major impact on the size of the heating unit.

The amount of heat that is allowed to pass through a single material when there is a temperature difference of one degree Fahrenheit is called the **conductivity (K)** of the material. Walls are built of a variety of materials. When the K value of each of these materials is known the **total heat loss** of the wall section, the **U value,** can be calculated, Fig. 15-5. This determines the total loss of heat through an entire wall. By determining the various U values of the ceiling, floor, walls, windows, and doors, you then can calculate the total heat loss for the entire structure. This figure is used to determine the size of the heating or cooling equipment that will be needed.

Insulating materials

By preventing or slowing down the heat transfer between buildings and the surrounding area, great energy savings can be achieved. This is why structures have **insulation** put into their floors, walls, and ceilings. Insulation helps keep buildings warmer in the winter and cooler in the summer.

Some materials are better insulators than others. As a rule, the lighter and more tiny air spaces a material has the better insulator it is. The ability to resist the flow of heat is called **resistivity (R).** The higher the R-value of a material the better it insulates. Fig. 15-6 shows the R-value of a variety of materials. Doubling the thickness of a material will roughly double the R-value of that material. Today, insulations have R-values clearly marked on their packages, Fig. 15-7. Examples of types of insulation are shown in Fig. 15-8. Insulation is discussed in more detail in Chapter 27.

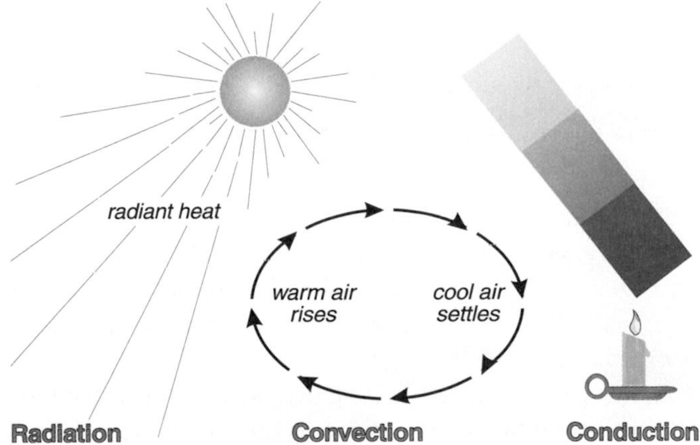

Fig. 15-4. There are three methods of heat transfer: radiation, convection, and conduction.

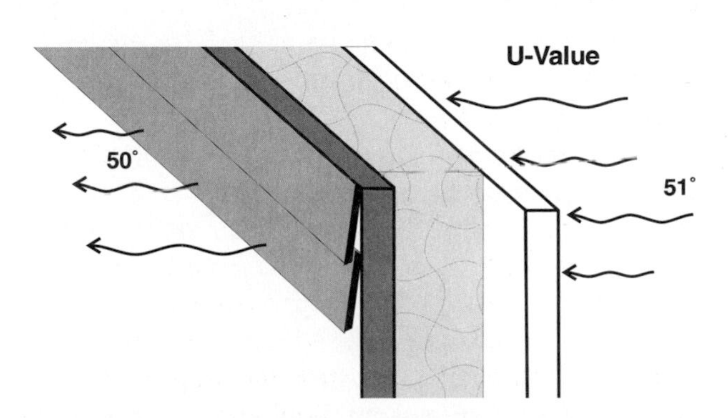

Fig. 15-5. U values are measures of the heat loss through a one square ft. section of wall with a one-degree temperature difference between the two sides for a period of one hour.

Infiltration

Another factor that must be considered when working with energy for heating is infiltration. **Infiltration** is the amount of air allowed to enter or escape the home. Any air that is heated and then allowed to exit the home or any cold air that enters the home adversely affects the heating load on the heating system, Fig. 15-9. During construction, great care needs to be taken to ensure that air infiltration is held to a minimum. In most homes, this is the single largest factor causing excessive heating costs. Yet, it is the most inexpensive problem to fix.

With currently available materials it is possible to construct a building so tightly that there may not be enough air exchange to maintain a healthy indoor environment. In super insulated structures an air to air heat exchanger is installed to ensure the proper air exchange. The exchanger moves polluted indoor air, via duct work, to the outside. At the same time, the exchanger moves fresh outdoor air into the structure.

The two air streams pass very near to one another in the exchanger. This allows some of the heat loss or gain to be exchanged between the air streams. This helps the exchanger to operate at a very high efficiency while providing healthy air for the structure. Years ago when buildings could not be built as tight or efficient, there was no need for this type of air handling equipment.

SOLAR ORIENTATION

Factors other than design also have a direct bearing on the energy needs of a structure. People have used the location and position of their homes to assist in heating and cooling for thousands of years. In the construction of homes and businesses, the free solar energy can be used by simply orientating the structure so that it absorbs some solar energy to provide some of its heating. Even roadways can take advantage of solar energy. Roads that are located on the southern side of hills and mountains require less energy to remove snow and ice, Fig. 15-10.

Material	Thickness in Inches	R-value
Exterior air film	---	.17
Interior air film	---	.68
Concrete block	8	1.11
	12	1.28
Light weight block	8	2.00
	12	2.13
Brick	4	.44
Concrete cast in place	8	.64
Wood sheathing	3/4	1.00
Fiberboard	1/2	1.75
Plywood	1/2	.63
	5/8	.79
Bevel siding	1/2 x 8	.81
Vertical siding	3/4	1.00
Drywall	1/2	.38
Interior plywood panel	1/4	.31
Building paper	---	.06
Vapor barrier	---	.00
Wood shingles	---	.87
Asphalt shingles	---	.44
Linoleum	---	.08
Carpet and pad	---	2.08
Hardwood floor	---	.71
Blanket glass wool insulation	1	3.70
	3 1/2	11.00
	6	19.00
Single pane window	---	1.00
Double pane window	---	2.00

Fig. 15-6. Listed are the R-values of commonly used construction materials.

Fig. 15-7. This fiberglass insulation has a stated R-value of 13. (Lowden, Lowden and Co.)

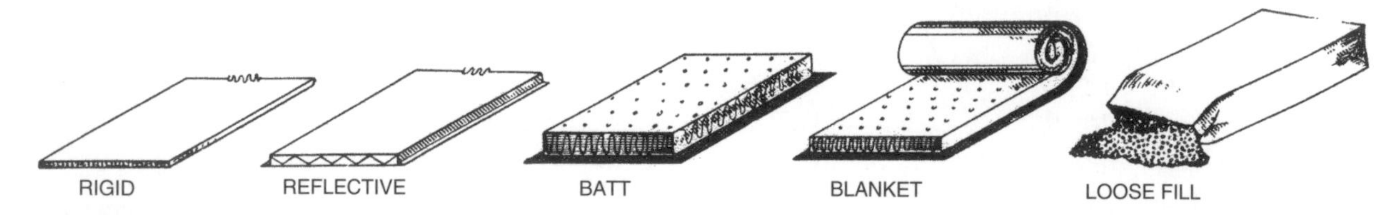

Fig. 15-8. Popular types of building insulation.

Fig. 15-9. Every door and window in a home is an opportunity for heated or cooled air to escape. These areas must be carefully insulated to keep utility bills at a reasonable level. (Velux-America Inc.)

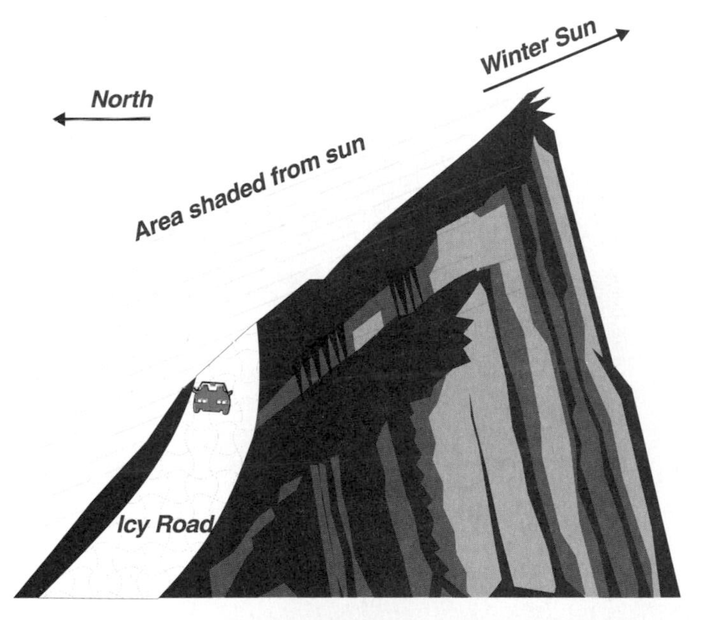

Fig. 15-10. A lack of solar radiation can leave roads icy and dangerous during the winter months.

Energy **conservation** must begin with the planning phase of the construction project in order to take the best advantage of the structure's location in relationship to the sun. Climate, wind direction, and the specific location on the site all have major impacts on the amount of energy a building will require to keep the occupants comfortable.

Generally, if a structure is being constructed in a northern climate, the structure should be placed in a location where it will receive the greatest amount of solar energy. This is to reduce the heating demands in the winter. If the structure is built in a warmer southern climate where there is a short heating season, the need to cool the building is more important. The location of these buildings should be where they will receive the least amount of solar radiation.

SUMMARY

Energy in some form is required for any construction activity. Without energy, no construction project could be started or completed. Depending on what is being built, a variety of energy sources will be required. Some of the energy is required for transportation of materials and labor. The construction job will also require energy to operate tools, equipment, and finally the finished construction site itself.

Depending on what materials are used, the amount of energy needed to convert raw materials into their finished products will vary. Wood, stone, and other naturally occurring materials require the least amount of energy for use in a construction project. On the other hand, materials such as steel, glass, and cement require a considerable amount of energy to be converted into a finished product.

Heat is transferred from one place to another by three methods: convection, radiation, and conduction. By using the correct materials and designing a structure properly, excessive heat loss or heat gain can be controlled. The proper use of insulation in addition to the restriction of infiltration keeps heat transfer under control.

The amount of energy the construction site uses once it is completed is very dependent upon how it is orientated in relationship to the sun. If it is carefully designed to work with the climate of the area, there can be significant savings in energy costs over the life of the structure.

KEY WORDS AND TERMS

All of the following words and terms have been used in this chapter. Do you know their meaning?

British thermal unit (Btu)
Conduction
Conductivity (K)
Convection
Conservation
Heat transfer
Infiltration
Insulation
Nonrenewable
Radiation
Renewable
Resistivity (R)
Solar
Total heat loss
U value

TEST YOUR KNOWLEDGE

Do not write in this book. Please write your answers on a separate sheet of paper.

1. Give two materials that are used in construction that do not require extensive processing before they can be used.
2. List three materials that require a great deal of energy before they are in a form that can be used for construction.
3. List the three methods of heat transfer.
4. If the thickness of a layer of insulation is halved, what happens to the R-value?
5. An exterior wall is made up of 1/2 in. of drywall, three in. of glass wool insulation, and four in. of brick. What is the R-value for the wall?
6. Air infiltration is one of the most difficult heat loss problems to fix. True or False.

APPLYING YOUR KNOWLEDGE

1. Create a drawing showing the different sun angles for your latitude.
2. Add a window with an overhanging roof to your drawing. Show how the roof affects the sun exposure during the four seasons.
3. Design three separate experiments that show how heat is transferred from one place to another.
4. Collect and display four or five different kinds of building insulating materials. Find out their R-values.
5. Research three construction materials. Estimate the amount of energy needed to convert each from its raw material form to the finished product.

MIA-COM, INC.

Jobs cannot always be performed on a site of our choosing. Thankfully, electricity can be transported anywhere.

NCR CORP.

Of the seven inputs to construction, people are the most important

TOM WOOD

MORRISON KNUDSEN CORP.

section 4

Types of Construction Enterprises

Construction activities are part of all technological systems. All of the transportation systems you use have to be constructed. All future transportation systems will have to be built by construction workers. In addition, the transportation infrastructure is maintained by construction workers. The same applies to manufacturing facilities, communications facilities, and structures used for shelter. All of these industrial enterprises rely on construction for their existence.

It is also true that the construction industry relies on the other industrial systems. These systems provide the construction industry with the needed communications links, the transportation of materials and workers, and the manufactured products needed in construction.

In this section, you will examine the enterprises of transportation construction, energy related systems construction, communications construction, and building construction. An understanding of the construction activities needed in each of these enterprises is important to providing a broad perspective of construction.

MARATHON

Industrial complexes must have easy access for materials and for large numbers of employees. Roadways and railways are common methods of transportation. This complex also has water access.

16

CHAPTER

TRANSPORTATION CONSTRUCTION

The information provided in this chapter will enable you to:

- *Identify transportation construction activities and assess their importance to a technological society.*
- *State the importance the transportation construction industry is to a technological society.*
- *Identify ten different types of transportation structures.*

Transportation systems are those systems used to move people and materials from place to place. A transportation system may be as simple as a path through the woods, Fig. 16-1, or as complicated as the interstate highway systems, Fig. 16-2. The interstate highway links the United States with a road system that allows cars, buses, and trucks to move material and people quickly, efficiently, and safely over long distances.

Fig. 16-1. Despite its simplicity, a dirt path is part of a transportation system.

WHAT ARE TRANSPORTATION SYSTEMS

Often, when people think of transportation systems, they limit their thinking to such structures as roads, bridges, and perhaps tunnels. Some people will also include the railroad system and perhaps airports or ships, Fig. 16-3. Seldom, however, do people think of such things as pipelines, sidewalks, smokestacks, elevators, canals, and waterways. Yet, all of these are part of our transportation system. They provide the means to move people and materials from one place to another.

Fig. 16-2. The interstate highway system crisscrosses the United States. It is a fairly complex transportation system. (Martin Marietta Corp.)

Fig. 16-3. An airport is a unique transportation system. Highway systems, air traffic systems, and very often railway systems meet in the same location. (Airport Public Affairs)

Fig. 16-4. Variances in length, climate, soil composition, and the expected traffic load make engineering every new bridge a challenge.

The size of our transportation system is enormous. In 1990 the United States Department of Transportaiton reported the fixed value of public transportation assets in highways, bridges, roads, aviation and transit facilities and vehicles, inland and coastal waterways, and ports and harbors at approximately 800 billion dollars. Included in these assets were 2.2 million miles of paved highways and roads, 150,000 miles of private railroad track, and 26,000 miles of commercially navigable inland and coastal waterways. Also included were 757 commercial ports, 140 million automobiles, 40 million trucks, 5300 commercial aircraft, 220,000 general aviation aircraft, 20,000 intercity buses, 80,000 local transit service buses, 10,000 subway cars and other commuter transit cars, 1.1 million miles of natural gas pipeline, 205,000 miles of oil pipeline, and four active launch pads for governmental and commercial spacecraft.

To create and maintain these transportation systems a great deal of construction activity must take place. As an example, look at the highway system. Then, tighten your focus to the need for the construction of a variety of bridges. The design of each bridge, Fig. 16-4, depends on such factors as the length that the bridge must be, the composition of the earth supporting the structure, the available materials, and the look of the completed structure. Those are just a *few* of the factors that go into the design of this one small part of the road system. A great amount of work is needed for the design and construction of each bridge. Of course, there are many other factors that deal with the final design of the road itself.

TRANSPORTATION SYSTEMS

The criteria that are important when transporting people or materials vary. The shape or special needs of the object being moved will affect the transportation system chosen. Very often, cost is the most important factor in choosing a path. Yet, other times the need for speed is even more important. Most travelers would pay the added expense to fly in an airplane across the country, saving the several days needed to travel over land. These varying needs have given rise to many different modes of transportation. Each transportation system constructed has a large number of components that must be properly designed and constructed to make the total system work properly.

PIPELINES

Pipelines are a very important transportation system. Pipelines take advantage of the fluid motion of liquids and gases. They transport materials such as water, oil, and natural gas in a speedy and economical method, Fig. 16-5.

Fig. 16-5. Pipelines are a very economical way to move fluids. Large pipelines bring oil and natural gas across much of the United States. (Montana Power)

The most common types of piplines are those that carry gas and oil throughout the North American continent. Most of these systems are underground and provide a primary transportation system that is very effective and safe.

Pipeline construction consists of supports for the pipe itself. These supports may be structures above the ground or just a ditch to contain the line. A pipeline may also have **pumping stations** built along its route. These stations help in the movement of the fluid from place to place.

Solid materials such as coal, wood chips, grain, sand, and gravel can be transported through pipelines if these materials are mixed with air or water. Often these materials are transported within a manufacturing facility by these methods. Engineering tests are being conducted to determine the feasibility of transporting coal from western states to eastern states by the use of pipelines.

A conveyor system acts like a pipeline in that it continuously moves material from place to place, Fig. 16-6. Examples of conveyor systems are the movement of gravel from the rock crusher to the stockpile, moving people on a "people mover" in an airport, and moving baggage in an airport terminal.

WATER TRANSPORTATION

Water serves as an excellent means to transport great quantities of goods and materials. However, to make it efficient, shipping ports, docks, and locks, Fig. 16-7, are required. Construction of these components of the shipping industry must be carefully designed and built.

Because of their great size, large ships and barges can be considered part of the construction industry even though they are also part of the transportation system. They are often constructed one at a time and require the skills of many construction workers, Fig. 16-8.

Canals and waterways can be very complex projects, Fig. 16-9. In addition, these waterways vary tremendously in scope. Large canal projects require extensive earthwork to be completed. On a smaller scale, fresh water canals used to supply cities and farms with water also require the use of construction knowledge and skill to be properly designed and built. In the Pacific Northwest, where salmon must migrate upstream, a fish ladder is a very important waterway. These ladders must be designed to allow the fish to travel upstream past dams to their spawning grounds. Without the fish ladders, the salmon would not be able to survive.

The best known example of a lock system is the **Panama Canal.** In this case, one set of **locks** is used for lifting the ships around one hundred feet up to the canal system. The canal provides a waterway for the ships to move across the isthmus of Panama. Another set of locks lowers the ships back down to sea level on the other side. This construction project was very extensive and took many years to complete.

RAILROADS

The railroad system in the United States is another example of a transportation system that has been

Fig. 16-6. While most conveyors are relatively short, it is not a requirement. This conveyor stretches for two miles.

Fig. 16-7. Numerous dams have been constructed to control waterways and generate power. This creates a demand for a system of locks to allow ships to pass.

Fig. 16-8. The construction of a large ship involves many of the same construction techniques and workers as does the construction of a large industrial building. (Morrison Knudsen Corp.)

Fig. 16-9. Waterways are an important transportation system. They may be used to transport goods by boat, or it may be the water itself that is to be transported. (Morrison Knudsen Corp.)

constructed over many years. The **railroad system** provides an excellent means of moving people and material over long distances in a safe, economical manner.

Today, railroads are meeting the needs of a variety of users. Specially designed rail cars allow semitrailers to be loaded aboard and transported long distances across the country. This allows local delivery via trucks combined with the efficiency of long distance rail travel. Also, shipping containers off-loaded from ocean-going ships are loaded onto special rail cars to be transported inland. Once these containers reach a major population center the entire container is loaded onto a semitrailer to be transported to the customer. The facilities to load and unload these large containers must be properly constructed to move the freight safely and practically.

Long trains transporting a single material, grain or coal, are becoming common sights. Unit coal trains may even be loaded without stopping by the construction of specialized loading equipment at the coal mine. This allows for the maximum use of the tracks, locomotives, and rail cars.

Like the highway system, the railroads consist of more than a simple roadbed. Bridges and tunnels must also be designed. Hills pose an additional complication. Unlike the highway system that may travel over hills with ease, railroads have to be very carefully designed. Gentle grades are necessary because of the limited traction the locomotive has on the steel rails. See Fig. 16-10.

Fig. 16-10. The mountains rise up around the train, yet the slope of the tracks remains gentle.

AIR TRANSPORTATION

Another transportation system that relies on heavy construction is air transportation. In this case, a carefully constructed airport system must first be built before airplanes can take off or land. Construction of an airport requires the design of runways that can withstand the heavy loads of large planes, Fig. 16-11. Also, once the airplane has landed, a facility must be available to load and unload people and goods.

Along with construction of the airport, there must also be a system of moving goods, materials, and people into and away from the airport. Adequate highway and railway service is vital to all airports. The construction industry is needed to create and build these projects.

Launch pads for space travel represent another complex construction venture. They require some work that is very similar to airport construction. Although in the case of launch pads, additional construction must be in place to handle the actual launching of the spacecraft. In addition, a special transportation system to move the craft from its assembly building to the launch site is required.

Fig. 16-11. Airport runways must be prepared for the loads from large airplanes landing. In addition to the weight of the plane itself, some aircraft may be carrying quite a payload. (Morrison Knudsen Corp.)

SMOKESTACKS

People seldom think of a smokestack as a transportation system. Yet, a **smokestack** transports hot gases from one place to another, Fig. 16-12. A material is moved from one place to another.

Building smokestacks requires specialized skills from a construction company. The stack must withstand the hot and sometimes caustic gases without deteriorating. Often we see large stacks associated with electrical power generating stations and heavy industry. Today, there are many regulations controlling the emissions placed into the atmosphere. This has changed the way that many smokestacks are built. In addition, fewer new stacks are being constructed today than in years past.

PEOPLE MOVERS

Where there are large concentrations of people, there are various types of people moving transportation systems.

Fig. 16-12. Smokestack regulations have been tightened in recent years. Pollutants must be filtered out of the stacks before the gases can be released into the air.

These systems consist of sidewalks, escalators (Fig. 16-13), elevators, and subways.

Based on the volume of people moved by a mechanical system, the **elevator** is by far the safest transportation system. Generally, these transportation devices are considered part of the building construction industry. However, they are very important in transporting people from one level to another and consequently are also part of the transportation system.

The **subway** is another way to move large numbers of people relatively short distances in a very efficient manner. Subways are a rail transportation system for effectively moving large numbers of people in large population centers. They are similar to railroads in that they operate on rails. They are different in that, for the most part, they operate underground in an extensive tunnel system. They are also electrically operated. The electricity may be supplied by electric rails or overhead lines. They have an advantage in that they do not use up surface space. Though, this has the related disadvantage of requiring expensive tunnel construction. Extensive engineering is needed to complete a modern subway system.

ROAD SYSTEM

The road system in the United States is vast. People have become totally dependent on a system of paved roads to move ourselves, our goods, and our services from place to place. Without our network of streets, roads , and highways we would not be able to move these great numbers. The building and maintenance of the road system has been, and will remain, an enormous construction activity.

SUMMARY

The materials or resources needed to complete a task are seldom in one location. Consequently, we rely on a transportation system to move materials, products, and people from one place to another. This transportation system is dependent on the construction system to design and construct safe and efficient methods to move these people and materials.

The transportation system consists of more than highways, railroads, and airports. The vast system of pipelines, waterways, sidewalks, and elevators also play an important role in society. Through the years, designers and builders have been able to refine the designs of these various transportation systems. This ensures the construction of safe and efficient means to move people and materials.

Fig. 16-13. Typically, escalators transport people only one or two floors.

KEY WORDS AND TERMS

All of the following words and terms have been used in this chapter. Do you know their meaning?

Elevator
Launch pad
Lock
Panama Canal
Pipeline
Pumping station
Railroad system
Smokestack
Subway
Transportation system

TEST YOUR KNOWLEDGE

Do not write in this book. Please write your answers on a separate sheet of paper.

1. What is the definition of a transportation system?
2. List seven separate transportation systems.
3. Explain why a pipeline is a transportation system.
4. It is vital for airports to have adequate highway systems connecting them to the surrounding area. True or False.
5. What transportation system moves the most people in the safest manner?

APPLYING YOUR KNOWLEDGE

1. Write a journal listing and describing all of the various transportation systems that you have relied on over the past week. Discuss your journal entries with others in your class.

2. Collect photos from magazines depicting the various transportation systems in your hometown. Turn these photos into a display for the classroom bulletin board.
3. Have a discussion with others in your class on how the transportation system has changed over the past 10, 25, and 50 years.
4. Create a scenario for the transportation system of the year 2050.
5. Review the public transportation system in your hometown. Write a list of suggestions for ways that it might be improved.

A complex roadway system covers the United States. This system has cut travel times, changing our society.

CATERPILLAR

After the site is cleared, a level road service is prepared.

17

CHAPTER

ENERGY RELATED CONSTRUCTION

The information provided in this chapter will enable you to:

- *Identify construction activities that are related to the production of energy.*
- *Assess the importance energy construction plays in the total field of construction.*
- *Identify ten different energy related construction projects.*

Energy related construction involves the construction of those projects that deal with the production of energy from its natural state. These construction activities include drilling for oil, mining coal and uranium, constructing hydroelectric dams, and harnessing renewable energy sources, Fig. 17-1. Without these energy sources, our modern civilization could not exist. Society depends on a plentiful supply of energy to heat and cool homes and office buildings, to operate transportation systems, to trans-

Fig. 17-1. The construction of windmills and wind turbines taps into a relatively low cost and pollution-free energy source.

mit messages over long distances, and to supply power for manufacturing.

The construction industry plays a critical part in the development of our energy sources. Without construction, few of the natural energy sources could be efficiently harnessed. In turn, the construction industry relies on a plentiful supply of energy. As with other technology systems, one system is dependent upon the other.

HYDROELECTRIC PROJECTS

Hydroelectric dams, Fig. 17-2, are some of the largest single structures ever built by humans. In total cubic yards of concrete, project dams such as Grand Coulee Dam and the Aswan High Dam have no rivals. There is enough concrete in Grand Coulee Dam to build a sidewalk four inches inches thick and eight feet wide that would extend from Washington DC to Seattle, Washington.

The design, construction, and management of a project of this type is a major undertaking. The output of such a project is useful to millions of people. Hydroelectric dams produce electrical power that is used to power our homes, offices, and industries.

In the northwest part of the United States and in the Rocky Mountains, where there are large fast flowing rivers, hydroelectric dams are very effective in producing cheap electrical power. In other parts of the United States, the lack of high mountains and the need to maintain open waterways for shipping makes it difficult and less cost effective to build large dams.

PETROLEUM INDUSTRIES

The petroleum industry requires a tremendous amount of construction labor. Construction locations are scattered across the land and some extend out into the oceans. In addition, the construction required in this field is as varied as it is specialized.

OIL RIGS

The building of large **offshore drilling platforms** as well as **land based drilling rigs** falls into the construction field. These large rigs are often moved from one location to another. Yet, due to their size, they are still considered construction projects. A land based drilling rig may be 150 ft. tall, Fig. 17-3.

Moving a land based drilling rig is no small task. To move the rigs and all of the associated equipment needed in a drilling operation, a large number of trucks and workers are required. The workers must disassemble the rig before it can be moved. Upon arrival at the new location, the rig must then be reassembled.

Offshore drilling platforms, on the other hand, can be towed by ships (tug boats) from one location to the next. In both cases, their construction requires large quantities of steel and workers as well as heavy construction techniques and methods.

REFINERIES

The building of **refineries,** Fig. 17-4, is another petroleum related project that requires specialized skills in construction fabrication. In this case, the designer has to determine the best location for the large quantity of required equipment. The construction workers then build and connect the various pieces of equipment into a maze of pipes, fittings, tanks, and pumps.

As with any industry, refineries require a great deal of maintenance and remodeling. Minor maintenance is con-

Fig. 17-2. Hydroelectric dams, like windmills, are relatively pollution-free energy sources.

Fig. 17-3. Large drilling rigs (right) are erected and taken down many times by construction crews in the search for oil. (The Coastal Corp.)

Fig. 17-4. Numerous plumbing systems dominate oil refineries.

ducted on a continual basis. Major remodeling and maintenance requires the refinery to shut down its entire operation for a short time to complete the scheduled procedures. When the refinery is completely shut down, no profits are made. Consequently, these shutdown times are held to a minimum. Scheduling work activities during the shutdown is very important in getting the refinery operating again in the shortest possible time. For the safety of the workers, when they are working close to highly combustible materials such as petroleum products, strict standards are followed. This may require specialized clothing such as shoes, suits, and breathing apparatus. Construction techniques and requirements as well as tools and equipment are also specifically designed to ensure the greatest safety for the worker.

OIL STORAGE TANKS

Because these petroleum products are used in large quantities in all cities and by most industries, there is also a need for large storage tanks to supply fuel for local customers. Some of these oil storage tanks are over 300 ft. in diameter and nearly 100 ft. tall containing many millions of gallons of petroleum.

To meet today's environmental laws and regulations, each tank must be contained within a pit that will hold the entire contents of the tank should the tank spring a leak. The tank is constructed from heavy steel. In some cases, the steel is over one in. thick. This requires expert welders to fabricate. Often, all of the welds on a tank are x-rayed to ensure there will be no leaks. Some oil storage tanks have a roof while others have a top that floats on the liquid contents in the tank.

OIL TANKERS

The United States uses much more petroleum than it produces. Thus, there is a need for large ships, Fig. 17-5, to transport the crude oil from foreign ports. Many ships are much larger than homes and in some cases larger than entire office buildings. Consequently, the building of these ships is considered a construction project more than a manufacturing job. The techniques and large equipment needs match the construction field perfectly.

Fig. 17-5. Oil tankers move slowly through rivers. Their tremendous size overshadows almost all other vessels.

MINING

Mining is another task used in power generation. Two common fuels dug out of the earth are coal and uranium. Coal and uranium mines provide fuel for plants that produce electricity.

COAL MINING

In the case of coal, Fig. 17-6, the process used to retrieve the ore depends on how close it lies to the surface of the earth. If the coal seam is close to the surface, then large earthmoving equipment is used to first remove the **overburden.** This exposes the coal seam. Next, large trucks, Fig. 17-7, and loading equipment are used to remove the coal and stockpile it in a location where it can be shipped to customers.

If the coal is deep in the ground, different types of equipment are used to mine the ore and transport it to the surface. In both cases, large specialized earthmoving equipment are needed to move large quantities of coal.

URANIUM MINING

Uranium is used in nuclear power plants, Fig. 17-8. It is also mined using these conventional mining methods.

Fig. 17-6. Coal is burned to drive steam turbines in electrical generating plants. (DuPont)

Fig. 17-7. Coal is loaded onto large trucks and stockpiled until it is needed.

This involves large earthmoving equipment to first remove the overburden off the mineral bearing rocks. The uranium ore is then hauled to a rock crusher where the ore is crunched into a fine powder. This allows the ore to be concentrated before it is further refined. Because the uranium ore is in such a low concentration in its raw form, there is little danger from its radiation when workers follow the proper safety procedures.

In the United States at this time there are very few uranium mines operating because of the small number of nuclear power plants in operation. As the demand for electricity continues to grow, and the cost of fossil fuels increases, wc may turn to using more nuclear power plants to generate electricity.

RENEWABLE SOURCES OF ENERGY

During the energy shortages of the seventies, people became aware of dwindling supplies of fossil fuels. The demand arose for the generation of energy from renewable sources. Solar power, wind power, and water power gained extensive attention. The construction industry was called upon to construct large numbers of these renewable energy facilities.

Fig. 17-8. The Palo Verde Nuclear Generating Station is the United State's largest nuclear plant. (Arizona Public Service Co.)

Solar panels, Fig. 17-9, were installed on homes and offices. Large **energy farms,** which concentrate the solar energy to produce heat and electrical energy, were constructed. Also, large **wind farms,** Fig. 17-10, were built to harness the energy of the wind. Various types of windmills have been developed to generate sustainable power. Although at this time, only a small percentage of our energy needs are supplied by the wind, in the future we may see greater use of this source of power. Studies have already been done to determine in what areas it is economical to produce electrical energy from the wind.

More recently, large gasohol plants have been designed and built. **Gasohol** plants use the excess production of grain. Through a chemical process, they convert the grain into alcohol, which is then blended with gasoline. This fuel burns well in automobiles and helps conserve our nonrenewable resources such as oil.

Fig. 17-9. Solar panels take advantage of an inexhaustible source of energy, the sun.

Fig. 17-10. Acres of land on the west coast of the United States are filled with wind turbines.

Another energy industry that relies on construction is **geothermal** energy. In this case, the earth's core serves as the source of heat that produces hot water and steam. This hot water and steam may then be used to generate electricity, Fig. 17-11, providing a source of energy for buildings. Again the construction industry is called upon to design and build the structures needed to harness this energy source.

One area that could be employing construction workers in the future is in drawing energy from waves and tides. With many of the major population centers on or near oceans, it may prove cost effective to produce electrical power by using wave action. Many ideas have been proposed to use the force of the waves to produce electricity. As the cost of fossil fuels increase, more research and development will be done in this field.

There are many other sources of energy being used or are being considered for use in the future. The construction worker will be a key player in the development of these new energy sources.

SUMMARY

As with all other technological systems, the construction industry is critical in the development of our energy sources. Without the construction industry, it is doubtful if any of the sources of energy could be developed.

Fig. 17-11. Geothermal energy often appears in the form of geysers on the earth's surface. (Cathy Stanford)

Energy sources can be roughly divided into fossil fuels and renewable energy sources. The construction worker is needed to build and repair all of the energy related projects. Also, the construction worker depends on the energy produced by these plants for the energy needed to continue to work in the construction field.

KEY WORDS AND TERMS

All of the following words and terms have been used in this chapter. Do you know their meaning?

Energy farm
Gasohol
Geothermal
Hydroelectric dam
Land based drilling rig
Offshore oil platform
Overburden
Refinery
Uranium
Wind farm

TEST YOUR KNOWLEDGE

Do not write in this book. Please write your answers on a separate sheet of paper.

1. List six energy related industries that rely on construction workers for their construction and maintenance.
2. Explain why it is necessary to have a reliable source of energy in a modern technological society.
3. Explain the relationship between energy related industries and the construction industry.
4. List eight different sources of energy and identify each as nonrenewable or renewable.

APPLYING YOUR KNOWLEDGE

1. Take a survey of the energy related industries in your local community. Discuss your list with others in your class.
2. List the careers you might consider as a construction worker who specializes in working on energy related construction projects.
3. Make two lists showing the interrelationship between energy and the construction field. Compare your list with your classmates.
4. Make a future forecast of what might happen to the energy picture in this country in the next 50 years.
5. Suggest ways at school and in your home on how you might be able to conserve energy by adjusting everyday life-styles and habits.

18
CHAPTER
COMMUNICATIONS CONSTRUCTION

The information provided in this chapter will enable you to:
- *Explain the importance of the construction industry to the communications field.*
- *List communications systems that rely on the construction industry for construction of their components.*

In order to establish a successful and rapid mass communications system, buildings and towers must be constructed. Without a construction system to install fiber-optic lines, telephone cables, Fig. 18-1, and satellite dishes, Fig. 18-2, communications systems would not be able to operate at the almost instantaneous rates at which they now run. The communications system relies on the construction field to build and install major components for all communications systems.

Modern communications systems consist of various **antennas,** Fig. 18-3, and land lines that must be installed. The installation of these communications components is part of the overall field of construction.

Fig. 18-2. Installing satellite dishes is a specialty of some construction crews.

Fig. 18-1. Large phone lines are buried underground to protect against accidental damage caused by the weather or other forces.

BASIC COMMUNICATIONS SYSTEMS

When people think of communications today, television, fax lines, satellite systems, and other electronic networks come to mind. Still, there are several other basic communications systems that affect our daily lives. Road signs and billboards are good examples.

Fig. 18-3. Radio transmission requires the construction of large antennas.

ROAD SIGNS

On first thought, we may not think of simple road signs, Fig. 18-4, such as mile markers, curve signs, no passing zones, and speed limit signs as part of the communications system. However, without these signs, driving would be much more dangerous. These systems all have to be installed and maintained by the construction workers.

Fig. 18-4. Road signs communicate very important information. The orange construction sign and the flashing yellow arrow give advance warning of a lane closure. This allows traffic to merge smoothly.

Within the vast road system in the United States, there are millions of road signs that must be attached to sign posts and planted along the road. These activities employ a large number of private and government construction workers. It is estimated that to just change the mile post markers on the interstate highway system to their metric equivalent, kilometers, it would cost over $200,000 in labor alone.

BILLBOARDS

Advertisement is a very important part of our society. A portion of the advertisement industry relies on visual communications provided by signs placed to attract the eye of travelers, Fig. 18-5. The installation and maintenance of these **billboards** is performed by a number of construction workers. Specialized companies have been established to provide this service to customers.

ELECTRONIC COMMUNICATIONS SYSTEMS

Our electronic communications travel in many ways. Some messages travel along copper or fiber-optic lines underground. Other messages are sent as electromagnetic waves through the air. Both methods of transportation require construction installations.

LAND LINES

Land lines are those communications links that rely on copper or fiber-optic lines to directly link one location to another. Construction workers must physically connect each of the components together. These components in-

Fig. 18-5. Billboards are a common sight along highways.

clude the telephone lines found in homes and offices, Fig. 18-6.

An electrical worker connects and routes cables from a main line to each individual communication device. Outside of the building, construction workers must run the wire or cable to take the signal to distant locations. These cables may be buried underground or held overhead on towers or poles.

ELECTROMAGNETIC WAVE COMMUNICATIONS

There are many different methods to send information through the air, and there is an even greater number of types of information transmitted using these methods. To set up all of these systems, towers and satellites must be produced and buildings of operation must be constructed.

Fig. 18-7. Cellular telephones have become extremely popular. These telephones are dependent on local transmitter towers constructed in their area. (J. Majka)

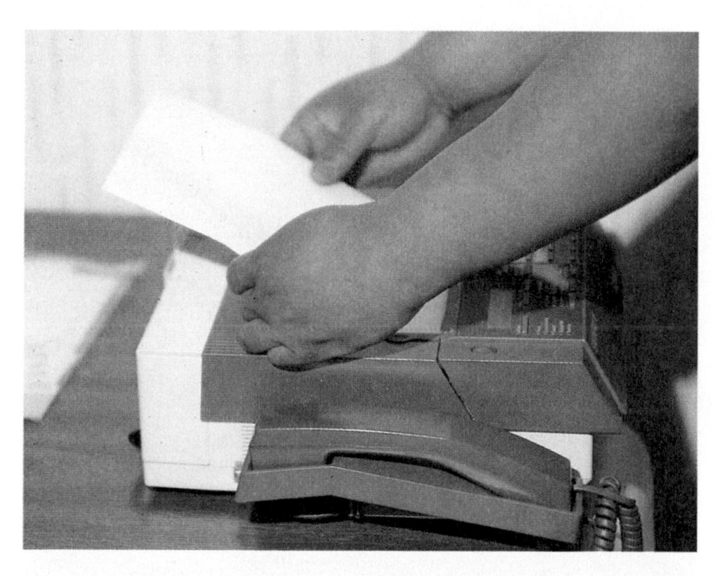

Fig. 18-6. FAX machines, like telephones, are connected with land lines.

Earth Based Transmitter Towers

Everybody listens to the radio or watches the television. Without the construction of the radio and television towers or the buildings that support this communication field, you would not be able to enjoy these communications systems.

The cellular telephone, Fig. 18-7, which is very popular in cars, depends on transmitter towers as well. These towers are called **local responders.** These towers pick up the low power signal from the car phone and relay it to more powerful transmitters. These, in turn, send the signal through standard communications networks. Construction plays an important part in the construction of these local transmitting sites.

Satellites

Communications systems also rely indirectly on construction systems. Today, many communications systems use distant **satellites** to relay communications signals, Fig. 18-8. Although the satellite itself is not part of the construction system, a launch pad was needed to place the satellite in orbit. Construction crews are responsible for building and maintaining these sites.

Many other aspects of satellite communications rely on construction systems. The facility in which the satellites are built, as well at those buildings that shelter the computers and communications networks, are important construction projects. Earth bound transmitters and receivers are installed by construction crews.

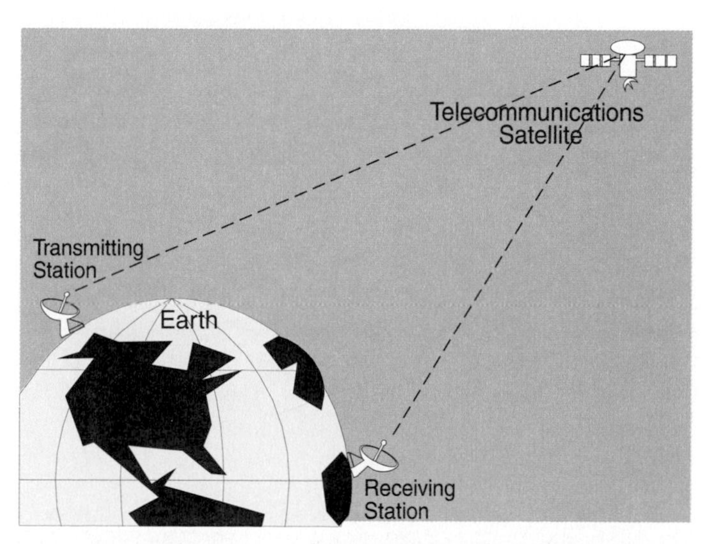

Fig. 18-8. Telecommunication satellites simplify reception from distant points on the earth. Without the satellites, many more transmitting and receiving towers would be necessary.

Satellite Dishes

The signals sent by the satellite are useless if there is no where for the signal to be received. Dishes that pick up satellite signals have become common sights. If one travels to rural areas beyond the cable television networks of the city, you will see more and more home satellite receiving dishes.

The satellite dish allows rural homes to receive many of the same programs available to city dwellers. Installation of the satellite dish requires holes to be dug, concrete and steel installed, and electrical lines installed. These activities are most often done by a construction worker who is familiar with these fixtures, Fig. 18-9.

Commercial satellite dishes are often installed on the tops of mountains, on towers, or on the tops of tall buildings. Again, all of these installations require the skills of construction workers.

Fig. 18-9. Skilled workers put the finishing touches on a dish they installed.

Radar Domes

Although radar signals do not penetrate your activities in the same fashion as television and radio waves, they affect your daily activities. Radar is associated with airports, the United States Defense system, the Weather Bureau, and shipping channels.

Radar is a sophisticated communications system that identifies the location of objects. This communication system relies on radar dishes that are in constant motion. Because the radar system must operate in all kinds of weather, it must be housed inside of a dome that is transparent to the radar waves. The design and construction of the dome and support structure are part of the construction industry. Without these structures, radar would not be usable during bad weather, when it is needed most.

COMMUNICATIONS CONTROL SYSTEMS

Traffic lights and railroad crossings are commonly used communications systems, Fig. 18-10. These systems regulate traffic flow. It is easy to imagine what would happen if there were no means of controlling traffic in our busy cities. In addition to the many delays and traffic jams, the loss of life would skyrocket with the number of accidents. Even though these signals are communications devices, construction crews do all of the installation. Again, without the construction industry we would not be able to be served by these communications devices.

AIRPORT RUNWAY LIGHTS

All commercial airports have runway lights. These lights mark the edges of the concrete or asphalt runways and allow the pilots to land their planes in the dark. Today's modern planes do have equipment on board to assist in the landing of the plane. However, standard airport regulations require minimum visibility in order to land the plane. Runway lights assist the pilots in the landing of the plane and also help the air traffic controller to identify where planes are situated on the runways. These lights are a vital part of the airport communications system.

LIGHTHOUSES

The shipping industry relies on the accurate location of reefs and small islands that are near shipping lanes so they can steer clear of these dangers. One recent notable mishap was the Exxon Valdez oil tanker. The ship got off course and ran into a reef in Alaskan waters.

Lighthouses and beacons are constructions that are built to mark and identify these dangers. They assist in the

Fig. 18-10. Railroad crossing signs are painted with a highly reflective paint. Many railroad crossings are also equipped with flashing red lights.

navigation of the ships. These structures, Fig. 18-11, are communications systems. They communicate the location of areas of danger and help to locate safe routes.

SUMMARY

The communications system relies on the construction system for the creation of its buildings and towers as well as the installation of other components. The systems are interrelated, and one cannot adjust any part without having an impact on the entire technological system.

Satellite systems depend on earth based transmitters and receivers that must be installed and maintained. Phone systems rely on land lines that connect into a network. These networks must be installed by construction workers either in the ground or overhead. Traffic signals must be properly installed and maintained. All of these communications activities depend on knowledgeable construction workers.

KEY WORDS AND TERMS

All of the following words and terms have been used in this chapter. Do you know their meaning?

Antenna
Billboard
Land line
Local responder
Satellite

TEST YOUR KNOWLEDGE

Do not write in this book. Please write your answers on a separate sheet of paper.

1. List five communications installations that rely on construction workers for their construction and maintenance.
2. Explain why it is important to have an extensive communications system in modern society.
3. What information does a traffic light relay?
4. Cellular telephones depend on local _____ to relay information.

APPLYING YOUR KNOWLEDGE

1. Keep track of the various communications devices you depend on each day to carry on your normal activities. Discuss your list with your classmates.
2. Make a poster showing the relationship between construction and communications systems. Be sure to point out how each system is dependent upon the other.
3. Make a diagram of the construction careers that are associated with the communications industry.
4. Invite a communications worker to visit with you about construction jobs in her/his field.

Fig. 18-11. Lighthouses communicate with shipping traffic. They ward boats away from islands and dangerous reefs.

Building systems are typically located in one central area.

19

CHAPTER

BUILDING CONSTRUCTION

The information provided in this chapter will enable you to:

- *Explain the importance buildings play in a modern technological society.*
- *Identify three types of building construction found in the construction industry.*
- *List several differences between the three types of building construction.*

Our society today is dependent on buildings. Without buildings, our civilization would be limited to those areas where the temperature was moderate. The climate would have to be one such that people could survive without shelter.

Buildings and shelters allow us to inhabit all parts of the world. People live and work in the Antarctic, Fig. 19-1, where the temperature may be –100 degrees Fahrenheit. People can live at the bottom of the ocean, where there are great pressures. People have lived for short periods of time on the moon, Fig. 19-2, and there are current plans to launch an inhabited space station. In the vacuum of space, unprotected exposure means instant death. Building construction technology allows us to live and work in these environments.

GENERAL BUILDING CONSTRUCTION

Construction of all buildings has many similar aspects. They all require a foundation system, a wall system, a floor system, a roof system, and utility systems. Although the specific details of each system will vary depending upon the type of building being constructed, they all follow the same principles of construction.

The skills required by the architects, engineers, supervisors, and workers involve the same basic understanding of construction systems. The construction materials may vary in type, size, and shape. The fastening system used on the materials may be different, but the principles do not change.

Fig. 19-1. In the Antarctic, buildings are not just for comfort, they are necessary for survival. (Morrison Knudsen Corp.)

Fig. 19-2. Artist's conception of a permanent lunar outpost. (NASA)

TYPES OF CONSTRUCTION

In general terms, the construction of buildings can be broken down into three types.

- Residential buildings, Fig. 19-3.
- Commercial buildings, Fig. 19-4.
- Industrial complexes, Fig. 19-5.

Each of the three types of buildings poses a different set of requirements upon a contractor.

RESIDENTIAL CONSTRUCTION

Residential construction is the construction of homes for people to live in. This includes single family homes as well as multi-family homes and apartment houses. Construction of these buildings most often involves the use of wood framing materials. These materials are put into place without the use of heavy construction equipment, Fig. 19-6. Consequently, residential construction is sometimes called **light construction.** Residential construction also includes mobile homes as well as modular homes, Fig. 19-7.

Fig. 9-3. Residential construction.

Fig. 19-4. Commercial construction.

History of residential construction

The first type of human construction effort was to provide some protection from the elements. Early dwellings

Fig. 19-5. Industrial construction.

Fig. 19-6. The construction of wooden-framed homes uses very little heavy construction equipment. (Lowden, Lowden and Co.)

Fig. 19-7. Residential construction also includes the construction of mobile homes, even though many mobile homes are produced in factories.

were natural caves or in the shelter of large trees. As humans began to travel away from natural shelters, it was important for them to design and construct shelters.

Early residential structures were made from simple natural materials such as animal skins, tree branches, and rocks. As they learned more and more about how to work with materials, their homes became more complex and provided better shelter. Thousands of years ago builders learned how to shape stone and wood to meet their specific construction needs and the modern age of residential construction began.

Current residential construction

Residential construction is the type of construction that is most familiar to most people. We all live in residential structures and, therefore, have a better understanding of this type of construction. In terms of the number of buildings constructed, residential construction is the most common type of structure built.

As the size of the residential structure becomes larger, heavier equipment and more workers must be used. Multifamily units, Fig. 19-8, or very large and expensive homes, Fig. 19-9, often require heavy equipment.

Fig. 19-8. For dwellings that house a multiple of families, contractors will make use of heavy construction equipment.

Fig. 19-9. Heavy equipment may be needed for complex landscaping on more expensive homes. (Lindal Cedar Homes, Inc. Seattle, Washington)

Multiple-dwelling units such as high-rise apartment houses, Fig. 19-10, use the same techniques as commercial construction. Building codes in the area have a great deal to say as to the type of construction required. The number of people who will occupy the building has a great effect on the requirements that must be followed in the construction of the structure.

COMMERCIAL BUILDINGS

Commercial buildings are those buildings in common use by the general public, Fig. 19-11. These buildings include schools, banks, stores, churches, office complexes, sports centers, malls, libraries, and government buildings.

For the most part, commercial building is multi-story. This requires stairs and/or elevators to move people and goods from the first floor to the other floors. Also, the commercial building may be required to have a fire sprinkler system, emergency lighting system, sound proofing, larger utility systems, and a system of signs that provide directions for the public who will be using the building.

Because of the increased foot traffic in the commercial building, materials used to construct the building have to be of a higher grade than found in residential construction. The maintenance and upkeep of the building after it is put into service are significant factors in selecting what finishing materials will be used for the flooring, walls, and finished utility systems.

One example of the differences in construction can be seen in something as simple as the hinges on doors. In commercial construction the hinges used are made much stronger than those in a residential building. The reason, of course, is that the door in a home is not opened and closed as many times a day as a door in a store.

Legal requirements for commercial buildings

Unlike most homes, commercial buildings must be designed to allow easy access for all people. This requires accommodations to be made for people with handicaps. These accommodations include the installation of ramps, elevators, braille signs, wide doors, and specialized plumbing fixtures. Exterior accommodations to streets for delivery of people and materials are also considerations when designing the building.

Although some of these commercial buildings use wood frame construction, they are more often constructed out of steel and masonry products. These materials allow the buildings to be made larger in size and assist in meeting the more stringent **fire codes.** Public safety is a major consideration in the construction of commercial buildings. Large numbers of people use these building every day. Consequently, the buildings have stricter building codes than residential construction projects.

Another difference in the construction of commercial buildings is in the location where the structures can be built. Commercial buildings are generally built on land that is specifically zoned for business, Fig 19-12. The commercial areas often have much better access for traffic.

With many stores and offices occupying spaces so close to each other, the risk of fire spreading from one business to the next is great, Fig. 19-13. Consequently, the need for the construction to be fireproof is even greater. This need affects the design as well as the material used in construction. If all of the structures were made of wood, the chance of fire spreading would be greater. Also, the exterior doors of all commercial buildings must swing out. This is to provide an easy exit for large numbers of people if they must leave the building in a hurry. In residential construction, the exterior doors generally swing in.

Fig. 19-10. Large apartment complexes involve extensive use of large construction equipment. Strict building codes govern their construction.

Fig. 19-11. Commercial buildings, like this parking structure, see heavy use on a daily basis. The construction materials used must be superior.

Fig. 19-12. Commercial buildings must be built on land that is zoned for commercial structures. In cities, large residential structures and commercial buildings often can be found side by side. (Cityfront Place/Photographer, David Clifton)

Fig. 19-13. Commercial buildings tend to be closely packed together. Fire is a great danger.

INDUSTRIAL COMPLEXES

Large commercial buildings are often difficult to distinguish from **industrial complexes,** where various types of manufacturing are taking place. The outside of an industrial complex may very closely resemble that of a commercial building. However, the interiors of industrial complexes are designed to house manufacturing equipment. The main design concern in industrial complexes is to make the interior space functional and efficient for production of some product.

Industrial complexes must meet most of the standards and codes required for commercial buildings plus the addition of any specialized needs. If, for example, the industrial complex is a manufacturing plant for steel, it will require specialized air handling equipment to remove dangerous fumes before the exhaust air can be vented from stacks. Also, if the major source of energy to melt the steel is electricity, the electric utility system will have to be very large to accommodate the electrical energy used.

Industrial complexes tend to use large quantities of materials. For this reason, the complexes are often located next to excellent transportation systems such as highways, railroads, or shipping channels, Fig. 19-14.

Often, industrial complexes will cover very large areas of ground. The buildings are more spread out. They are only a few stories high. This makes the movement of large quantities of materials, equipment, and people easier. Industrial complexes may cover many acres of land and be thousands of feet long. The largest building in the world, based on the enclosed volume, is the Boeing airplane assembly building in Everett, Washington, Fig. 19-15. This building covers 98 acres and contains 472,000,000 cubic feet of space, which is more than 50,000 average sized houses.

SUMMARY

Building construction is the area of construction systems that includes the construction of residential, commercial,

Fig. 19-14. The Vehicle Assembly Building in Cape Canaveral, Florida is where the Space Shuttle is remounted. Notice the main transportation lines allowing easy access to the complex. (Morrison Knudsen Corp.)

A

B

Fig. 19-15. The assembly of large vehicles requires a large industrial worksite. A–The Boeing assembly plant in Washington. B–Three aircraft under construction. (Boeing Commercial Airplane Group)

and industrial structures. These structures are designed to house people, materials, and equipment so they can be protected from harsh environments.

Residential construction, primarily, deals with light materials. The buildings are constructed without the use of heavy equipment.

Commercial construction deals will those buildings that are used often by the public. These buildings must meet strict safety standards to protect the public's health and safety.

Industrial complexes are generally very large buildings. Industrial complexes may be spread out on large amounts of land. Their primary use is to provide a quality environment for workers as well as the materials and equipment they are manufacturing.

The construction materials may change from residential to commercial to industrial, but basic principles apply to all building construction.

KEY WORDS AND TERMS

All of the following words and terms have been used in this chapter. Do you know their meaning?

Commercial building
Fire code
Industrial complex
Light construction
Residential construction

TEST YOUR KNOWLEDGE

Do not write in this book. Please write your answers on a separate sheet of paper.

1. List three different types of building construction.
2. There are very few similarities in the construction of the different types of buildings. True or False.
3. Schools, post offices, and restaurants are examples of what type of building?
4. List three differences between the three types of buildings.
5. Industrial complexes tend to be only one or two stories high. True or False.

APPLYING YOUR KNOWLEDGE

1. Along with your classmates, make a list of the commercial and industrial buildings in your community.
2. Develop a chart showing how the three different types of building construction are similar and dissimilar to each other.
3. Discuss how the factory (industrial building) has changed over the past 50 years and how it might change in the next 20 years.

20

CHAPTER

SITE EVALUATION AND PREPARATION

The information provided in this chapter will enable you to:

- *Explain the processes and activities that must be completed before a site is ready for construction.*
- *Locate the position of the construction on the actual construction site.*
- *Explain how batter boards are used.*
- *Identify several pieces of earthmoving equipment commonly used in preparing the site for construction.*

A construction site must be evaluated before any project is built or even bid upon. The contractor who will construct the project must have an accurate idea of what the site looks like.

If the project is a road, the contractor must know how much land must be cleared and how much earth must be removed or added to make the road bed. Perhaps road gravel will have to be hauled long distances. Possibly the rock will have to be dynamited before it can be moved. The contractor must also know about the soil and sub-soil conditions. An understanding of **geology,** the study of the rocks and soil, is vital to a contractor. Bad soil conditions make some projects impossible, Fig. 20-1.

If the construction project is a building the contractor will have to know many similar things. Once again, is more soil needed at the site, or does some of the earth have to be hauled away. If the building is very large, detailed studies will have to be done to determine how much weight the ground can support. The makeup of the soil (the quantity of dirt, clay, or rock) varies from location to location. Some locations can bear much greater loads than others.

Fig. 20-1. Clearing large stones from a project site may be necessary. If the soil is too rocky, this cost may be prohibitively expensive.

SITE EVALUATION

The evaluation of a site often begins with a **survey** of the construction site, Fig. 20-2. The information gathered in the survey bears on all aspects of the project construction. The survey establishes the precise location of the project so that all individuals are certain they are designing and biding on the same thing. Look back to Chapter 4 to review surveying.

Once these surveys are completed, the location of the construction project can be described. All structures, whether they are homes, office buildings, roads, dams, or bridges must first be located on the site very accurately. **Locating** a structure on a site means establishing the distance that specific points on the structure are from the boundaries of the property. A crew marks were the structure is to be established. You have probably seen a number

Fig. 20-2. Readings taken by this surveyor will affect all aspects of the construction project that follow.

of survey crews with surveying instruments locating roads or buildings.

GROUND WATER

An important consideration in site evaluation is the quantity of **ground water** at the location. Water below the surface has an effect on the load carrying capacity of the ground. Sometimes, the water exists in the soil throughout the year. In other cases, the water is seasonal and is present only in the spring, Fig. 20-3. In all cases, the foundation design for the construction project must take this water into account. Otherwise, a stable project cannot be built.

Fig. 20-3. Excess ground water can pose hazards during and after construction. (Caterpillar)

The problem of having a quantity of water in the soil is compounded if the construction project is located in a cold climate. In cold weather the water may freeze, causing it to expand. If not properly dealt with, this can cause serious damage to the projects.

ENVIRONMENTAL IMPACT STUDIES

In recent years, people have become more conscientious about the long range impacts major construction projects have on the environment, Fig. 20-4. Laws have been passed that ensure careful **environmental impact studies** are completed on the potential impact of any proposed construction.

These studies take considerable time to complete. Often, there are hearings conducted to assure that all individuals affected by the construction have the opportunity to give input into the project well before the construction begins. In a number of cases, construction projects were not allowed where the environmental impact study showed that local impacts would have been too great.

OTHER FACTORS THAT AFFECT THE CONSTRUCTION PROJECT

There are many factors that a contractor must research when evaluating a site. Are the roads strong enough and wide enough to handle heavy loads? Are there underpasses that equipment will have to clear when moving to the site? The best ways to move materials to the site must be determined by the contractor, Fig. 20-5.

The contractor must also examine the land *surrounding* the site, Fig. 20-6. If there are large buildings nearby then some method of stabilizing those structures will have to be used. If the project is an addition to an existing structure, the existing structure is part of the surroundings and must be stabilized, Fig. 20-7.

The availability of services needs to be explored. Contractors need to make sure that the local electrical lines are large enough to handle the equipment used during construction. Water, gas, and sewer lines must be available on the site.

The contractor should also examine the local building codes. In some areas, the building codes are not as modern as others. There is also the possibility that the building codes could be revised before construction begins.

Labor is another important consideration. Before tackling a project a contractor should be acquainted with local supply of labor. Are there enough of the proper types of skilled workers? Will the contractor have to rely on work-

Fig. 20-4. Building a road through this forested area has an environmental impact. (Washington Construction)

Fig. 20-5. If heavy machinery is brought to a construction site, a route must be planned that can handle the load.

Fig. 20-6. Construction on this site may affect the buildings nearby. The surrounding buildings need some form of protection.

Fig. 20-7. During remodeling, an existing structure may need to be stabilized as shown.

Fig. 20-8. Contractors need to protect their sites. In addition to guarding against theft, fences stop individuals from wandering into dangerous areas.

ers from other towns and communities to complete the project? If the climate is harsh, special care may have to be taken to ensure safe working conditions. Special products and tools might be necessary to withstand the forces of the climate in the region.

Careful and successful contractors take all of these factors into consideration long before they submit a bid. The more a builder learns about the site, the fewer the number of surprises encountered. This makes a much more successful contractor.

CLEARING THE SITE

It is seldom that a new construction site is in shape to begin the actual construction of the project. The site has to be **cleared** of undesired trees, large rocks, fences, and even other buildings. Sometimes unwanted buildings are moved to new locations. Often the buildings are simply demolished. If the site is in a populated area, there is some danger to individuals who pass by. These sites must be fenced, or guard rails should be installed to protect the public from the hazards of construction, Fig. 20-8.

Up unto this point, no permanent work on the project itself has taken place. However, on some projects, preparing the site is the most difficult part of the entire construction project, Fig. 20-9. In some cases, clearing the site is subcontracted out to contractors who specialize in this difficult type of work.

Fig. 20-9. For construction on this site, many tons of earth were removed. Clearing the site was a major portion of the entire project. (Caterpillar)

SITE LAYOUT AND PREPARATION

Once the location of the site has been determined, the next step is to carefully locate the construction project on the surveyed site. Often, this will require further removal of unwanted trees, brush, and large rocks. After these are removed, and salvaged where possible, the actual construction layout can begin.

The first step is to carefully lay out the boundaries of the project being constructed, Fig. 20-10. If it is a road, the road is marked by **staking.** Staking a road is the process where the edges of the road are first located and elevations are taken. These figures are then used to determine the amount of material needed to be cut or filled to produce the rough grade of the road. The stakes are used to inform the equipment operators how much to cut or fill at particular places along the road.

If the project is a building, its location in relation to the boundaries of the lot must be laid out even more carefully.

Fig. 20-10. Stakes mark the boundaries for the construction workers. (Tom Wood)

In most cities, these distances are carefully controlled in accordance with local building codes. The actual layout of the building is begun by locating the corners of the building. Stakes are put at the corners. These corner stakes will be lost during excavation for the foundation. Thus, it is necessary to be able to relocate these points after the dirt is removed for the foundation. This relocation is done by **offsetting** these locations. Offsetting is accomplished by the use of batter boards. Batter boards temporarily establish the outside dimensions of the structure. The **batter board** consists of a horizontal board supported by posts or stakes securely anchored to the ground, Fig. 20-11. Batter boards are often set several feet from the actual edge of the building so they are not disturbed during excavation. A chalk line is run between the batter boards. This line is adjusted to exactly line up with the outside of the proposed building. These lines are then carefully marked on the batter boards.

Before excavation begins the line is removed to make way for the earth moving equipment. When the excavation is complete the lines are restrung on the batter boards. These lines establish the original corners of the building so construction can proceed, Fig. 20-12. They allow the removal of earth without damaging the building layout. After the foundation is complete the batter boards are removed.

Fig. 20-11. Batter boards are set several feet away from where the ground excavation will be.

EXCAVATION AND EXCAVATION EQUIPMENT

The top layers of soil must be removed, or **excavated,** before any structure is built. This is to ensure that the structure rests on a stable surface. The top layers of earth contain a considerable amount of organic material, Fig. 20-13. This material decays in time. This decay can cause the structure to settle, doing serious damage in the process. Also, water and frost action must be dealt with, both of which will cause the structure to shift. If the structure is a very large, such as a dam or large building, it may be necessary to excavate to bedrock, Fig. 20-14. **Bedrock** is the hard layer of rock under the looser surface material.

EXCAVATING EQUIPMENT

Excavating equipment can be as simple as a hand shovel or as complicated and costly as a crane. The contractor

Fig. 20-12. After excavation, the lines that mark the outer edges of the building are restrung.

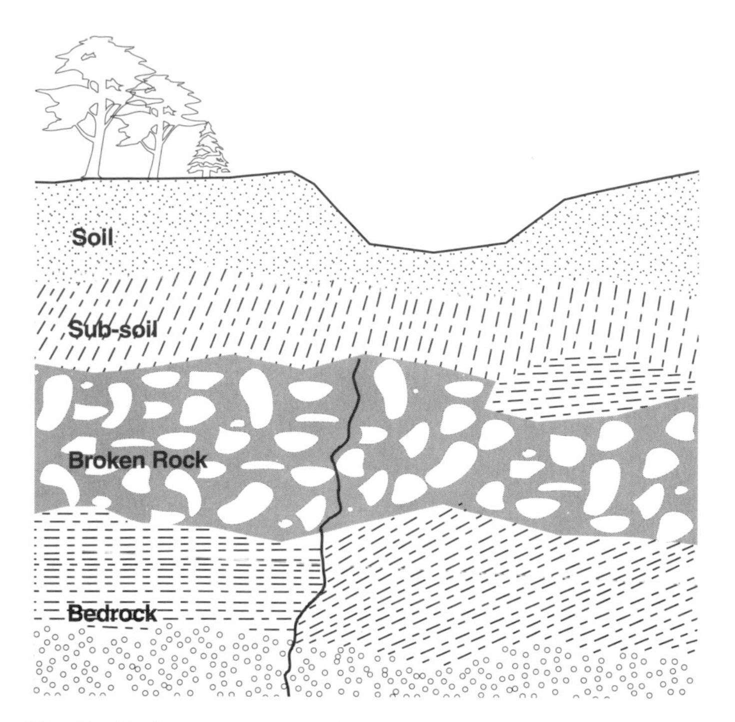

Fig. 20-13. Large amounts of organic material are worked into the top layers of soil.

Fig. 20-14. When constructing projects as massive as this dam, the ground must be excavated to bedrock.

must carefully select the pieces of equipment to be used during excavation to make sure they are economical in both time and money. The amount and type of material to be removed, as well as the distance the material has to be transported are important considerations in determining the best equipment for the job.

For example, if there is a need to dig a shallow ditch, a tractor mounted backhoe, Fig. 20-15, is appropriate. On the other hand, if millions of yards of earth must be transported several miles, large trucks and loading equipment might be used, Fig. 20-16. If large quantities of sand must be removed from under water to make a new shipping channel, some type of dredging equipment may be selected. The successful contractor is knowledgeable about the excavation equipment available and uses it to solve the excavations problems that come up on the construction site. Figs.

Fig. 20-15. This construction worker is digging a trench with a backhoe. What other tasks is the vehicle pictured designed to do?

Fig. 20-16. The vehicle on the right not only excavates the dirt, but conveys it into a dump truck for removal.
(Holland Construction)

20-17 through 20-20 show several other common pieces of excavation equipment used in the various construction areas.

SUMMARY

Long before construction begins the builder must review the site and determine what must be done in order to begin work. This may be as simple as locating exactly where the construction project will be erected. It could also be complicated and involve the removal of large previously existing structures and natural obstacles. The site evaluation provides the contractor with information to arrive at a reasonable bid for completing a proposed project.

A good understanding of geology is important to any large contracting firm. Without this understanding the company will not be able to assess the soil, rocks, and water that may cause problems and additional expenses. Today, another consideration is the environmental impact the project will have on the surrounding area. This is a critical review to determine if the project should be built or if alternatives should be considered.

The successful contractor will also take into consideration the various utility systems in the area as well as access to the site by heavy equipment.

Once all of the above considerations have been satisfied, then the construction site must be cleared. Sometimes certain natural features need to be preserved. Other times the land is completely cleared, as is the case with a large interstate highway.

The actual layout of the site begins with survey stakes that locate the boundaries of the project. Depending on the type of structure to be built, there may be a need to offset

Fig. 20-19. This scraper is rough grading the ground for a roadway. The bulldozer to the right is being used for added traction. (Caterpillar)

Fig. 20-17. Large excavators are digging to bedrock. This will provide the foundation for a large building.

Fig. 20-18. A power shovel is loading large amounts of soil into a truck. (Washington Construction)

Fig. 20-20. Giant drills are often used when creating a strong foundation. The shafts that are produced may be filled with concrete.

the stakes and use batter boards. This provides a reference point for relocating boundary lines after excavation.

Excavation equipment comes in a great variety of sizes and types. The equipment includes the simple shovel and complex earthmoving machines.

KEY WORDS AND TERMS

All of the following words and terms have been used in this chapter. Do you know their meaning?

Batter board
Bedrock
Clear
Environmental impact study
Excavate
Geology
Ground water
Locate
Offset
Stake
Survey

TEST YOUR KNOWLEDGE

Do not write in this book. Please write your answers on a separate sheet of paper.

1. Why is it important for the contractor to understand geology?
2. List two ways in which ground water affects the construction site.
3. List three utilities and services that will be needed during the construction of a large commercial building.
4. What does the term "bedrock" mean?
5. What are batter boards?

APPLYING YOUR KNOWLEDGE

1. Design an experiment that will show how the load carrying capacity of soil is different when it is wet or dry.
2. Visit your county courthouse and ask to review an environmental impact statement for a large project being considered.
3. Dig a hole at least two feet deep. See if you can identify the different layers of soil in the soil profile you have exposed. When your studies are complete make sure to back fill your hole.
4. Observe a construction site that is being prepared for actual construction. Make a list of the various pieces of construction equipment being used at the site.
5. Layout the batter boards for a building of your design. Make sure the batter boards are level before stretching your line. Also, check to make sure opposite lines are parallel to each other and the diagonals are equal.

Accuracy is vital in surveying. Surveyors are well-trained construction workers.

section 5

Application of Construction Systems

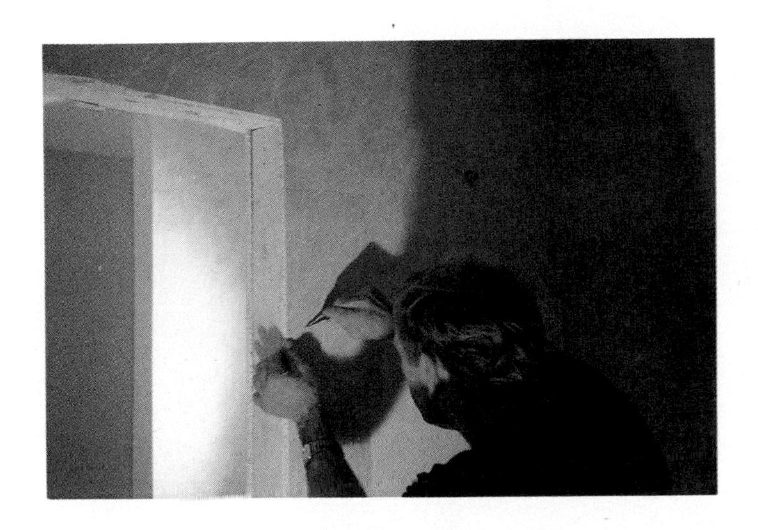

Construction projects, both large and small, can be intimidating efforts. Many complex tasks are involved. However, when a project is divided into multiple smaller goals, the project as a whole becomes less difficult. Consequently, the construction project is completed quickly and efficiently. There are many subsystems into which a construction project may be divided. In this section, you will examine each of these subsystems.

Most construction efforts will require one or more of the following systems: foundation, floor, wall, and roof. If the project is to be used for shelter, it will contain all of those systems. Each system can be created in many different forms using a variety of materials. This allows contractors to produce the perfect foundation, floor, wall, or roof for their customers.

After the exterior systems have been erected, there are a number of interior systems that must be installed. Electrical systems and plumbing systems, in addition to heating, ventilating, and air conditioning systems, are all very important. In industrial complexes, these systems may make up the most difficult parts of the project. Often, these systems are bid out to subcontractors.

The concluding stages of a construction project involve finishing tasks, cleanup procedures, and service work. After all the heavy construction machinery has left the site, construction workers can go through the area polishing the project as well as the surrounding land. Finally, service and maintenance work should be considered. An entire section of the construction industry makes a living through remodeling, repair work, and preventative maintenance.

Complex systems of batter boards are used when creating some foundations.

21

CHAPTER

FOUNDATION SYSTEMS

The information provided in this chapter will enable you to:

- *Distinguish between substructures and super structures as they pertain to construction projects.*
- *Discuss the purpose of the foundation on construction projects.*
- *List and discuss five different types of foundation systems that might be used in the construction of buildings.*

All construction projects can be divided into two major components: the substructure and the superstructure. The **substructure** is that part of construction that ties the construction project to the earth. Often this substructure is called the **foundation.** The foundation is generally out of sight of the casual observer. The **superstructure** is that part of the constructed project that is above the foundation or substructure. Fig. 21-1 shows the various types of substructures and superstructures.

The foundation of a construction project is a critical part of the construction. Without a strong and stable foundation, the structure built on top will soon fail. You have seen roads deteriorate and buildings that have fallen. Often, the reason for this collapse is a foundation that was not strong enough to properly support the structure.

The main purpose of the foundation is to spread out the weight of a structure, such as a house, large building, road, dam, or bridge. This allows the soil or rock supporting the structure to do so without deforming, Fig. 21-2. If the structure rests on solid bedrock, the foundation can be quite small. On the other hand, if the structure rests on clay soil that may get wet, the foundation must be very large. This large foundation will spread the weight out uniformly over a wide area. The type of soil at the construction site helps determine what type of foundation needs to be designed and built.

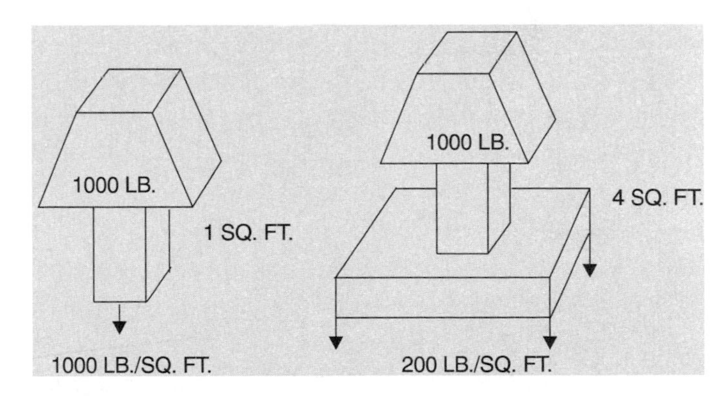

Fig. 21-2. Foundations spread out the weight of a structure. At the top, the structure exerts a force of 1000 lb. per square ft. At the bottom, the concrete base reduces the force to 200 lb. per square ft.

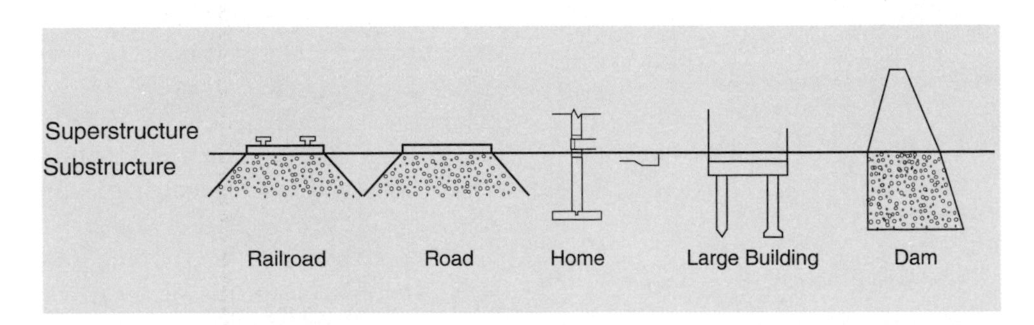

Fig. 21-1. Substructures and superstructures vary dramatically depending upon intended use.

MATERIALS OF FOUNDATION SYSTEMS

Architects choose foundations and materials carefully. Concrete is the most commonly used foundation material. It is strong and versatile. Wood is another common foundation material. Stones and the earth's bedrock are also used when they suit the construction project.

CONCRETE FOUNDATIONS

By far, the most common type of material that is used for foundation work in all types of construction is concrete. Concrete resists decay, moisture, and corrosion from acids. Concrete not only has great compressive strength, but it is also easy to handle and form to fit the specific needs of the construction project. It is commonly used in the construction of homes, dams, bridges, roads, and tunnels.

For concrete to be properly placed, a **form** must be built. As an example, the footing forms for a house, as shown in Fig. 21-3, contain the wet concrete while it sets. The location of this form is very important. Once the concrete has set, it is not possible to move the footing without breaking the concrete and destroying the foundation.

Depending on the size of the construction project, concrete may be mixed near the site and hauled in dump trucks. This is common in projects such as road construction. For longer distances, concrete is hauled and mixed in ready-mix trucks. Fig. 21-4 shows a ready-mix truck. When only a small amount of concrete is needed, people can purchase dry mixed concrete in sacks. The user need only add water and mix, Fig. 21-5.

Fig. 21-4. The rolling barrel on a ready-mix concrete truck keeps the concrete from setting during delivery.

Fig. 21-5. Bags of premixed concrete allow anyone to produce high quality concrete quickly and effectively.

Form work

For some structures the footing and foundation merge with the rest of the structure. In these cases, it is often difficult to tell the difference between the substructure and the superstructure. Smokestacks and dams are two examples. Form work is important to quality construction with these structures.

Fig. 21-6 shows the types of forms commonly used in construction of large projects. The construction of a concrete foundation wall commonly used in residential construction is shown in Fig. 21-7.

A **slip form** allows the construction of a large smokestack in one continuous pour. The form is moved up the stack at the rate at which the concrete cures. This requires very careful calculations regarding the amount of water in the concrete as well as careful control of the curing process. The advantage of slip casting a smokestack is that, when completed, the stack is without seams. This makes a very strong, as well as economical, smokestack. Some large stacks require continuous pours over several days or weeks.

Another example of slip forming is shown in Fig. 21-8. Here a guard rail for a new highway is being slip formed. It is a strong and efficient method used in concrete rail construction.

Fig. 21-3. Wooden forms hold the wet concrete in place. When the concrete sets the forms are removed.

Fig. 21-6. Forms used in construction are built in as many different shapes as there are projects. (Lowden, Lowden and Co.)

Fig. 21-7. This worker is examining the form for the foundation wall prior to filling it with concrete.

Fig. 21-8. The pouring machine slides right over the slip form. Long and strong guard rails are produced in a very short time.

Placing concrete

Placing concrete in forms is much more complicated than just pouring the concrete into the form. If the concrete falls too great a height, the large gravel particles will separate from the fine particles when the concrete reaches the bottom. This produces a very weak concrete. Concrete can only be dropped a maximum of around eight feet before the separation becomes a problem.

Another consideration is the strength of the forms. The outward force that concrete exerts on the forms is great. If the forms are not built extremely strong they may burst.

Modern day construction practices requires that concrete be delivered to the exact location in an efficient manner. To do this, various types of equipment have been developed. These methods may consist of the standard wheelbarrow, concrete shoots, cranes with large buckets, conveyor belts, concrete pumps, or spraying equipment. Fig. 21-9 shows several of these methods of delivering the concrete to the forms.

Fig. 21-10. Consolidating concrete produces a stronger end product.

Working concrete

In order for concrete to dry strong, several conditions must be met after the concrete is placed. First, the concrete must be consolidated. **Consolidating** is accomplished with the use of a concrete vibrator. A **vibrator** works out the air that is trapped in the concrete. It also works the smaller aggregate particles in between the larger aggregate, Fig. 21-10.

On concrete work where the end product will be flat (or nearly flat), the concrete must first be leveled off with a screed board, Fig. 21-11. Next, the surface rocks are forced into the concrete so the surface can be smoothed. After the surface is level, it is floated with a **bull float,** Fig. 21-12. This smooths and further consolidates the surface of the concrete. After the concrete has firmed up a little, it is then finished with a steel **trowel,** Fig. 21-13. This leaves a smooth finish.

Fig. 21-11. This patch of concrete is being leveled by two workers with a screed board.

Care must be exercised while the concrete is curing. The concrete must not be allowed to freeze or cure too fast. This will produce concrete that is weak and will not hold up to

Fig. 21-9. The quantity of concrete and the area in which it is needed help determine the method of delivery. Large funnels and hoses are just two systems.

Fig. 21-12. A bull float (center) consolidates the surface of this concrete sidewalk.

Fig. 21-13. Trowels are finishing tools. The concrete is allowed to begin hardening, then it is troweled to a smooth finish.

traffic or weather conditions. The best temperatures to work concrete fall between 60°F and 80°F. Special care and methods should be used when working concrete outside of these temperature limits.

Reinforcement steel

Concrete can hold up under very heavy loads, it has great **compressive strength.** But, as discussed in previous chapters, concrete does not have great strength when it is in tension. In order to make a strong foundation, the concrete is often reinforced with steel bars, Fig. 21-14. The bars provide the **tensile strength** for the concrete. Most structures made of concrete use these reinforcement bars so the entire structure will be strong and have a long life expectancy.

BUILDING WITH PRECAST MATERIALS

Sometimes it is difficult or simply not economical to build a structure with the use of poured-in-place concrete. Concrete block, rock, clay bricks, and large blocks of cut stone, Fig. 21-15, are examples of alternate foundation materials. The term used to describe this type of construction material is **masonry.**

Fig. 21-14. After the concrete has been poured over reinforcing steel bars, the crew pushes the concrete around. This forces the concrete into all the gaps around the steel.

The strength of the foundation is based on the shape of the individual units in addition to the material used to bond the units together. In ancient times it was important to make sure that stones fit with the stones next to them. The fit and shape of the stones was very important for the entire strength of the structure, see Fig. 21-16. Today a strong masonry mortar is used to bond the units together. Properly installed, a masonry structure is nearly as strong as one that is cast in place.

OTHER FOUNDATION MATERIALS

Concrete is the most common foundation material, but other materials are handy for special cases. Wood, gravel, and bedrock serve select needs in construction projects.

Fig. 21-15. Masonry products can be purchased in a variety of sizes, colors, and shapes.

Fig. 21-16. Back when this aqueduct was constructed, stones had to be carefully matched and fit together. (Esther McLatchy)

Wood foundations

Wood is often used for foundations in small structures. Some homes, bridges, and docks have wood foundations. Wood can be treated to keep it from rotting or decaying. The foundation must last as many years as the rest of the structures. Early water lines were wood structures. In some cases, the pipes have served for well over 100 years. Wood foundations for homes have been used for over 25 years with excellent results.

Gravel foundations

Roads also need some type of foundation just as houses and other buildings do. As Fig. 21-17 shows, this foundation is often heavy gravel with large rocks. This gravel layer helps to spread out the weight of the road itself as well as the weight of the vehicle traffic.

Gravel foundations are used in railroads for the same reasons, Fig. 21-18. In this case, the gravel base supports the railroad ties. These in turn spread out the weight of the tracks and the trains.

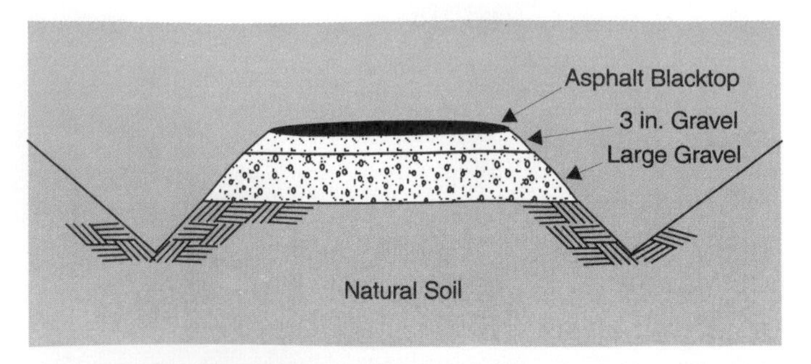

Fig. 21-17. Cross section of a typical poured road.

Fig. 21-18. Gravel produces a strong and economical base for railroad ties. (Morrison Knudsen Corp.)

Bedrock foundations

When large hydroelectric dams are constructed there is a need for a very strong foundation. This foundation must be the bedrock of the earth. With a large concrete dam, the many thousands of tons of concrete used in the construction would cause any other type of foundation to move. Construction workers must remove all of the loose rock from the bedrock. This is so construction of the dam can begin with a solid foundation.

TYPES OF FOUNDATIONS

As discussed in Chapter 20, the design of the foundation depends on the weight of the structure and the load bearing ability of the soil. To determine the weight of the structure the engineer or architect determines the total load of the structure. Also, they determine the load carrying ability of the soil by performing a variety of tests. The combination of these two factors help to determine the size and type of foundation designed for the construction project. There are numerous types of foundation systems.

PILE TYPE FOUNDATIONS

Sometimes it is not possible or it is too expensive to dig all the way through to bedrock for a foundation. In these cases, long **piles** may be driven into the earth to provide a solid foundation upon which to build. Piles can be made from wood, steel, concrete, or combinations of these materials. The type of pile an engineer selects for the construction is determined by the place at which the piles are driven and the conditions the pile will be put under. The weight of the structure and the climate are two important conditions.

The piles driven into the ground firm up the soil and keep the structure from settling unevenly. The piles support the structure using the friction between the pile and the soil. The piles can also be driven until they strike bedrock. In both cases, the pile transfers the load of the structure to the ground. Fig. 21-19 shows one type of pile used in construction.

Piles are driven into the soil by the use of various types of **pile drivers,** Fig. 21-20. The most common pile driver is an engine-driven device. It is held in place on top of the pile with a heavy crane. The device drives the pile into the ground much as you would drive a stake into the ground by striking with a hammer.

At the completion of the pile driving, the top of the piles are cut evenly and concrete is pored over the tops. This links the piles together. This is called **grade beam construction,** Fig. 21-21, because the beam is formed on grade, at ground level. The structure will rest on top of the grade beams which in turn rests on the piles.

DRILLED SHAFT FOUNDATIONS

A **drilled shaft foundation,** Fig. 21-22, is simply a round hole drilled into the earth and then filled with concrete. This type of foundation system is quick to complete, versatile, and cost-effective. A single drilled shaft can replace a group of driven piles. This results in faster construction.

Fig. 21-19. This massive steel pile is one of many that will create a stable foundation.

Fig. 21-20. Pile driver in action.

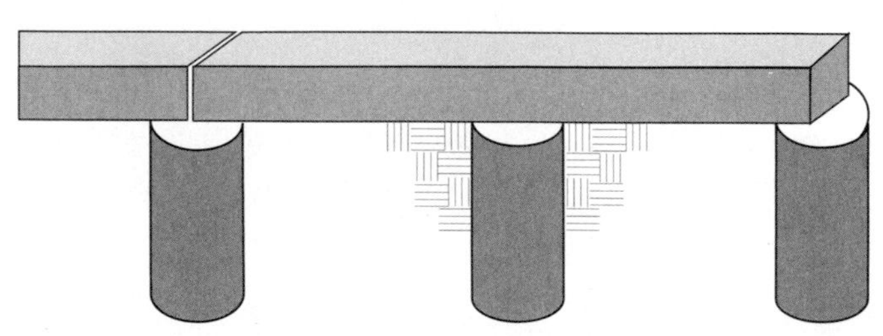

Fig. 21-21. Grade beams are the concrete girders that are poured on top of concrete pilings. Grade beam construction is one form of pile type foundation.

When the hole is being bored, Fig. 21-23, the engineer has the opportunity to visually inspect the sub soil. This inspection helps the engineer determine the load bearing ability of the ground. The shaft drilling equipment is large enough to bore holes several feet in diameter, even into rock formations. Often steel reinforcing is used to strengthen the foundation before the concrete is placed in

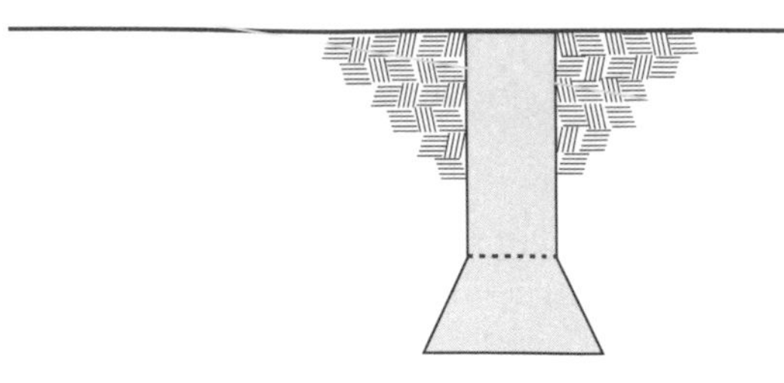

Fig. 21-22. Cross section of drilled shaft foundation.

the hole. Also, the bottom of the shaft may be flared out to make a bell bottom pier. A bell bottom pier will spread the weight over a greater area. This type of foundation system is called **pier foundation.**

MONOLITHIC SLAB FOUNDATIONS

A **monolithic slab foundation** loads the ground beneath it very little. In this case, the monolithic slab simply "floats" on top of the ground. The term *monolithic* is used because the foundation is very solid and rigid. On a monolithic slab foundation, if you raise one corner of the slab the whole slab will tilt.

This type of foundation allows the construction of structures in areas where it may not be economical to build other types of foundations. By spreading the weight of the entire structure over such a large area, the load per square foot is very small, Fig. 21-24.

FOUNDATIONS IN THE ARCTIC

Foundation systems built in the far north have an additional challenge, permafrost. **Permafrost** is a layer of permanently frozen earth. In this case, the foundation

Fig. 21-23. Special boring equipment is used in drilled shaft foundations. The holes bored are several feet in diameter. (International Association of Foundation Drilling)

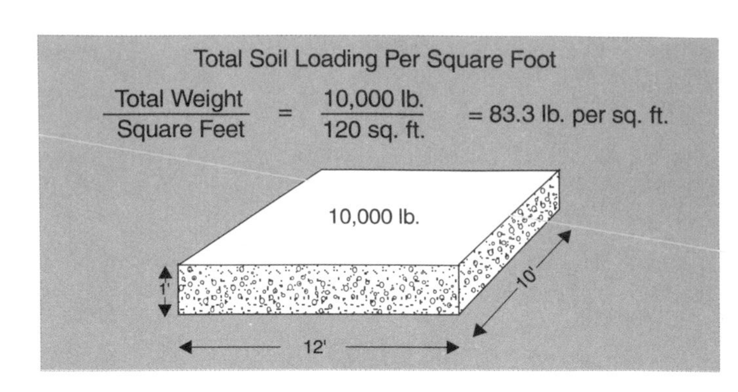

Fig. 21-24. Loading forces on the soil of a monolithic slab foun-

system relies on this permanently frozen ground to support the structure, Fig. 21-25. Care must be exercised to assure that the frozen ground below the support walls of the structure remains frozen.

CONTROLLING WATER AND FROST ACTION

All substructures must have a means of controlling the deterioration caused by water or frost. In some cases, the water is controlled by the use of a drainage system, Fig. 21-26. No large quantities of water are allowed to sit near the foundation. In other cases, water is sealed off from the foundation by the use of waterproofing materials, Fig. 21-27. These materials put a thin barrier over the foundation materials that prevent moisture from entering.

In some cases, such as docks, the water is not kept from the structure. With a dock, the structure must be in contact with the water. Consequently, the materials used are treated with preservatives. These chemicals prevent decay.

Fig. 21-25. Areas that stay cold year round have a layer o permafrost. Structures such as this one can be built on this soli earth.

Fig. 21-26. If water is not drained well away from a foundation, extensive damage will occur. (Lowden, Lowden and Co.)

Fig. 21-27. A layer of tar is often used to prevent moisture from reaching the foundation.

In colder climates, structures must be designed to drain away the water for an additional reason. Great damage can occur when water freezes in a substructure. Water expands when frozen. This expansion can cause the structure to shift. Another alternative is to place the footing deep enough in the ground so it is below the frost line.

SPECIAL CASES

In some specialized construction projects, waterproofing is very important. In gold mines toxic chemicals are used to leach the gold out of the ore. Waterproofing is necessary to protect the environment.

In this type of construction, large sheets of heavy plastic are laid over a layer of sand, Fig. 21-28. Then, several layers of asphalt are placed on top of the plastic membrane. This contains the toxic chemicals. Fig. 21-29 shows the completed pad. The pad becomes the foundation that holds the weight of the ore to be processed.

Earthen dams are another structure that must control the flow and seepage of water. In this case, a layer of clay placed within the dam is used to seal off the seepage of water. When the seal does not work properly, serious problems can arise. The Teton Dam demonstrated just what can happen. One June 5, 1976, the Teton Dam gave way. Over 3000 homes were lost or severely damaged by the flooding that followed. Approximately 99,252 acres of land were affected.

Fig. 21-28. Large sheets of plastic are laid across the sand in this gold mine. The chemicals used in mining will not be able to penetrate the plastic.

Fig. 21-29. The unique foundation for the processing area of a gold mine stands out from among its surroundings.

SUMMARY

The foundation of any construction project is critical to the completed project. Without a well-designed and constructed foundation, a structure will not withstand the forces of nature. All construction projects have foundations. Sometimes these foundations are very extensive such as those for large high rise buildings. Other times, the foundation is very simple such as that of a road that is built on gravel.

The purpose of the foundation is to support the weight of the structure above. The type and size of the completed structure as well as the load carrying ability of the soil determine the design and size of the foundation.

The most common foundation material used for buildings is concrete. Concrete is easy to put in place and it has great compressive strength. Other foundation materials for buildings are wood, metal, and masonry. Foundations for civil construction projects, such as roads, often consist of a layer of heavy gravel that is spread over the earth.

Foundations, because they connect the construction project to the earth, must deal with water and frost action. Either water must be drained away from the foundation and sealed out, or the materials used in the foundation must be designed to work effectively in the water.

KEY WORDS AND TERMS

All of the following words and terms have been used in this chapter. Do you know their meaning?

Bull float
Compressive strength
Consolidate
Drilled shaft foundation
Form
Foundation
Grade beam construction
Masonry
Monolithic slab foundation
Permafrost
Pier foundation
Pile
Pile driver
Slip form
Substructure
Superstructure
Tensile strength
Trowel
Vibrator

TEST YOUR KNOWLEDGE

Do not write in this book. Please write your answers on a separate sheet of paper.

1. Describe the difference between substructures and superstructures.
2. Explain why the foundation is so important to a quality construction project.
3. What two factors determine the size and type of foundation a structure will require?
4. What material is most commonly used for the foundations of buildings?
5. Reinforcement steel adds what kind of strength to a completed concrete structure?

APPLYING YOUR KNOWLEDGE

1. Calculate the amount of weight a footing must carry on a 1200 square ft. house assuming that the weight per square ft. of the house is 85 lb.
2. If all of the weight of the house is carried by just the foundation wall footings that are 16 in. wide, how much weight does each square ft. of earth under the footing hold?
3. If a monolithic slab was used for the same house, how much does each square ft. of earth have to carry?
4. Collect 1/2 cubic ft. of a clay type soil and the same amount of a gravel type of soil. Place the soils in plastic dishpans. Set a concrete block on end in the pan. Measure the height of the top of each block and record this measurement. Next, add an equal amount of water to each pan. Observe any sinking of the block over several days. Add more water and continue to observe the blocks. Was there any difference in the sinking of the two blocks? Which soil provides the best foundation and why?
5. Arrange a field trip to a construction site that is pouring concrete. Observe how the concrete is placed in the forms and how it is finished.

Construction activities may take place in remote locations. Preventative maintenance of equipment is important to avoid breakdowns in the field.

22

CHAPTER

FLOOR SYSTEMS

The information provided in this chapter will enable you to:
- *List several support structures for flooring systems.*
- *Calculate the dead load for simple floor systems.*
- *Describe four different floor systems used in construction.*
- *Distinguish between support columns, beams, girders, and bearing walls.*

The simplest type of floor system is the earth beneath our feet. Although we seldom think of the earth as a floor system, it meets all of the criteria for a floor. There are two major criteria for floors. They must be relatively flat, and they must be strong enough to support the loads they will be used for. In many places the earth is fairly flat and can support itself as well as additional loads. The floor of the homes of early humans was the dirt upon which they erected their tents and other structures. The floor of early cave dwellers was the dirt in the bottom of the cave.

Today, we have advanced to designing and building very elaborate floor systems that better meet the needs of our modern civilization. Though, in some cases, dirt floors still find use, Fig. 22-1. Some large sports arenas, such as field houses, rodeo arenas, and indoor football and baseball stadiums, make use of dirt floors.

TYPES OF FLOOR SYSTEMS

Floor systems exist in a variety of types. This variety extends from the carpeted floors in buildings to the paved streets and roads. A paved road is little more than a garage floor extended over long distances. Generally, this type of floor is placed on or near the surface of the earth. However, at times there is the need to place the highway over rivers or other roads, Fig. 22-2. In this case, more extensive construction must be completed. The construction must support the weight of the road as well as all of the traffic it will carry.

Additional support is also needed for buildings that have more than one floor. The floors that are above grade must be supported in some manner from below. Adequate room between the supporting members has to be provided if the structure is to be useful. Through the years, architects and engineers have been very successful in achieving this. The result allows large buildings to be erected in cities where the available land is costly. It is much more cost efficient to build tall buildings with many floors in large cities, Fig. 22-3, rather than purchase enough land to provide the needed floor space on a single level.

FLOOR SUPPORT SYSTEMS

The first step in building a floor system it to build the support system upon which the floor will rest. These sup-

Fig. 22-1. Dirt roads do not handle wear like paved roadways. But in areas with light traffic, dirt roads provide cheap transportation. (Bureau of Land Management)

Fig. 22-2. Every bridge introduces its own intriguing challenges for construction engineers.

port systems vary in form to match the intended uses for the floor. The weight the floor must carry, the makeup of the earth underneath, and the floor's elevation are just three of many considerations for the architect. You will look at the supporting systems for roads, bridges, high-rise buildings, and residential buildings.

Fig. 22-3. Expensive land in the city forces construction upward. (Montana Power)

ROAD CONSTRUCTION

In the case of a road, Fig. 22-4, the underlying earth is the support system. Steps must be taken to prepare this ground.

Generally, the first step with roads is to remove the top soil. It is stockpiled to be reused later. After the removal of the top soil, additional soil is removed to provide the slope of the road. Then, large gravel is hauled in to make the substructure. Succeeding layers of crushed rock are placed and carefully packed onto the gravel. The drawing in Fig. 22-5 shows the various layers of a paved road.

Finally, the top surface of the road is placed into position. This surface is usually made of concrete or asphalt. They provide a smooth surface that is easy to maintain. This surface also keeps the water out of the base of the road, which helps prevent deterioration.

BRIDGES

When a roadway system is required to cross rivers or other obstacles, the road must be supported by heavy substructures and superstructures. There are a variety of methods used to support the floor of a bridge. The bridge floor is called the **decking.** It is a solid horizontal covering that is supported underneath by columns, girders, and beams. The deck itself may be constructed of wood, steel, or concrete. The main purpose of the superstructure of the bridge is to support this decking. The decking, in turn, supports the traffic that the bridge will carry. Fig. 22-6 shows a support structure.

The simplest type of bridge places heavy beams on top of columns or support walls. The bridge beams are strong enough to carry the traffic load as well as the weight of the structure itself. This type of support system works well for short spans, but it is not efficient for longer bridge spans.

As the length of the bridge becomes greater, other designs must be used. Fig. 22-7 shows a long span being supported by heavy cables suspended from high towers. This bridge is called a **suspension bridge.** Smaller cables are attached to large main cables that support the bridge deck. Another method uses trusses to support the decking of the bridge as is shown in Fig. 22-8. A **truss** is a group of beams formed into a rigid frame. This type of bridge is called a **truss bridge.**

Fig. 22-4. Trees and shrubs are removed from the dirt road base. (John Deere)

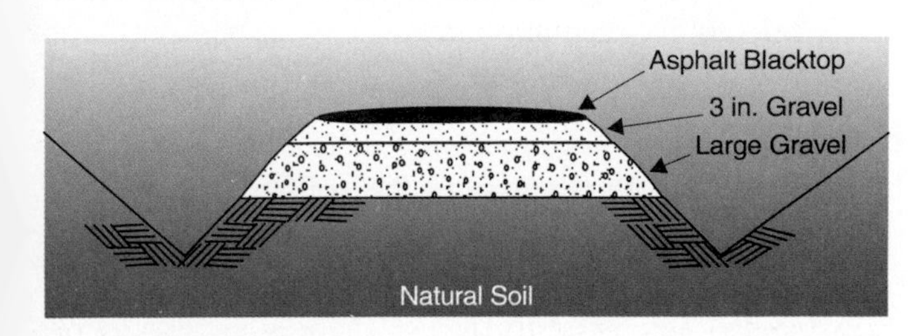

Fig. 22-5. Cross section of a road.

Fig. 22-6. Bridges with short spans are used to cross over intersecting roads.

Fig. 22-7. Suspension bridges are a popular support system for crossing very large rivers and bays. (Morrison Knudsen Corp.)

Fig. 22-8. The triangles beneath this truss bridge provide structural support.

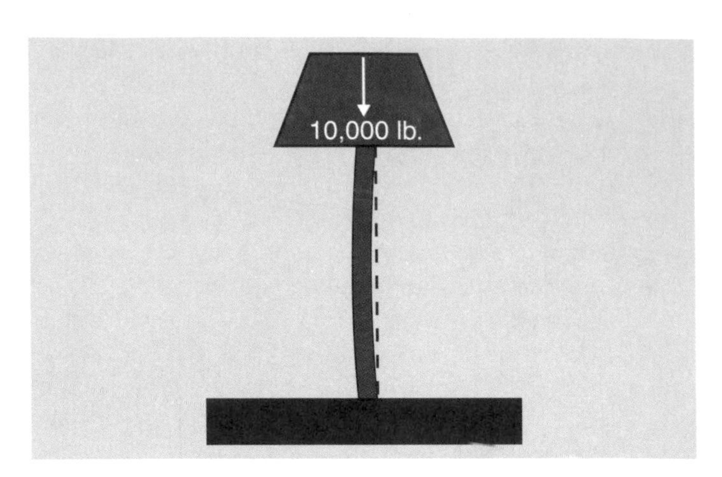

Fig. 22-10. A downward force will bend a column.

Fig. 22-9. Before this building was erected, engineers calculated the load that the foundation and each floor must carry. This factor was then multiplied several times to ensure this apartment is safe.

HIGH-RISE BUILDINGS

The decking on high-rise buildings uses a support system similar to the simple beams used in the construction of a bridge. In this case, each floor is supported by a series of support columns. The support columns hold up the heavy beams for each floor, see Fig. 22-9. Between these heavy support beams are lighter beams that hold up the flooring material itself.

In all cases, an engineer must carefully calculate the load limits of the materials to be used in the beams and columns. For safety, this load factor is estimated to be *several times* the anticipated load and is called the safety factor. This ensures that the structure will still be sound after many years of service and even some deterioration.

RESIDENTIAL CONSTRUCTION

The primary differences between home construction and commercial construction are the size of the project and the materials used in that construction. The floor system in homes and small apartment houses is supported by systems very similar to those of the high-rise buildings. In homes, though, lumber is most often used instead of concrete and steel.

Once the foundation of the home is in place, support columns and beams are used to make the distances between foundation walls into shorter spans. This allows the use of standard horizontal wood framing members to support the load of the wood floor as well as the future loads of furniture and occupants. Successive stories are similarly supported with the use of interior walls. The interior walls serve as support members for the framing members overhead.

SUPPORTING COLUMNS, WALLS, AND BEAMS

Support columns are the part of a structure that transfers the load from above to the foundation or substructure below. The shape and length of the column is critical in determining how great a load it can safely carry. If the column is thin in relation to its length, it will begin to bend as heavy loads are applied, Fig. 22-10. On the other hand, if the column is very short, it can support heavier loads

Fig. 22-11. Workers pouring a reinforced concrete column. The concrete is poured into a form with the rebar in place.

A

B

Fig. 22-12. A–Form for producing a round reinforced concrete column. B–Round reinforced column. (Morrison Knudsen Corp.)

without bending. Also, if the column is produced in the shape of an H, an I, or an O, it will not bend as easily. These shapes help distribute the stress from the weight of the load. Careful calculations must be made to ensure that the shape of the columns will support the desired load.

CONCRETE COLUMNS

Reinforced concrete is used often in support columns in heavy building construction and on some highway construction, Fig. 22-11. In this instance, concrete has been formed in a rectangular shape to support the structure above. Concrete has a great load carrying capacity and it does not fail under the heat of a fire.

Many commercial buildings use round concrete support columns for the main support structure. These round concrete columns are formed with large round tubes, Fig. 22-12, similar to the ones used for rolls of paper. These tubes are several feet in diameter. They have heavy walls that can withstand the outward pressure from the wet concrete. The tubes are waxed on the inside. This prevents the concrete from wetting the paper and makes the tubes easy to remove once the concrete has set.

STEEL COLUMNS

High-rise buildings often use heavy steel columns and beams, Fig. 22-13. The columns nearest the ground are the heaviest. Lighter columns are used for the upper stories. The lower stories, of course, carry a great deal more weight than the upper stories.

I beams are the most common shape of support columns in large buildings. Their load carrying capacity can be easily adjusted by making the material thicker. I beams come in a variety of standard sizes and weights. Also, additional beams are easily attached to each other.

Steel beams do have a weakness. Steel beams loose much of their strength when subjected to a great amount of heat as in a fire. To remedy this problem, building codes require steel beams to be sprayed with a fireproof insulation to protect them from the intense heat of a fire. See Fig. 22-14.

WOOD COLUMNS

The most common building material for residential and light construction is wood, Fig. 22-15. Not only is wood very strong for its weight, but wood is very easy to work with. When properly protected from moisture and insects,

Fig. 22-13. Steel beams support a heavy load for their weight. I beams are the most common.

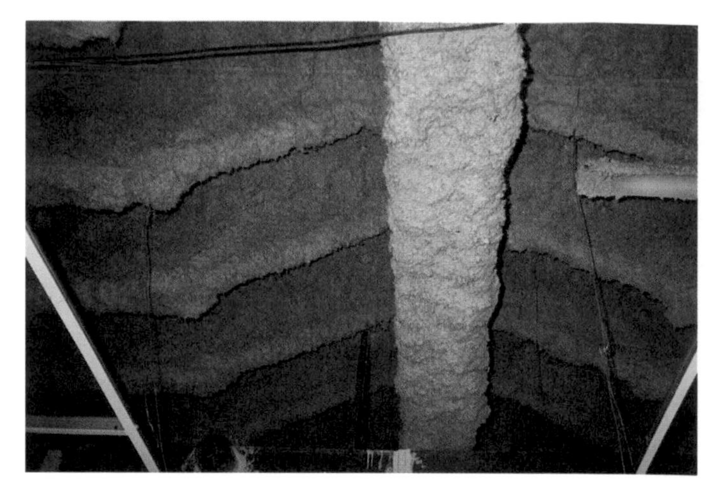

Fig. 22-14. These beams and girders have been covered with a heavy layer of fireproof insulation. This insulation allows the building to remain structurally sound in the event of a fire.

wood structures have lasted hundreds of years. It takes a relatively small amount of energy to produce lumber from the raw material. Additionally, wood is a renewable resource.

The shape of most wood columns is square or rectangular. Sometimes, however, the support columns are turned round. They also may be used in their natural condition. Log homes, wood piling for foundations of buildings and docks, and telephone poles are examples, Fig. 22-16.

BEARING WALLS

Another way to support a load from below is to build a **bearing wall** that will support the beams above, Fig. 22-17. The wall can be made from a variety of materials. Wood is common in residential structures. Concrete and steel are often used in commercial structures.

A bearing wall can be thought of as a series of support columns that are placed close together. A concrete foundation wall in the basement of a home is an example of a bearing wall you may see frequently.

Fig. 22-15. Wood framing was chosen for this three-story complex.

GIRDERS AND BEAMS

Once the support columns or walls are in place, the next phase of construction of a floor system begins. Horizontal members are placed from one support to the next. **Girders** are the main horizontal support structures. **Beams** are generally smaller. They may be connected to the girder or directly to columns and bearing walls. Because these horizontal members carry the load to the supporting structure below, their design must be carefully calculated.

Fig. 22-18 demonstrates the forces that operate on any horizontal member. Note, the forces on top of the beam are towards each other. These are **compression forces.** The forces at the lower edge of the girder pull the girder apart. These are **tensive forces.** The forces in the center of the beam are neutral. This allows you to better understand why I beams are often used. The tops and bottoms of beams must resist compression and tension forces. The center portion does not have to be as strong.

Concrete girders and beams

Concrete girders can be designed to carry heavy loads. Fig. 22-19 shows a small bridge that is supported by

A

B

Fig. 22-16. A–Logs give homes a very distinctive character. B–Above ground telephone lines rely on wood columns for support.

Fig. 22-17. This bearing wall is part of a residential structure. Wood was chosen to support the beams above.

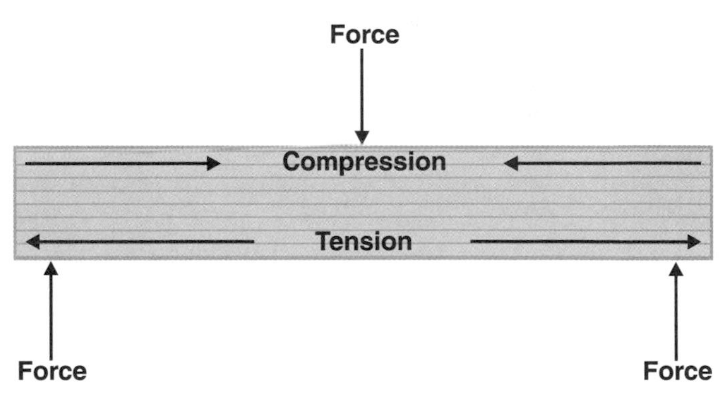

Fig. 22-18. Forces operating on a girder.

Fig. 22-19. Concrete girders withstand the tremendous stresses applied by the trucks that pass daily.

A

B

Fig. 22-20. A–Hoisting a prefabricated beam for delivery. B–Beams have arrived on site and await use.
(Elk River Concrete Products)

Fig. 22-21. These workers are applying the finishing touches to a massive beam. (Elk River Concrete Products)

concrete girders. Sometimes concrete girders are manufactured in a plant and then transported to the construction site, Fig. 22-20. With these girders, quality during manufacturing can be carefully controlled ensuring a high quality product. Fig. 22-21 shows one of these facilities.

Fig. 22-22. Bulldozer loads concrete into steel reinforcing cage.

Fig. 22-23. Steel cable is used for reinforcement when producing prestressed concrete. (Elk River Concrete Products)

Steel reinforcement is placed in concrete girders to provide the strength needed to make the entire girder strong, Fig. 22-22. If the reinforcement steel is placed under tension before the concrete is set, Fig. 22-23, the girder is further increased in strength. This type of construction is called prestressed concrete and is often used in the construction of bridges and large buildings. Review Chapter 11 on steel reinforcement.

A **ribbed slab,** Fig. 22-24, is a combination of a concrete girder and a floor slab that is cast into one unit. These units can have one or more ribs depending on the manufacturer.

The advantage of this type of floor system is that the various components can be manufactured off site.

Once the supporting structure for the building is in place, the concrete flooring sections may be added. A large crane can set in precast flooring sections, Fig. 22-25. This allows construction to move at a rapid pace. An alternative is to form a **waffle pattern** with specialized metal pans. Reinforcement steel is placed between the pans. Then concrete is placed in the forms. Once the forms and pans are moved, a waffle pattern exists as viewed from under the floor, Fig. 22-26. The ribs in this pattern provide the strength for the floor because they are deeper.

Steel girders and beams

Steel is used for beams in high-rise buildings for several important reasons. Steel can be assembled quickly. Steel is very strong for its weight. Steel, also, handles tensive and compressive forces well. Just as with concrete, the engineer must carefully calculate the loads that a structure will carry. The design and size of the steel beams are based on these calculations.

Structural steel comes in a large variety of shapes, sizes, and weights per lineal foot. This flexibility provides for a virtually limitless number of options in the construction of structures using steel.

Connecting the steel is most often done by welding or bolting. Each method has certain advantages and disadvantages. Bolts make quick and easy connections. Very little special equipment is needed to make connections with bolts. Also, workers do not need to learn any specialized skills. However, bolts do not produce connections as strong as those produced by welding. In addition, bolt holes are drilled in the fabrication plant. This does not provide for easy adjustment in the field.

Welding provides very strong connections and adjustments are easily made in the field. However, welding does require special skills and tools. Welding is also a more time consuming process.

Fig. 22-27 shows the use of structural steel in a high-rise building. With its versatility, many smaller commercial buildings are now using steel as the major component in their construction. Steel buildings have been in common use for agricultural buildings for many years, see Fig. 22-28.

Fig. 22-24. Shown is a ribbed concrete slab. The ribs give this flooring material significant structural strength.

Fig. 22-25. Heavy construction equipment, such as this crane, is needed when working with precast concrete products. All of the floor slabs are put in position with this crane.

Fig. 22-26. The waffle pattern may be used for its structural strength or for interesting architectural effects. (Morrison Knudsen Corp.)

Fig. 22-27. Steel beams provide quick assembly. In high-rise buildings, where there are thousands of beams to assemble, the time saved is tremendous.

Fig. 22-28. Steel provides marvelous versatility. (Butler Manufacturing Co.)

Wood girders and beams

Wood, like steel, is another versatile material. Wood is very strong for its weight and has excellent strength under compression and tension. The earliest wood buildings used the whole logs for the construction members, Fig. 22-29. More recently, the log has simply provided the raw materials for standardized forms of framing and finishing materials. Just as with the concrete and steel floor framing members, wood materials must be carefully selected and loads calculated to ensure an economical, safe, and useful structure.

Because wood is a natural product, its production cannot be tightly controlled like steel or concrete. The strength of a board of wood is not uniform throughout the material. To overcome the problems that weak areas can produce, two or more wood members are bonded together to make larger and stronger framing members.

Fig. 22-29. This historical site is an example of early whole log construction.

A truss joist is a wood framing member that resembles a steel I beam. The top and bottom parts of the beam are made up of many laminations of wood. The center portion is a piece of plywood.

Wood beams can also be made in the shape of a box. These are called **box beams.** Box beams are very strong for the amount of material used. Beams, also, may be built up by the use of successive layers of lumber as shown in Fig. 22-30. These beams are called **gluelams.**

With the advent of new technology in glues and adhesives as well as wood cutting procedures, the usefulness of wood as the structural supports for floors has greatly expanded. Fig. 22-31 shows a structure that has taken advantage of wood as a commercial construction material.

SYSTEMS FOR SMALL BUILDINGS

When constructing small buildings and homes, contractors must deal with the same forces that exist in large high-rise buildings, bridges, dams, roads, and commercial structures. The main difference is that the forces are smaller.

In residential construction the columns and girders are often cut to length on site. Because of their small size, the girders can be erected by hand. Crossing the top of the girder are smaller framing members called **floor joists.** These floor joists support the floor above as shown in Fig. 22-32.

Just as with the commercial structures, the architect must select floor framing materials based on load calculations. Architects run into many similar structural situations when designing homes. Rather than hand calculating the strength of materials needed for each job, architects often use tables or computer programs to aid them in designing the size and location of floor framing members. Fig. 22-33 is a table for determining the size and spacing of floor joists for a home.

Because wood framing members will vary in quality from piece to piece, the carpenter must select the various framing members to ensure a strong structure. Longer framing members tend to bow slightly. The carpenter must be careful to make sure the bow, or **crown**, is up. As weight is put on the joist, it will tend to straighten out. The carpenter must also position any knots so they will have the least adverse effect on the strength of the joist.

The construction industry in this country is based on the four foot module. This applies to residential construction in general and to framing in particular. The joists are spaced either 24 in., 16 in., or 12 in. on center. All of these are modules of 48 in. or 4 ft. The module allows for the most efficient use of materials, and it sets some standards so load carrying capacities can be easily calculated from prepared charts.

Fig. 22-30. Gluelams produce a very solid beam from smaller cuts of wood.

Fig. 22-31. This entire apartment building is framed in wood. With proper maintenance, it will provide secure housing for many years.

Fig. 22-32. In this two-story structure, these floor joists will support the ceiling for the first floor and the floor for the second.

PERMISSIBLE SPANS FOR SOUTHERN PINE FLOOR JOISTS

NOMINAL SIZE		SPACING OC		40 lb./sq. ft. (195 kg/m^2) LIVE LOAD GRADE NO. 2 DENSE SPAN		30 lb./sq. ft. (146 kg/m^2) LIVE LOAD GRADE NO. 2 DENSE SPAN	
In.	cm	In.	cm	Ft.	m	Ft.	m
		12	30	11–5	3.5	12–5	3.8
2 x 6	5 x 15	16	41	10–5	3.2	11–4	3.4
		24	61	8–11	2.7	10–0	3.1
		12	30	14–9	4.5	16–1	4.9
2 x 8	5 x 20	16	41	13–6	4.1	14–9	4.5
		24	61	11–11	3.6	13–1	4.0
		12	30	18–3	5.6	19–11	6.1
2 x 10	5 x 25	16	41	16–9	5.1	18–3	5.6
		24	61	14–10	4.5	16–2	4.9
		12	30	21–9	6.6	23–9	7.2
2 x 12	5 x 30	16	41	19–11	6.1	21–9	6.6
		24	61	17–7	5.4	19–2	5.8
		12	30	17–0	5.2	18–7	5.6
3 x 8	7.5 x 20	16	41	15–7	4.8	17–0	5.2
		24	61	13–10	4.2	15–1	4.6
		12	30	21–1	6.4	23–1	7.0
3 x 10	7.5 x 25	16	41	19–4	5.9	21–2	6.4
		24	61	17–1	5.2	18–8	5.7

Fig. 22-33. Permissible spans are charted for southern pine floor joists.

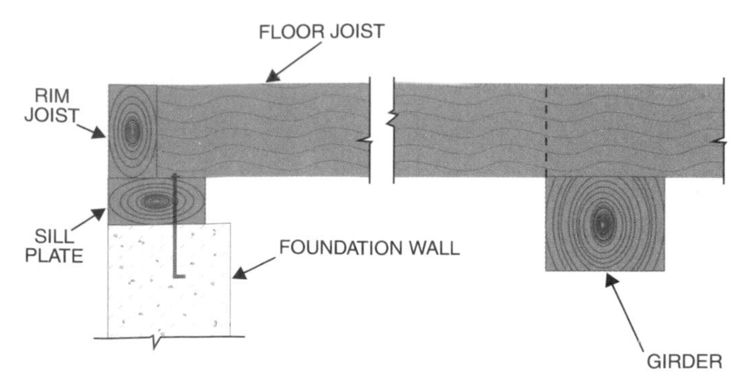

Fig. 22-34. Cross section of a residential floor framing system.

Fig. 22-35. Without special framing, the floor surrounding a staircase will be weak and unsafe. (Lowden, Lowden and Co.)

One end of the floor joist rests on top of the girder, and the other end rests on top of the **sill plate** over the foundation wall, as shown in Fig. 22-34. The sill plate is anchored to the foundation wall with bolts. The joists are, in turn, nailed to the plate. This makes for a strong flooring system.

Where there are openings through the floor for items such as stairs or fireplaces, the standard framing members must be interrupted. Special framing procedures are used to ensure that the floor is strong around the opening, see Fig. 22-35.

Additional strength can be achieved by using a system that more evenly distributes the load on the joists. These framing members are called **bridging.** Bridging helps to distribute the load placed on one joist to the joists on each side of it. See Fig. 22-36.

Once the floor joists and bridging are installed, **sheathing** material is placed on top of the joists. For residential construction this material is most often plywood, Fig. 22-37. This is called the **subfloor.** This subfloor material serves as the base for the finish floor that is installed later in construction. The thickness of the subfloor will vary from 1/2 in. to 3/4 in. depending on joist spacing, local building codes, and the thickness of the finish floor.

Fig. 22-36. These straps help distribute the forces on each joist to the surrounding joists.

Fig. 22-37. Sheets of waferwood are used as the subflooring in this residential construction.

SUMMARY

A floor system may consist simply of the earth itself such as a dirt road. Generally, however, the floor system requires additional construction activity to meet certain needs. Roads will require drainage and surfacing, and buildings will require a solid surface that is easy to maintain and clean.

A sturdy support system must be utilized in order for the flooring system to meet the demands of the completed project. For road construction this requires the selecting, placement, and compaction of gravel fill material. For a bridge or a high-rise building, columns and beams must be installed to support the load. In residential construction the careful selection of framing methods and materials ensure a long lasting floor.

Builders have a large variety of materials from which to choose for floor systems. Concrete, steel, and wood are the most common materials used for columns, girders, and beams. Concrete has good compressive strength and is very fire resistant. Steel has good compressive and tensive strength. Wood is easy to work with and strong for its weight. Each of these materials is available in a very large variety of products.

KEY WORDS AND TERMS

All of the following words and terms have been used in this chapter. Do you know their meaning?

Beam
Bearing wall
Box beam
Bridging
Compression force
Crown
Decking
Floor joist
Girder
Gluelam
I beam
Ribbed slab
Sheathing
Sill plate
Subfloor
Support column
Suspension bridge
Tensive force
Truss
Truss bridge
Waffle pattern

TEST YOUR KNOWLEDGE

Do not write in this book. Please write your answers on a separate sheet of paper.

1. List three methods that can be used to support the flooring system.
2. What three materials are most commonly used in the construction of flooring systems?
3. What kind of force squeezes a framing member?
4. What kind of force pulls a framing member apart?
5. List the two primary differences between the floor systems used in home construction and commercial construction.
6. There is no difference between a column and a girder. True or False.

APPLYING YOUR KNOWLEDGE

1. Draw a cross section of support columns and girders for a bridge and the foundation wall and floor joists of a home. Compare the two. List similarities and differences.
2. Arrange a visit to a construction site to observe how the flooring system is being installed.
3. Calculate the total dead load for a 1200 square ft. house if the weight per square ft. is 40 lb.
4. Design an experiment that demonstrates the compression and tensive forces exerted in a horizontal support member
5. Design and construct a bulletin board showing different types of flooring systems.
6. Build a model of the different types of flooring systems used in civil, commercial, and residential construction.

MORRISON KNUDSEN CORP.

The ridges in this preformed floor slab significantly strengthen the material.

23

CHAPTER

WALL SYSTEMS

> *The information provided in this chapter will enable you to:*
> - *Describe the function of wall systems.*
> - *List seven different types of wall systems.*
> - *State major differences between three different wall systems.*
> - *Identify the major components of standard wall framing.*
> - *Discuss the importance of shape as it relates to the strength of a structure.*

The main purpose of a wall system is to enclose the interior of the structure from its surrounding environment, Fig. 23-1. Walls, however, have many purposes. In our homes, walls shield us from the elements, provide us with privacy, and protect our belongings. In a smokestack, walls contain the toxic gases until they can be filtered. The walls of a canal contain the water so it can be directed to some other location. On some highways, retaining walls are used to hold back the earth from the road, or walls are set up to keep road noise from nearby homes. Dams are essentially great walls that hold back the water of a river to make a storage reservoir.

Some types of walls are used to support a floor or roof system above. In residential and commercial construction, interior walls divide the total space of one floor of a building into smaller units.

Walls can be made of a large variety of materials. Some of these materials are wood, metal, concrete, clay, glass, and plastic.

STANDARD WALL TYPES

The function of the structure is the most influential factor in determining how to design parts of a structure. This information determines what types of wall systems might be used to accomplish the job.

Understanding the forces that will be applied to a wall system helps the designers determine what materials to use. If the structure is to be a simple one-story building, the walls may be used to support only the roof system. On the other hand, if the structure is to be a high-rise office building of many stories, the walls will likely only be used to fill the spaces between the supporting columns and girders. If the structure is a dam then the wall must be very massive. The dam wall must support not only its own weight but also that of the water behind the dam. There are many considerations taken into account when determining what type of wall system is the best and most economical to use for each particular project under construction.

EARLY WALL CONSTRUCTION

Early pioneers constructed their shelters out of what materials they had available. Tents were used as temporary housing. Some early permanent buildings had walls built out of sod, Fig. 23-2 or logs, Fig. 23-3. Both of these wall systems are still used to some extent in modern construction.

Rammed earth structures have soil and other binding materials mixed together and then packed into forms. They are being researched as an economical means to construct walls of buildings. These structures are energy efficient and durable.

Log homes are gaining in popularity for aesthetic reasons. With modern equipment and tools, Fig. 24-4, the labor costs of log construction can be considerably reduced. This makes the price of log construction competitive with other structures.

STANDARD WOOD FRAME CONSTRUCTION

The most common type of construction used to build homes is **wood frame construction,** as shown in Fig. 23-5.

Fig. 23-1. Wall structures are common construction projects. A,B–Providing privacy and shelter. C–Containing agricultural and industrial supplies. D–Channeling water. E–Separating road noise and pollution from the surrounding area. (Alpine)

Fig. 23-2. Pioneers did not have the tools or materials that are available today. Homes with sod roofs, such as this structure, were common.

Fig. 23-3. In areas where there was enough wood, simple log homes were erected. (Bureau of Land Management)

With a wood frame construction system, the builders use standard lumber sizes to erect frames made of wood framing members. Fig. 23-6 shows the walls of a home that have been framed using this system. The vertical members are called **studs,** and the horizontal members are called **plates.**

Because this system of wall framing supports the floor or roof above, where there are openings in the walls for doors or windows, a horizontal member is needed to

Fig. 23-4. A rustic look may be achieved through log construction. (Alpine)

Fig. 23-5. Wood framing was the choice for this condominium complex. Wood is the most popular choice for residential structures.

Fig. 23-6. The use of standard-sized studs and plates speeds up the construction process.

support the structure above. This member serves the same purpose as a small beam or girder. It is called a **header.** Without this header, the roof system or floor above the opening would not have enough support. The structure would be unsafe.

Like the wood floor framing from the previous chapter, most wall frame construction is based on the four foot module, Fig. 23-7. The studs are spaced either 12 in., 16 in., or 24 in. on center. All of these distances are evenly devisable into 48 in. and make the most efficient use of the available materials.

Interior walls that are *not* required to support the structure above the wall are made similar to load-bearing partitions. However, these walls do not require the heavy header construction, see Fig. 23-8.

The tops of all walls are tied together to make a strong wood frame. The plate that the tops of the studs are connected to is called the **top plate.** The **double plate** is attached on top of the top plate and crosses all of the joints that are in the top plate, Fig. 23-9.

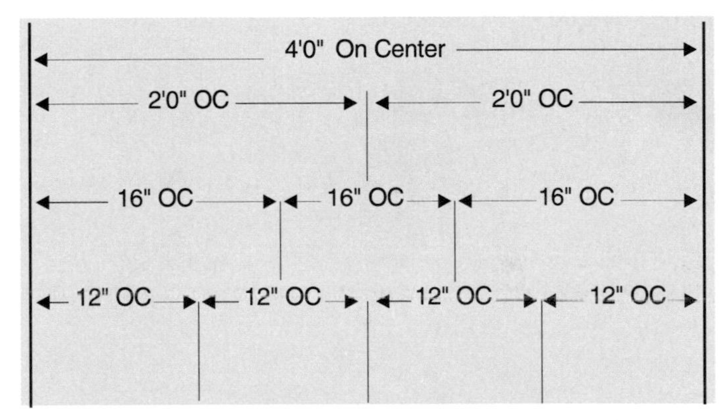

Fig. 23-7. The four ft. module allows for uniform spacing of framing members on two ft., 16 in., or 12 in. centers.

Fig. 23-8. Shown is a framed opening on a wall that is not a load-bearing partition.

Fig. 23-9. The double plate crosses the tops of all the joints, strengthening the wall.

To add additional strength to the wall, **diagonal braces** are used. Diagonal braces ensure that the wall will remain plumb. If the walls are **plumb** it means that they are absolutely vertical. The wall has to be plumb for doors and windows to operate properly.

TIMBER FRAME CONSTRUCTION

With **timber frame construction,** a heavy timber frame is constructed, Fig. 23-10. The frame consists of columns and girders that are used to support the walls and roof of the structure. Once the frame is constructed, the spaces between the columns are filled with wall sections, Fig. 23-11.

Fig. 23-10. Timber frame construction uses thick, strong wooden boards. Notice the size of each framing member used. (Big Timber Works)

Fig. 23-11. Sections are affixed to the walls and ceiling to complete the structure. The wall sections do not need to be very strong. They will not be bearing the weight of the building. (Big Timber Works)

Because the wall sections are not required to support any structural members of the building, they do not have to be made of heavy materials. Fig. 23-12 shows the wall sections during the manufacturing process. Notice that the sections have sheathing material on the outside and inside. The sheathing is separated by a foam core. This makes a wall that is very energy efficient, easy to construct, easy to erect, and light in weight. These walls are manufactured in a plant with specialized equipment. This greatly reduces the labor costs per square foot for this type of wall.

There are fewer framing members in the timber frame construction system than in conventional wood framing. Consequently, the total labor cost of the construction project is smaller.

LIGHTWEIGHT STEEL FRAMING

Lightweight steel framing, Fig. 23-13, uses beams of steel in place of wood framing members. This framing has some advantages over wood frame construction, particularly in commercial buildings where there are strict fire codes. Steel is a human-made product. Thus, steel can be produced in consistent qualities, sizes, and shapes. Also, steel is not subject to rotting or insect infestation.

Fig. 23-12. Foam-filled walls are stacked and waiting on a conveyor. Mass producing these walls in a factory creates walls of a consistent quality.

Fig. 23-13. Lightweight steel framing is very common in small commercial construction.

Fig. 23-14. The parts for this steel building arrived on the site precut and preformed.

Depending on the cross-sectional shape of the framing material, a steel frame can be designed to support the same loads as the wood frame wall system and more. With modern tools and joining methods, it is very easy to erect a structure using steel, and it requires few specialized tools. Holes are pre-punched in the metal studs and plates when they are manufactured. This way the builder does not have to drill holes to accommodate wiring and plumbing.

Some steel buildings, as shown in Fig. 23-14, come in **kit form.** The builder simply erects the structure from plans provided by the manufacturer. This construction is similar to building a model, just on a larger scale. The labor costs on this type of structure are greatly reduced. The components are manufactured in a factory with labor-saving equipment. Field assembly of the components is quick and easy. Little time or material is wasted in cutting and fitting the parts together.

HEAVY STEEL FRAME CONSTRUCTION

The wall system used for large buildings is similar to those systems discussed previously for smaller buildings. The difference is in the size of the columns and girders. Fig. 23-15 shows **heavy steel framing** used to construct a large office building. Heavy duty cranes are used to lift the steel I beams into place. Workers secure them with bolts or by welding, Fig. 23-16.

Fig. 23-15. Larger and heavier buildings require stronger steel beams and girders.

Fig. 23-16. Workers high in the air guide the I beam into place. Once the beam is positioned correctly, they will begin to bolt it down. (Montana Power)

Fig. 23-17. Masonry materials come in the widest variety of styles.

This system is similar to all building construction in that it involves support columns used to support heavy girders. These, in turn, support the floors of the structure.

As the floors are supported by the columns, the wall sections of each floor can be made relatively light. They are held in place by hanging them on the heavy girders that are already in place, hence the term **curtain walls.** The curtain wall may be made from a variety of materials such as concrete, steel, glass, aluminum, or combinations of these materials.

MASONRY WALLS

Masonry walls may be used on all types of structures to provide a good-looking and low-maintenance wall system, Fig. 23-17. Masonry walls can be support walls for the structure above or simply function as coverings for the wall system used to support the structure. A very large variety of clay and concrete blocks as well as native stone are available for the contractor to use.

One type of masonry construction is tilt up construction. With **tilt up construction,** the walls are either built in a plant and transported to the construction site, or they are poured flat on the construction site and then lifted into place by cranes. Tilt up construction uses a smaller amount of labor than other masonry constructions. It is very common in Europe and is gaining favor in the United States.

To begin tilt up construction on the site, concrete forms are set up on a prepared surface. Door and window locations must be carefully formed. Reinforcement steel is placed in the forms. Next, concrete is poured and tamped or vibrated into place and allowed to cure. Steel plates are embedded into the curing concrete. When the concrete cures, the sections will be attached to each other with the use of these steel plates. After the concrete has cured the, slabs are lifted into place and welded to each other to secure their position.

SPECIAL PURPOSE WALLS

Although most of the walls erected are for homes or commercial buildings, there is also an assortment of special purpose walls. Numerous construction workers each year are employed to erect and maintain storage tanks, dams, and retaining walls. Each of these walls serves a special need.

TANK CONSTRUCTION

Large tanks used to store fuels and other liquids are usually made of steel plate, Fig. 23-18. To make these tanks, heavy sheets of steel are shaped in arcs. These arcs are then welded together on top of steel plates that make up the bottom, Fig. 23-19. Because of their circular shape, the thin walls of the tanks can withstand the pressure from stored liquids. Circular walls distribute the pressure from the liquid evenly about their entire surface. In addition, circular walls allow for a greater volume with a smaller surface area. This reduces the quantity of materials needed for a job. This is an excellent example of how the shape of a structure can be used to make walls the most efficient.

Fig. 23-18. Large steel storage tanks are a necessity in industry and agriculture.

Fig. 23-19. Steel storage tanks under construction.

DAM CONSTRUCTION

Another wall structure that is used to contain liquids is the dam. A **dam** is simply a structure that contains water in a particular location for storage purposes or sometimes to produce power. Sizes of dams vary widely. A simple wood structure, Fig. 23-20, can function as a dam. Earth structures and large concrete dams fill larger needs.

Straight gravity dam

The dam wall rests on a foundation. With large concrete dams, the foundation must be the bedrock of the earth to support the tremendous weight of the concrete. The Grand Coulee dam is one such dam. The Grand Coulee dam is a straight gravity dam. **Straight gravity dams** hold back the water simply by their massive weight. This type of dam can be identified by its straightness, see Fig. 23-21.

Fig. 23-20. Dams are not all massive construction projects. Sometimes a simple dam structure is all that is needed

Dams in mountains

In mountainous areas, a dam can be designed to take advantage of steep canyon walls. These dams are curved, Fig. 23-22. This allows the canyon walls to be used as supports for the ends of the dam. This dam construction does not require the massive weight of the straight gravity dams to support itself and the water behind it.

Earth filled dam

Another type of dam is an earth filled dam. **Earth filled dams** use very large amounts of earth to build a large bank

Fig. 23-22. The foundation for a dam in a mountain pass is under construction. Notice the curvature of the structure. (U.S. Department of Interior, Bureau of Reclamation)

Fig. 23-21. Dams provide a cheap source of electricity with minimal pollution. (U.S. Department of Interior, Bureau of Reclamation)

of earth and stones, Fig. 23-23. This bank holds back the water. This type of dam is relatively inexpensive to construct. However, earth filled dams are not as durable or as long lasting as concrete dams.

RETAINING WALLS

Water is not the only material that is held back with walls. Wall structures are also used to hold back or support the earth. These structures are called **retaining walls.** They allow us to change the landscape. Often, along the sides of the road in steep country, you will see structures used to hold back the earth. This allows the road to be built without the removal of large amounts of earth, Fig. 23-24.

The design of a retaining wall is determined by the type of earth to be retained. For example, if the soil is a clay type, it will exert a great deal of lateral force as it becomes saturated with water. On the other hand, if the earth is made up of mostly rocks and gravel, there will be less lateral force exerted on the wall. Another design consideration is the slope of the retaining wall. If the wall is to be made near vertical, it must be constructed of very strong materials. The wall must also be well anchored to the earth. If the retaining wall can be made with a slope (more in line with the natural terrain), the wall does not need to be as heavy duty.

Landscape architects use retaining walls when designing landscapes in and around buildings. The designer must have knowledge of forces of the earth and be able to determine how the retaining wall should be constructed. Often large treated timbers are used to make retaining walls several feet in height, Fig. 23-25. Higher walls use steel, rock, or concrete as the main building components.

In road construction, it is sometimes more economical to tunnel *through* a mountain rather than construct the road around or over the mountain. A **tunnel,** Fig. 23-26, consists of a wall with two sides as well as a top and bottom. This creates a retaining wall that totally encases the road. A

Fig. 23-23. Earth filled dams are less expensive to construct than concrete dams. (U.S. Department of the Interior, Bureau of Reclamation)

Fig. 23-25. Here, a landscape architect has used a timber retaining wall to beautify the outside of the building.

Fig. 23-24. Retaining walls control sections of earth

Fig. 23-26. Railroads frequently tunnel through the mountains in the western United States.

smokestack is essentially a vertical tunnel and is used to carry away gases in an industrial site. The round shape of the stack allows it to be made of relatively thin materials and yet be very strong.

SUMMARY

Wall systems are used in construction to enclose, protect, and define a space to meet our needs. Some walls bear the load of the structure above. Other walls cannot support loads and are hung on frames. These frames bear the load of the structure.

Early construction used readily available materials such as stones, logs, and dirt. Today's construction projects utilize more manufactured products that allow us to build more complicated structures.

Wall systems may be constructed from a large variety of materials. The selection of the material depends upon the type of structure being built and the intended use of the wall. In residential construction, wood, metal, and masonry are the most common materials. In commercial construction, concrete and steel are generally used.

The function of the wall may be to separate an area or to support the floor system above. It may also be used to contain or control liquids or gases as in canals, dams, smokestacks, and storage tanks.

KEY WORDS AND TERMS

All of the following words and terms have been used in this chapter. Do you know their meaning?

Curtain wall
Dam
Diagonal brace
Double plate
Earth filled dam
Header
Heavy steel framing
Kit form
Lightweight steel framing
Masonry wall
Plate
Plumb
Rammed earth structure
Retaining wall
Straight gravity dam
Stud
Tilt up construction
Timber frame construction
Tunnel
Wood frame construction

TEST YOUR KNOWLEDGE

Do not write in this book. Please write your answers on a separate sheet of paper.

1. List six different materials that can be used to construct wall systems.
2. Identify two reasons why walls are constructed.
3. In standard wood framing, what are horizontal framing members called?
4. In standard wood framing, what are vertical framing members called?
5. Why is a header used in a wood frame wall?
6. What is a curtain wall?
7. Why can the walls of a tank or a smokestack be relatively thin?

APPLYING YOUR KNOWLEDGE

1. Build models for four different types of wall systems.
2. Visit a construction site and determine what type of wall system is being installed and why.
3. Build a small storage shed using wood frame construction.
4. Using cardboard, construct a small smokestack a few inches in diameter and several feet tall. Explain why the stack is so much stronger than a flat piece of material you used in its construction.
5. Locate and observe a retaining wall in your community. Determine its current condition. Give reasons why it is in this condition.

24

CHAPTER

ROOF SYSTEMS

The information provided in this chapter will enable you to:

- *Describe the function of the roof system.*
- *Identify various types of roof systems.*
- *Describe the function of roof framing members.*
- *Explain the difference between framed and trussed roofs.*
- *List five types of roof systems and briefly describe each.*

Roof systems are similar to the floor systems studied in Chapter 22. **Roof systems** are often horizontal, they connect the wall structure together, and they provide protection from the elements. Some structures do not have a roof system. Examples of these are dams, bridges, canals, and some tanks. Some structures have a single roof that provides weather protection for a number of floors below. This is the case in high-rise buildings. Other structures have little more than a roof structure that is supported by columns. Picnic shelters and covered sidewalks are designed this way. It is the function of the roof to protect the occupants and contents of the structure from rain, snow, wind, and sun.

In some structures it is not possible to distinguish between the walls and the roof system. This is true in geodesic domes and some air-supported structures, Fig. 24-1. In **A-frames,** Fig. 24-2, the roof system serves as the walls on two sides. Conventional walls are placed on the other two ends. In large commercial buildings, there are many similarities between the floors on each story of the building and the roof system. The major difference is the roof is covered with a roofing material that prevents water from seeping into the interior of the structure. The floors are simply covered with carpet, wood, or tile.

CONVENTIONAL ROOF DESIGN

Just as with a floor system, a roof must be designed to withstand the forces that loads may apply upon it. A roof must be able to hold up under the loads of heavy snows

Fig. 24-1. Roof and wall systems are not always distinct entities. Can you pick out where the walls end and the roof begins on this structure?

Fig. 24-2. The walls to the left and right on this A-frame home make up the roof.

common in some parts of the world. They also must bear the weight of the materials used to construct them and mechanical equipment placed on them.

If a roof is very steep, forces applied by the wind must also be considered during design. The wind load can be quite large on roof structures such as A-frames.

TYPES OF ROOFS

Roofs may be designed flat, or they may slope in one or more directions. Fig. 24-3 shows several basic roof designs commonly used on buildings. When designing the roof, the architect takes into consideration the climate, the costs of various roof systems, and the finished appearance of the completed structure.

Flat roof systems

The simplest roof system to construct is the **flat roof**. Even though this type of roof is called a *flat roof,* it often will have a slope of a few inches per foot. This slope is to ensure that water does not collect on the roof. An important advantage of the flat roof is that all of the space beneath it is usable. The flat roof also covers less area, offering a savings in construction materials. Most large commercial buildings use a flat roof system. In addition to its efficiency, it provides a platform for mechanical equipment on the roof tops, Fig. 24-4. However, in cold climates, where there is considerable frost action and snow, flat roofs require considerably more maintenance than a sloped roof.

The supporting framework of the flat roof is very similar to the frame ork that supports the floors below. On large buildings the roof is constructed of concrete, steel, or wooden materials. This frame is covered with water and moisture resistant roofing materials. Because the roof is directly exposed, it must withstand temperatures as high as 150° F to temperatures well below 0° F. On large roofs this variation in temperature requires the installation of an expansion joint system. An **expansion joint system** will allow the roofing material to expand and contract without breaking.

Fig. 24-4. A flat roof provides a stable base for a building's mechanical equipment.

The flat roof can be supported by the box and pan system. This system was discussed in Chapter 22. The box and pan system for constructing floors and roofs consists of a series of boxes or pans placed on a temporary floor supported by scaffolding. The pans are laid out in rows separated by several inches. This space allows for the addition of reinforcement steel to add strength to the completed system. When the form work is completed concrete is poured over the entire area. After the curing of the concrete, the pans are removed along with the scaffolding.

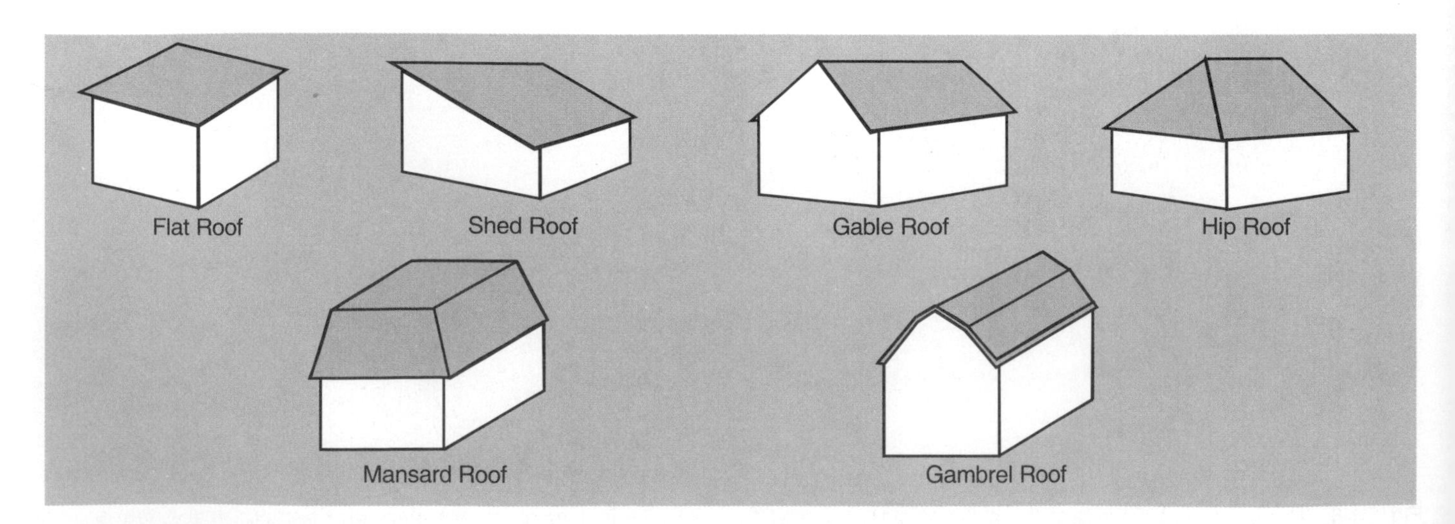

Fig. 24-3. Six common roof types.

When viewed from underneath, the box and pan system resembles a large waffle pattern.

A flat roof can also be supported by an extensive system of flat trusses as shown in Fig. 24-5. The trusses must be designed to withstand the weight of the structure and any accumulation of water and snow.

Sloped roofs

There are several different types of sloped roofs. The simplest type of sloped roof is the shed roof. With the **shed roof,** the entire roof is sloped in one direction in a single plane as shown in Fig. 24-6. If the roof slopes in two directions, as shown in Fig. 24-7, it is called a **gable roof.** The location where the two slopes meet is called the **ridge.** The triangle formed at the end of the structure by the meeting of the two slopes is called the **gable.** This is what gives this type of roof its name.

Some roofs slope in four directions, Fig. 24-8. These roofs are called **hip roofs.** These roofs do not have a gable end. The line of the roof is uniform entirely around the structure. An advantage of this roof is its easy maintenance. It is often difficult to maintain the gable ends of a gable roof.

There are variations on these basic types of roofs. One variation uses two different slopes on each side of a gable roof. This is called a **gambrel roof,** Fig. 24-9. The first slope is quite steep, while the upper slope is more gentle.

Fig. 24-5. This truss construction is designed to support a flat roof for a condominium.

Fig. 24-6. The entire shed roof falls into a single plane. The roof of this car port is a shed roof.

Fig. 24-7. Gable roofs are very common in residential structures. The peak of the roof is called the ridge.

Fig. 24-8. Hip roofs are also very popular in residential construction.

Fig. 24-9. This barn has a gambrel roof. Notice that the steep slope starts about one-third of the way down the roof.

This design allows more usable space in the attic. The **Mansard roof** is very similar to the hip roof. The Mansard roof varies in that the slope of the roof is much steeper and the top is flattened off, Fig. 24-10.

ROOF FRAMING TERMS

The framing members that support the roofing materials need to be strong in order to withstand the forces applied. The main framing members of all sloping roof systems are called the **rafters.** The rafters must be carefully cut so they match up with the connecting framing members. This requires an understanding of the various components that make up the roof framing system. If a rafter is cut too long or too short, the rafter will be too high or too low. This gives a wavy appearance to a roof. It also weakens the roof structure.

Sloping roof systems are made up of framing members and cuts that are horizontal, vertical, and sloping. See Fig. 24-11. For example, on the standard gable roof, Fig. 24-12, the **ridge board** is a framing member that runs horizontal near the center of the structure. The rafters connect to the ridge board and form the sloping portion of the roof. The lower end of the rafters are often attached together with a **fascia board.** This board also runs horizontal between the ends of the rafters.

The cuts on the ends of the rafters must be vertical, or plumb. In order to make the proper plumb cut on the end of the rafter, a framing square is used, Fig. 24-13. The following terms are applied when using the framing square to lay out a roof.

- **Span** is the total width of the building that is to be covered by the roof. This distance is usually the distance measured from the outside of one wall to the outside of the opposite wall.
- **Run** is the horizontal component of the width covered by a rafter. On a simple gable roof this distance is equal to one half of the building span.
- **Unit run** is always equal to 12 in. This is the standard unit used to determine the slope of a roof.
- **Rise** is the vertical distance from the top of the wall to the top of the roof.
- **Unit rise** is a number in inches that determines the slope of the roof. Unit rise is used in conjunction with the unit run. For example, a 4-12 slope roof is one that would rise 4 in. for each 12 in. of run. Therefore, a building that had a run of 5 ft. would have 20 in. of rise. (4 in. multiplied by five units of run).
- **Birds mouth** is the vertical and horizontal cut made near the bottom of a rafter so it has full bearing on top of the wall.

The length of the rafter is a critical feature when trying to ensure a strong, well constructed roof. The rafter length must be calculated since it is very difficult to accurately measure the rafter length on the actual construction site. Mathematics, rafter tables, or the framing square may be used to determine the exact length of the rafter. Figs. 24-14

Fig. 24-10. Roofs have a dramatic effect on the appearance of a home. An architect chose the Mansard roof for this house.

Fig. 24-11. Framing members in a sloping roof system. (Lowden, Lowden and Co.)

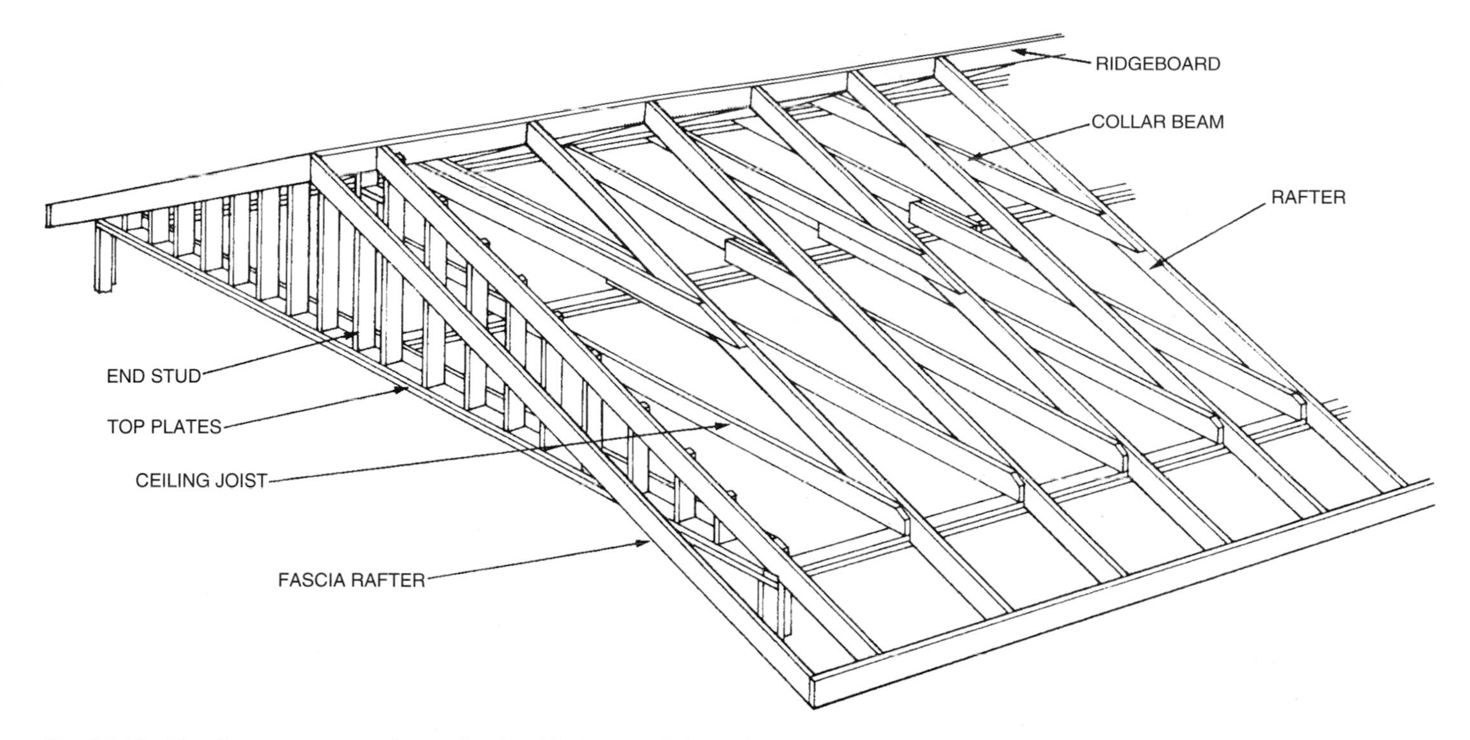

Fig. 24-12. Fascia rafters cover the ends of a ridgeboard. This roof is the gable type.

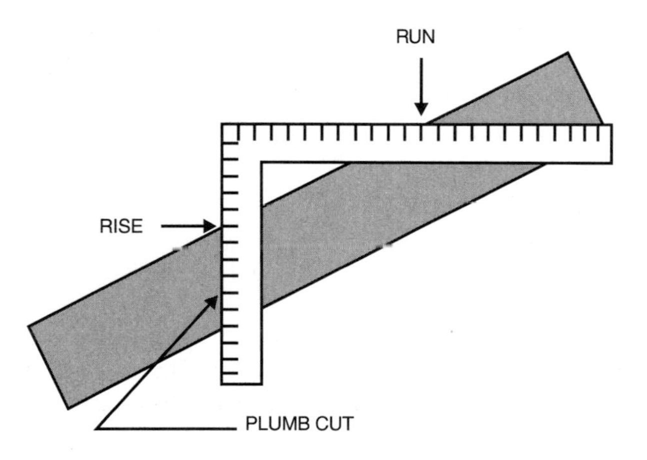

Fig. 24-13. The framing square can be used to find the length of a rafter or to make the proper plumb cut.

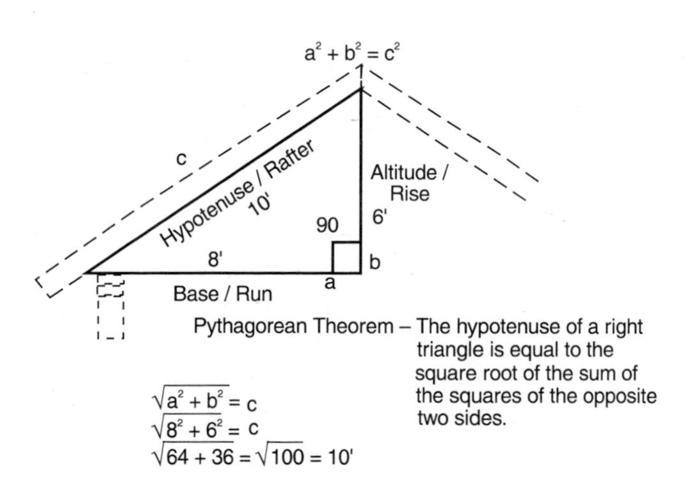

Fig. 24-14. Rafter length is based on mathematical principles.

and 24-15 show how math is important in these methods for determining rafter length.

TRUSSES AND TRUSS DESIGN

If the rafters are cut and assembled into large triangles and installed in units, it is called **truss framing.** See Fig. 24-16. This type of roof construction has several advantages. One major advantage of the truss system is a savings of construction time. The trusses can be ordered from truss manufacturing plants and delivered to the job site ready to install.

RAFTER TABLE FOR 4/12 SLOPE = 1/6 PITCH ROOF

Span	Rafter Length
1′	6-5/16"
2′	1′ 0-5/8"
3′	1′ 6-15/16"
4′	2′ 1-5/16"
5′	2′ 7-5/8"
6′	3′ 1-15/16"
7′	3′ 8-1/4"
8′	4′ 2-5/8"
9′	4′ 8-15/16"
10′	5′ 3-1/4"

Fig. 24-15. Section of a rafter table. It is used for calculating the rafter length for different spans.

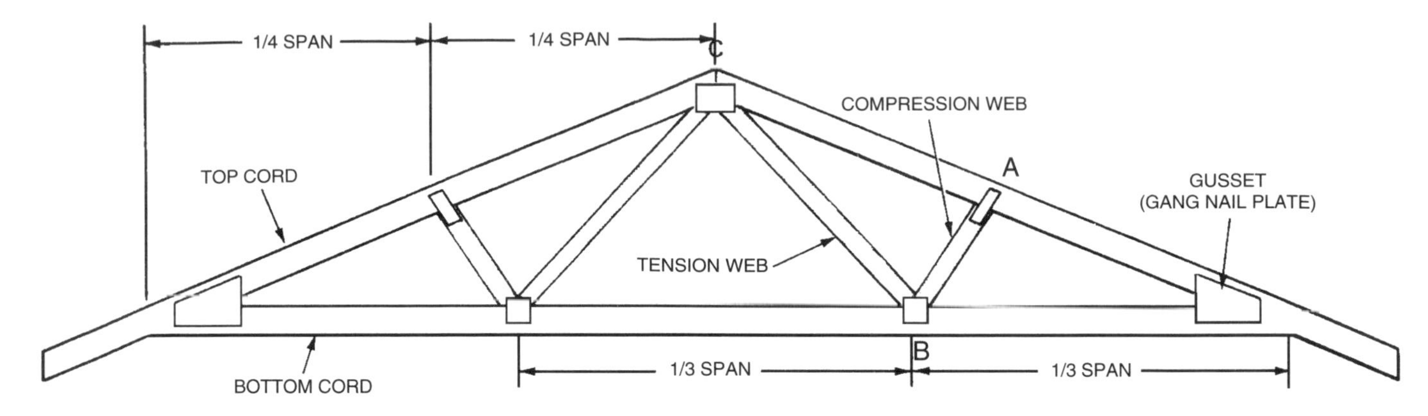

Fig. 24-16. Sketch of a typical truss construction.

Other advantages of truss framing include the stronger and more precise members that may be produced in a truss manufacturing plant. In addition, smaller framing lumber may be used. This is because of the bracing characteristics of the triangles contained in the truss.

Fig. 24-17 shows a few of the most common types of trusses. Truss systems are the most common type of roof system used in residential construction today. This widespread usage is due to the efficiency of the system.

NONCONVENTIONAL ROOF SYSTEMS

Roof systems can also be constructed in a number of other ways. A **geodesic dome** is a special type of truss system that transfers the load of the roof equally and uniformly to all members of the structure. In this way, individual components can be made smaller, yet the structure as a whole remains very strong. A simple demonstration geodesic dome can be built by assembling equal lengths of small pieces of wood as shown in Fig. 24-18.

Dome roofs, are designed to transfer the roof load to a high strength outer ring. This design allows large areas to be enclosed without the need for supporting columns in the middle area of the structure. Athletic arenas often use this type of construction, Fig. 24-19.

Cable-supported roofs also are used in buildings that require a large open area such stadiums, arenas, and airplane hangers. A cable-supported roof system gets its name from the strong steel cables used to support the roof. The

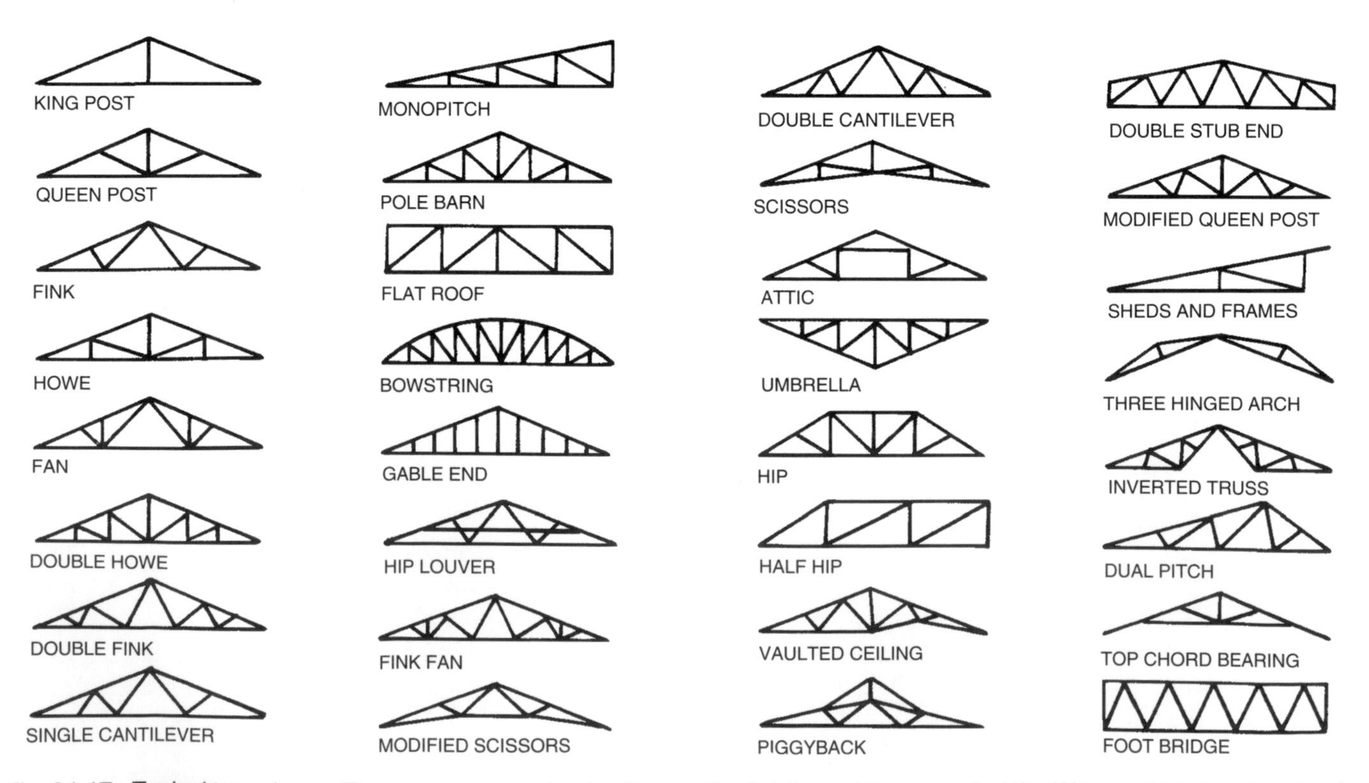

Fig. 24-17. Typical truss forms. The most common for dwellings is the fink truss. It is also called the W truss. Note how bracing forms triangles in each truss.

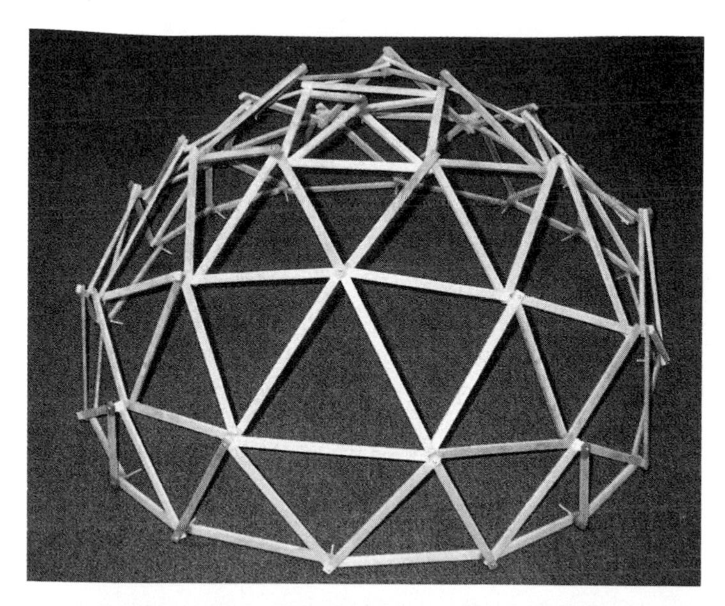

Fig. 24-18. A simple geodesic dome is easy to construct.

Fig. 24-19. Domed roofs have become the popular choice when enclosing a sports stadium. (American Plywood Assoc.)

cables are attached to a compression ring built into the walls of the buildings. Unlike the dome roof system that generally requires a circular shape, the cable-supported roof can be any shape the architect may desire. During construction, the cables are placed in multiple directions between supports. These cables, in turn, support the roof covering as show in Fig. 24-20.

AIR-SUPPORTED STRUCTURES

Another way to support a roof over a very large area is by using air pressure, Fig. 24-21. An **air-supported roof** is similar to a dome structure, however, the roof is supported by a few pounds per square foot of air pressure.

After the walls have been completed, cables are hung loose between opposite points on the wall. The cables are attached to an air-proof membrane. The building is sealed

Fig. 24-20. An architect has constructed this model to pitch his idea. Notice the elaborate setup that can be created with a cable-supported roof.

Fig. 24-21. A slight difference in air pressure keeps this dome aloft. The air pressure under the roof is greater than the outside air pressure.

and large fans are used to pump air into the structure. As the internal air pressure increases, the roof moves up into its final position. Air locks, in the form of double doors, allow people to move in and out of the building without decreasing the internal air pressure.

ROOF COVERINGS

After the supporting structure of the roof has been erected, sheathing material is placed over the framing members. The sheathing will support the final weatherproof roof covering.

Several different materials are used as sheathing. Wood, plywood, and wafer board are the most common for residential structures, Fig. 24-22. Industrial buildings often use corrugated sheet metal or concrete. Fabric is used as sheathing for inflatable structures. Prestressed wood, metal, or plastic foams may be used on other buildings.

The materials are carefully selected by the engineers and architect to meet specific building codes as well as to provide an economical and safe roofing system. The weatherproof roof covering completes the roof system. This covering on residential buildings may be wood shingles, asphalt shingles, tile, slate, or colored metal. On large roofs, various types of waterproof membranes are used to seal out the water.

WATERPROOF MEMBRANES

Waterproof membranes make excellent roofing material for flat roofs. They can be made up of a number of layers of building felt and bituminous tar covered with gravel. This is called a built-up roof.

Rubberized membranes are now available that can be installed on the roof in large sheets. These sheets are then chemically vulcanized at the joints. The roofing is held in place by placing gravel on top of the membrane. When this roof system is completed, the roof will not have any seams that may leak at a later date.

SHINGLES

Shingles are small, flat or curved pieces of waterproof material. They are installed individually on a sloped roof. They are often made from wood, concrete, fired clay, metal, slated, and asphalt impregnated fiberglass. The asphalt shingle is the most common shingle used in residential construction today.

Fig. 24-22. This home is having sheathing attached. Plywood is being used for this residence. (American Plywood Assoc.)

METAL ROOFING CONSTRUCTION

Sheet metal with a baked-on finish is another common roofing material used today. The metal is available in two and four foot widths. Both sizes come in very long lengths. This material provides an excellent roof with good fireproofing qualities.

SUMMARY

The roof system connects the wall system together and protects the space under the roof from rain, snow, and the sun. There are two important types of conventional roofs, flat and sloping. There are many variations on these two major types of roof.

The design of the roof must withstand forces exerted by water, snow, and wind as well as the pull of its own weight. The designer also considers the appearance of the structure when determining what type of roof system to use.

Flat roofs are simpler to build and provide a good surface for mechanical equipment located on the roof. The space directly beneath the roof can be fully utilized. However, it is much more difficult to keep a flat roof from leaking.

Sloped roofs shed water well, but the space directly below the roof is taken up by framing members. This space is not convenient for use. Sloped roofs can be constructed with rafters or they can be built using trusses. Trusses provide reduced labor costs and the opportunity to use lighter framing materials.

KEY WORDS AND TERMS

All of the followings words and terms have been used in this chapter. Do you know their meaning?

A-frame
Air-supported roof
Birds mouth
Cable-supported roof
Dome roof
Expansion joint system
Fascia board
Flat roof
Gable
Gable roof
Gambrel roof
Geodesic dome
Hip roof
Mansard roof

Rafter
Ridge
Ridge board
Rise
Roof system
Run
Shed roof
Span
Truss framing
Unit rise
Unit run

TEST YOUR KNOWLEDGE

Do not write in this book. Please write your answers on a separate sheet of paper.

1. Conventional roofs can be classified as either ______ or _______.
2. List four different kinds of sloped roofs.
3. Give one advantage and one disadvantage for both flat and sloped roofs.
4. What is the relationship between slope and run?
5. Explain how an air-supported roof stays up.

APPLYING YOUR KNOWLEDGE

1. Construct a model of a small building and then design and construct three different types of roof systems for the model.
2. Build an air-supported structure.
3. Design and construct a two frequency geodesic dome with a diameter of three feet.
4. Build a model of a sports arena with a cable supported roof system.
5. Build a scale model of a "W" truss and test its strength.

A well constructed roof helps produce a weather tight structure.

25

CHAPTER

PLUMBING UTILITY SYSTEMS

The information provided in this chapter will enable you to:

- *Describe two types of plumbing systems common to all construction.*
- *Identify the materials used in plumbing systems.*
- *Diagram a simple water supply system for a home.*
- *Diagram a simple wastewater system commonly used in a home.*

Plumbing systems handle the control of liquids and compressed gases. The most common fluid that plumbing systems are designed to move is water. However, oil, gasoline, compressed air, oxygen, and even vacuum lines fall under the plumbing codes and are part of the plumbing system.

A plumbing system is relatively simple in a home, Fig. 25-1. The plumbing systems for refineries, factories, and power plants may be very complex, Fig. 25-2. These industrial plumbing systems include both the supply and wastewater lines, venting for many systems, vast storage tanks and/or disposal systems, and additional plumbing fixtures for unique tasks.

Fig. 25-1. Plumbing systems in a home are fairly straightforward. With the right tools, many homeowners will work on their own plumbing. (DuPont)

Fig. 25-2. Plumbing systems for industrial sites can be complicated. Water is just one of many fluids running through the pipelines. Special training is needed for this work.

HOME PLUMBING SYSTEMS

The plumbing system in the house or apartment consists of two major parts. The **supply system** supplies water to the various fixtures such as the sink, washing machine, and water closet. This system includes both hot and cold water, Fig. 25-3. The **waste system,** Fig. 25-4, includes those pipes that carry the wastewater away to the sewer system. This system consists of the drain and waste pipes as well as a venting system to allow the waste plumbing system to work safely. The waste system is often called the **drain, waste, and vent system** or **DWV.**

The materials used to manufacture plumbing equipment have changed a great deal. Early systems required extensive skills and specialized equipment to properly join the various sections of plumbing, Fig. 25-5. Today, much of the complexity has been eliminated, yet some technical skill is still required to do a quality plumbing job. Because of these changes, many people do their own plumbing. Although, problems can arise if they do not follow proper plumbing procedures.

When it took extensive training and tools, most plumbing work was done by professionals. These people had the knowledge and experience to do a quality job of plumbing. Today plumbing systems do not have to be installed by professionals, but a quality job is still important. Plumbing systems have to be installed according to a set of rules or codes. The **Uniform Plumbing Code** is one such code, Fig. 25-6. This code is published by the International Association of Plumbing and Mechanical Officials. By following a code, the person installing the system ensures the finished plumbing meets the needs of the customer and provides a safe and long lasting system.

WATER SUPPLY SYSTEM

Water must be supplied to each fixture in the proper quality and volume. The size of water lines, fittings, and

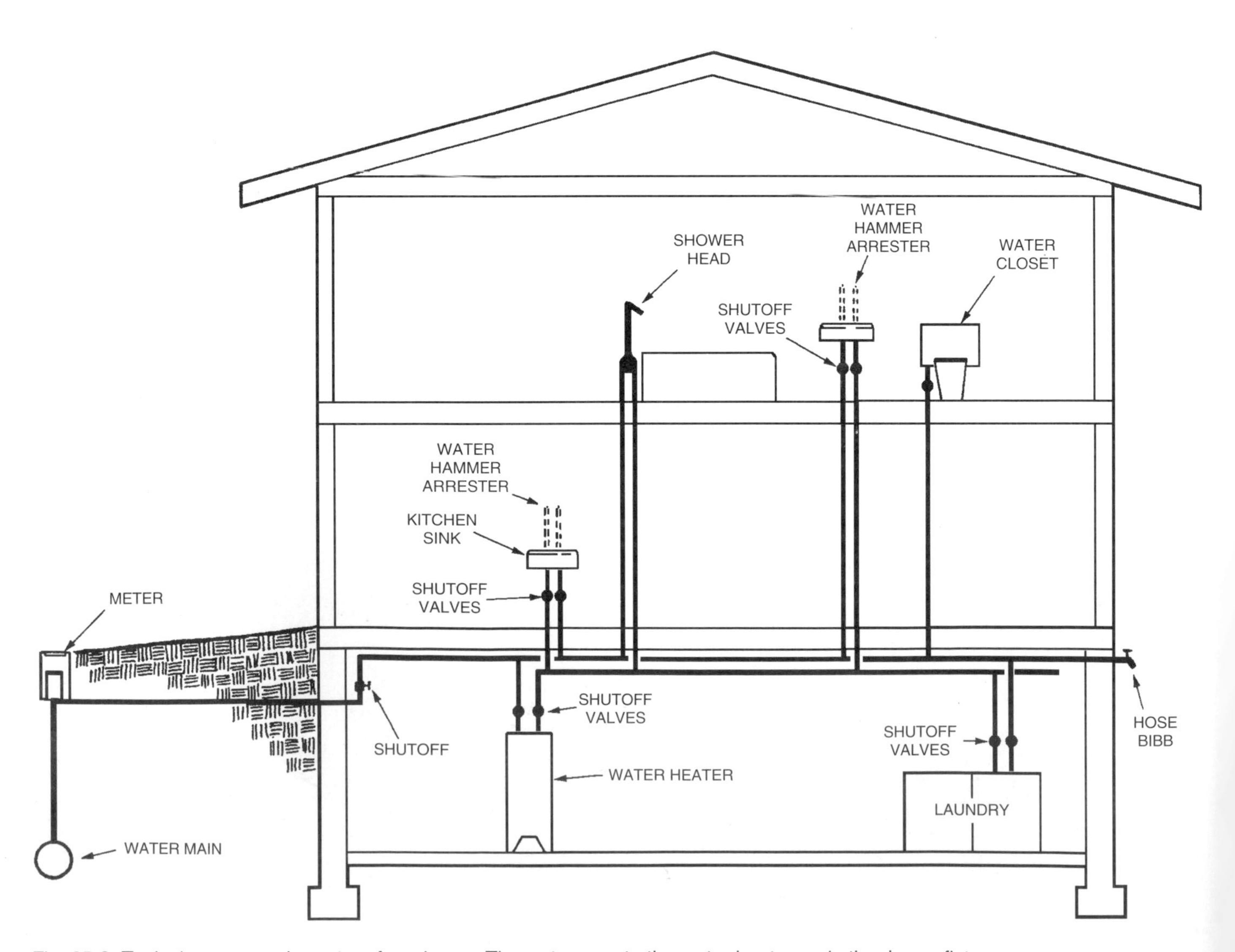

Fig. 25-3. Typical water supply system for a home. The water runs to the water heater and other home fixtures.

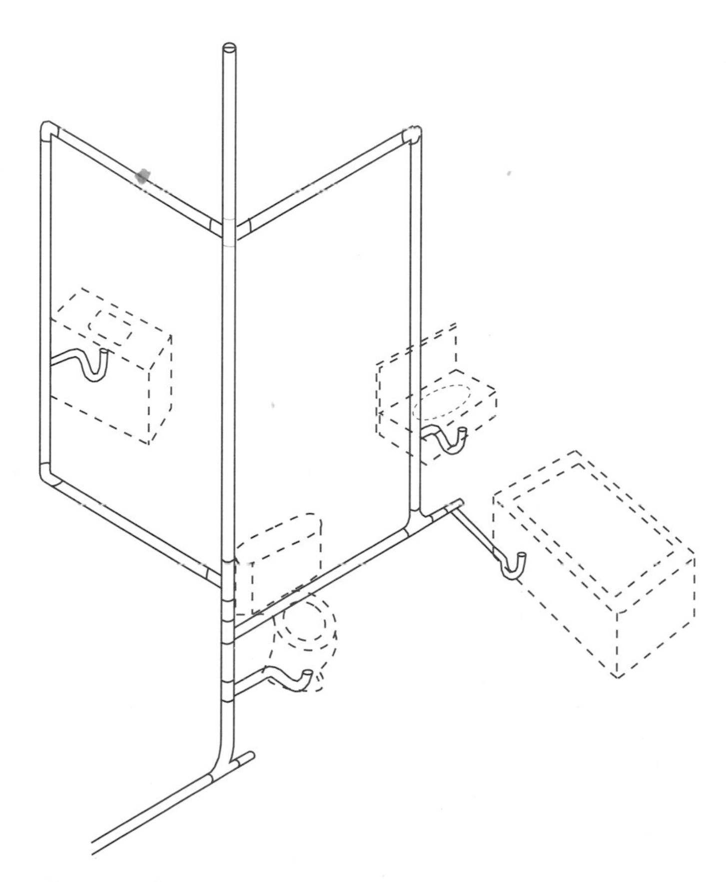

Fig. 25-4. Typical waste system for a home.

Fig. 25-6. The Uniform Plumbing Code book is one of several plumbing code books published. Different communities have different plumbing codes.

FIXTURE	FLOW RATE IN GAL./MIN.
Kitchen Sink Faucet	4.5
Bathroom Faucet	2.0
Bathtub Faucet	6.0
Clothes Washing Machine	5.0
Shower	5.0
Water Closet	3.0
Drinking Fountain	.75
Hose Bib	5.0

Fig. 25-7. Water demands for standard plumbing fixtures.

the meter are determined by the water demand of each fixture, Fig. 25-7. The *total water demand* for the residence is the sum of all of the individual demands for each fixture.

The designer plans the size of the piping carefully. The designer calculates the length of the supply lines, the elevation of the supply lines from the water meter to the fixture, the type of pipe used, and the water demand at the fixture, Fig. 25-8. All of this information is calculated with information from the building blueprints and using the help of formulas and tables.

Since it is the function of the supply system to provide clean water to the fixture, the supply lines must be installed correctly, and they must be made from the proper materials.

Fig. 25-5. Early home plumbing systems could not be fixed by most homeowners. Special tools to melt solder (left) and to thread pipe (right) were needed. Threaded pipe is now sold, eliminating the need for expensive tools.

These materials will withstand the water pressure and resist the corrosive action of the water flowing through them.

Common materials used for supply pipes are galvanized steel, copper, and plastic. All of these materials are available in hardware and plumbing supply stores. Large numbers of specialized fittings and valves are made from these materials as well. They come in a variety of sizes.

Water hammer

When water is flowing through supply pipes at pressures in excess of 70 pounds per square inch and the valve is suddenly closed, the plumbing system makes a pounding noise. This is called a **water hammer.** Water hammer is caused by quickly stopping the flow of water. This causes a sudden buildup in pressure in the pipeline, creating the hammer-like sound. This condition is damaging to the plumbing system and, if not fixed, may cause future maintenance problems.

Water hammer can be fixed easily. Installing an air chamber close to the fixture, as shown in Fig. 25-9, prevents the problem. An **air chamber** is a piece of the supply system that holds a small amount of air. The air pocket serves as a cushion for the water, easing the pressure on the pipes when the valve is closed suddenly.

Shutoff valves

Supply pipes should have shutoff valves installed in the plumbing near each fixture, Fig. 25-10. **Shutoff valves** allow the repair of the fixture without shutting off the entire supply system. Water can be shut off to each individual fixture. During the finish work on a structure, the plumber first attaches the shutoff valves. Then the plumber installs the fixtures and checks for proper operation.

Shutoff valves also help protect the rest of the household from water damage. If a fixture begins to leak or overflow, the fixture's water supply can be quickly stopped.

Fig. 25-8. Designers use the water demand of each fixture to calculate the width of the pipes needed.

Fig. 25-9. Air chambers are installed near water fixtures to prevent water hammer. Water hammer does more than just make noise. With time, it will damage a plumbing system.

Fig. 25-10. Shown are the hot and cold water leads for a washing machine. Each of the pipes has its own shutoff valve. (Tom Wood)

DRAINAGE SYSTEMS

Every sink, tub, and water closet in a system has a soil or waste stack, a vent system, and a trap, Fig. 25-11. The waste stack is made up of large pipes that carry both solid and liquid wastes out of the building. The waste stack connects to another large sewer line that carries the wastes to a sewage disposal plant or to a private septic system.

Vents

The **vertical vent** is a pipe that provides air circulation throughout the drainage system. It picks up any sewer gases that might be present and releases them into the air above the roof. This vent also is important for equalizing pressures inside the drainage system.

Water traps

Each fixture in a home needs a water trap. **Water traps** stop the flow of sewer gases back into the fixture where they could be released into the home. A trap is a U-shaped bend in a waste pipe. This bend holds water while the fixture is not in use. The water seals out gases that might come up through the pipe. If you look under the sink in the bathroom or kitchen, there will be a curved drain pipe. You can see how some of the water draining out of the fixture will remain and fill the lower part of the trap. Fig. 25-12 shows a sink trap.

Clean-outs

Another feature that a waste plumbing system needs is a **clean-out.** Sections of the waste system, on occasion, get clogged with solids. These sections must have the solids cleaned out. Clean-outs are capped openings that allow easy access to the insides of the pipeline. Without clean-outs spaced at strategic locations, a section of the plumbing would have to be removed and replaced.

Clean-outs are capped with threaded plugs that can be easily removed when cleaning becomes necessary, Fig.

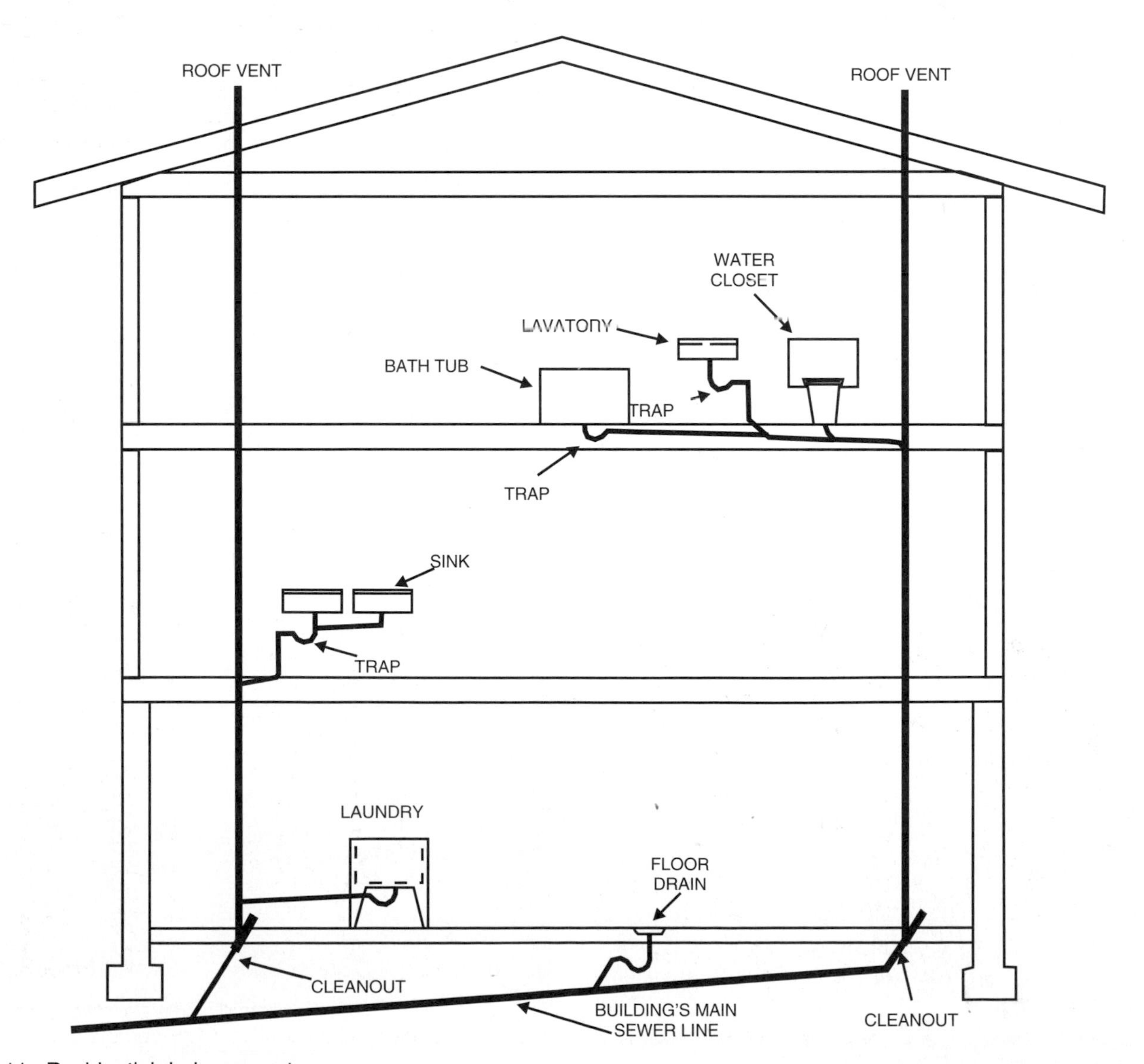

Fig. 25-11. Residential drainage system.

Fig. 25-12. Sink traps are a simple way of preventing sewer gases from escaping from the wastewater system.

25-13. Usually a clean-out is placed at the upper end of a pipeline each place the line changes direction. When a pipe becomes clogged, the cap is removed so a clean-out tool can easily remove the blockage in the line.

DWV materials

Drain, waste, and vent piping is manufactured from a variety of materials. Cast iron pipe has been very popular for many years and is still common in some locations. Fig. 25-14 shows two methods used to connect the various lengths together. Plastic pipe has replaced cast iron pipe in most areas. Plastic pipe, Fig. 25-15, is not affected by the waste it carries nor do soil conditions cause any problems with this material. Copper piping is more expensive, but copper is still installed inside some buildings. However, copper pipe is not used for drain pipes that are to be buried in the soil. This is primarily because other materials are less costly and work equally as well.

The joints on the sewer pipes must be tightly sealed in order to keep the sewer water from leaking out. It is also important to keep roots from entering and plugging the sewer system.

GRAY WATER SYSTEMS

Numerous authorities feel that there will be serious water shortages in many parts of the world. Some areas are feeling it even now. With these coming water shortages, a need exists to make better use of our available water. One method to save some of this water begins with the installation of a storage tank with a filtering system. Water from the clothes washer and bathroom sinks would be directed to this tank, Fig. 25-16. This water, once filtered and chemically treated, is then used in the flush tank of the toilet.

This type of water conservation system is called a **gray water system.** Gray water systems are being used in a number of homes. They add efficiency to our water use by reusing a portion of the household water.

There are many other conservation measures that can be installed in homes along with gray water systems. These measures include low volume shower heads and water closets that require less water to flush.

PLUMBING SYSTEMS IN COMMERCIAL CONSTRUCTION

Plumbing systems in large buildings are basically the same as the systems found in homes. The most important

Fig. 25-13. The cap of this clean-out simply twists off to give easy access to the pipe. The work area around a clean-out should be clear. A good deal of water may be trapped in the pipe.

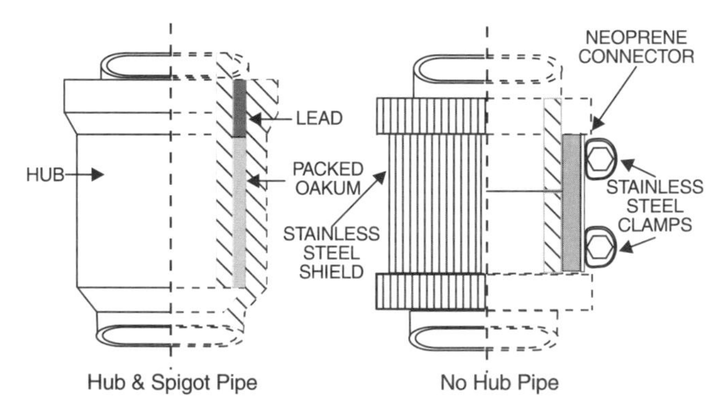

Fig. 25-14. Two methods for connecting and sealing sections of pipe in waste systems.

Fig. 25-15. Building supply stores carry a wide variety of PVC (polyvinyl chloride) pipes and fittings.

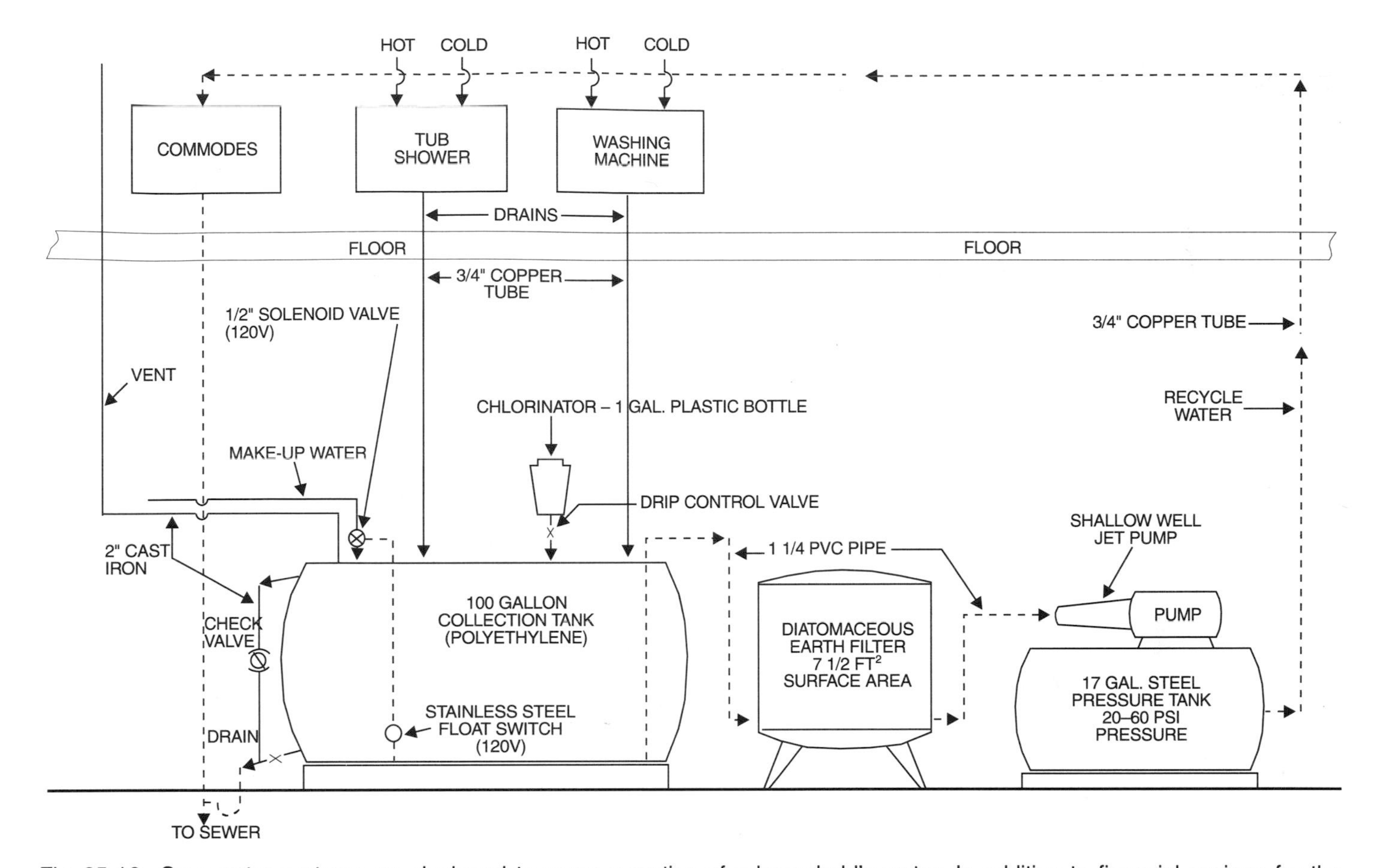

Fig. 25-16. Gray water systems are designed to reuse a portion of a household's water. In addition to financial savings for the consumer, gray water systems can help prevent future water shortages.

difference is the size of the supply and waste pipes, Fig. 25-17. An engineer must be careful when sizing the pipe for a large commercial system. Many fixtures may be operating at the same time. If the main pipes are not large enough, some fixtures will not get the correct water volume or pressure to function properly.

Fig. 25-17. Larger buildings have more plumbing fixtures and much larger total water demand. Larger supply and drainage pipes are required. In addition, powerful pumps may be necessary to deliver water to upper floors.

FIRE SPRINKLER SYSTEMS

Larger buildings do have some systems that are not common in smaller homes. A plumbing system that is becoming very routine in large buildings is a fire sprinkler system, Fig. 25-18. In many cities, the building codes

Fig. 25-18. Large buildings under construction have sprinkler systems included in the design. The finishing work on this office building will hide all but the sprinkler heads.

require sprinklers. A **fire sprinkler** system contains a series of sprinkler heads that are designed to turn on when the temperature at the ceiling level reaches 170 degrees. Once activated, the sprinkler sprays water until it is shut off manually by the fire personnel responding to the fire.

Where sprinkler systems have been installed, records show that they have saved many lives and considerable property damage. Most new commercial constructions require sprinklers, especially high-rise buildings where there are large concentrations of people. Some home owners are also installing residential sprinklers.

SPECIALIZED PLUMBING SYSTEMS

In large factories, research facilities, and manufacturing plants, the plumbing systems can become very complex, Fig. 25-19. The person installing these systems must be very experienced. These plumbing systems carry gas, oil, steam, compressed air, and a variety of other fluids.

A service station, Fig. 25-20, is a good example of one of these specialized systems. Gasoline is highly flammable. Extreme care must be exercised because a leak in the above ground fixtures could cause an explosion. A leak in an underground tank may contaminate the ground water. Naturally, there are many regulations that govern the installation of this type of plumbing. Only licensed individuals have the authority to work on these systems.

GAS AND OIL PIPELINES

Throughout the United States there are many networks of pipelines that carry gas and oil. Perhaps the most famous pipeline is the Alaskan pipeline. It carries crude oil from Prudhoe Bay to Valdez, a distance of about 800 miles (1300 km). Fig. 25-21 shows part of this pipeline. Some of the construction problems encountered with this pipeline included working with permanently frozen ground as well as the need to allow areas for migrating caribou to cross the path of the pipeline.

Fig. 25-19. The red pipes in this industrial building make up pre-action plumbing for a fire sprinkler system. The blue pumps and piping supply water from an on site storage tank to cooling towers in the event of a failure of the city supplied water system.

Fig. 25-20. Ground work in progress at the site of a future service station. Gasoline tanks are secure stored deep in the ground.

PLUMBING SYSTEMS IN CIVIL CONSTRUCTION

City, state, and federal governments provide plumbing systems for their residents. These large plumbing systems are used by everyone in the area. Water treatment plants, main sewer lines, and drainage systems for roadways are now necessities for our society.

Fig. 25-21. The Alaskan pipeline stretches 800 miles across the state. Sections of the pipeline are raised to allow animals to pass. (Morrison Knudsen Corp.)

WATER TREATMENT PLANTS

Water treatment plants are another important piece in civil plumbing structures. Cities require tremendous quantities of clean water. The water must be free from bacteria and other organisms. A **water treatment plant** is required to provide this water. The treatment facility is part of the plumbing system.

Treatment plants take water that is not fit to drink and run it through a series of filters. Additional chemical treatment produces water that is perfectly safe for human consumption. Due to potential water shortages in the future, the treatment of water from rivers and dams is becoming more and more important. Other water sources will not supply all of our needs.

CITY WATER SUPPLIES AND SEWER LINES

Populations continue to grow and in some areas exceed the capacity of the local water supply system. Cities have had to transport water many miles to bring an adequate supply to the growing population. This requires the construction of large **water mains,** Fig. 25-22. In some parts of the United States these pipelines are hundreds of miles long. They function the same as a main supply pipe in a home. They provide the city with the extra water needed to function properly.

Some cities and towns are located near large dams. These dams supply the community with water for both domestic use as well as fire fighting. All of these cities have a water treatment plant that carefully monitors the water quality.

All cities must have a system to remove the wastewater. Sewer lines from homes, businesses, and factories connect to the main sewer lines buried deep under the streets. These main lines pipe the sewage to a **sewage treatment plant,** Fig. 25-23. This treatment plant removes enough of the impurities that the water is safe to be returned to a river or lake. All cities and rural areas have strict enforcement regulations pertaining to the treatment of sewage. Detailed information can be obtained from the county health office.

A

B

Fig. 25-22. Many cities do not have an adequate supply of local water. Water mains are vital supply lines for these cities. A–Construction of one segment of a water main. B–Assembly of the water main.

RURAL WATER SUPPLIES AND WASTE HANDLING

Many rural homes, and some cities, obtain their water from **wells,** Fig. 25-24. The well water is pumped into large elevated tanks that supply the city with the needed water pressure. If these tanks were not present, the water supply would rely on electricity to operate pumps to supply pressure. If the electricity fails, there would be no water for homes and offices. There would be no water available to fight fires. For these reasons, you will see large backup water tanks in many cities, Fig. 25-25.

Many homes in rural areas will have their own tanks. Well water is pumped into a **water pressure storage tank,** Fig. 25-26. The air in the tank is compressed by the incoming water. When the pressure reaches a predetermined level, a pressure switch shuts off the pump. As water

Fig. 25-23. This treatment plant purifies wastewater. The water must be cleaned to such a degree that it may be safely returned to local rivers or lakes.

Fig. 25-24. Rural areas often rely on wells. Shown is a crew drilling a new well opening.

is used, the pressure drops. When a second predetermined setting is reached, the pump is turned on again, Fig. 25-27. This type of system stores from 30 to 90 gallons between pump cycles. This is adequate for single homes.

Sewage from the home is treated in a **septic tank** system, Fig. 25-28, in rural areas. The sewage from the home enters the septic tank where **anaerobic** (without air) bacteria decompose the sewage. The clear liquid from the septic tank is

Fig. 25-25. Water towers are a familiar sight. They provide water storage as well as water pressure when needed.

Fig. 25-26. Water pressure tanks, such as the one shown here, provide adequate water storage for single homes.

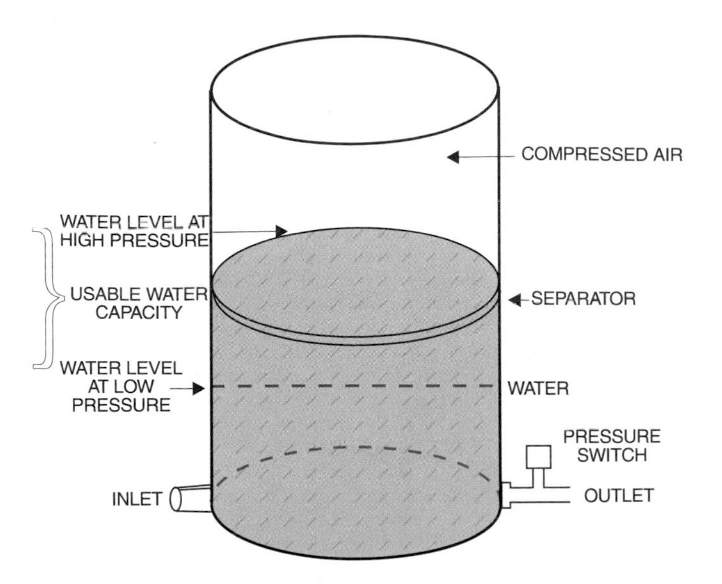

Fig. 25-27. Diagram of a water pressure tank.

Fig 25-29. Where there is plenty of land surrounding the roadways, ditches and culverts are used to remove the excess water.

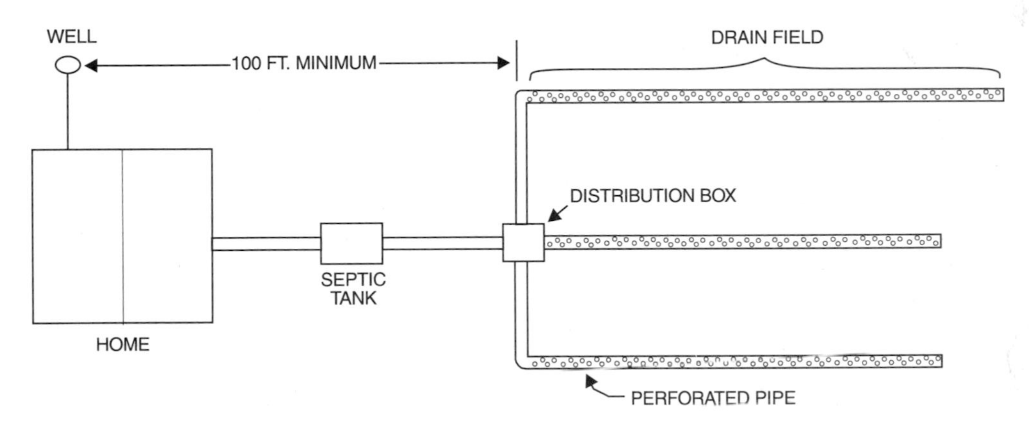

Fig. 25-28. Typical rural drainage system for a home. Note that there must be a minimum of 100 ft. between the well and the drainage field.

drained off underground to a drain field. It is disposed there. The solid waste (sludge) that remains settles to the bottom of the septic tank. There it remains until it is pumped out every few years on a regular maintenance schedule. This system provides many years of service if properly maintained.

ROAD AND STREET DRAINAGE

If water sits for a long time on or near the road surface, the roadbed softens and the road deteriorates. Water standing on a road also makes the roadway hazardous. However, controlling water on roads and railroads is generally straightforward. The designer makes sure that there is a place for excess water to flow during heavy rainfalls or during spring runoff. The water must be moved away from the roadbed.

On most roads water runoff is handled with the use of ditches, gutters, and culverts, Fig. 25-29. In the city, though, it is not practical to have open ditches. The ditches take up a considerable amount of space. Land in the city is too valuable. Thus, cities use a system of **storm sewers.** The drains for the sewers, Fig. 25-30, drain off the water into large underground pipes, Fig. 25-31. These pipes are connected to the sewage treatment plant where the water is treated and returned to the river.

SUMMARY

Plumbing systems play an important role in residential, commercial, and civil construction. All types of construction require the control and containment of gases or liquids. In residential and commercial construction, where people reside or are employed, the plumbing system includes both a supply system and a waste disposal system. In civil construction, a waste system is needed to control the runoff of water from storms. If the water is not controlled properly, road damage may result.

Fig. 25-30. This drain will take in all of the excess water produced by the melting snow.

The supply plumbing system consists of piping that supplies potable water to the various fixtures. This system must be carefully designed so that it provides safe water to each fixture.

The waste system removes the used water from the fixture and transfers it safely to the proper treatment facility. The design of these plumbing systems must follow building codes.

In filling stations, refineries, and industrial plants, a large variety of different liquids and gases are used. This requires a very extensive and complicated plumbing system to properly control these materials. Safety problems can be tough when working with caustic or flammable fluids.

One of the services a city furnishes is providing main supply lines and sewer lines for individuals, companies, and industries. In civil construction projects, such as dams and roads, the plumbing system must be larger than residential and commercial systems. Also, civil projects are often in the open. Civil systems must be able to control enormous quantities of water.

KEY WORDS AND TERMS

Each of the following key words and terms have been used in this chapter. Do you know their meaning?

Air chamber
Anaerobic
Clean-out
Drain, waste, and vent system (DWV)
Fire sprinkler
Gray water system
Septic tank
Sewage treatment plant
Shutoff valve
Storm sewer
Supply system
Uniform Plumbing Code
Vertical vent
Waste system
Water hammer
Water main
Water pressure storage tank
Water trap
Water treatment plant
Well

Fig. 25-31. Workers installing a storm sewer. Storm sewers prevent the flooding of local businesses and residences as well as the streets.

TEST YOUR KNOWLEDGE

Do not write in this book. Please write your answers on a separate sheet of paper.

1. The two major types of plumbing systems are _______ and _______.
2. What three materials are commonly used in household plumbing systems?
3. Gray water systems are becoming (less/more) common in households.
4. DWV stands for _______.
5. Water treatment plants are part of _______ construction plumbing systems.

APPLYING YOUR KNOWLEDGE

1. Construct a simple plumbing system to deliver water to a specified location using valves and fittings. When complete, test your system for leaks using compressed air and a soap solution.
2. Research the pressure produced by water per foot of head. Design an experiment using a garden hose and pressure gauge to determine if your mathematical calculations are correct.
3. Make a display showing various types of plumbing fittings that may be used in the plumbing system of a home.
4. Using a piece of clear plastic tubing, a circulating pump, and an old faucet, construct a display as shown in the drawing.
5. Research how you can locate underground pipe lines that are near your community. Make a bulletin board showing these pipelines and what material they are transporting.

Plumbing systems come in all shapes and sizes.

26

CHAPTER

ELECTRICAL UTILITY SYSTEMS

The information provided in this chapter will enable you to:

- *Briefly explain an electrical circuit.*
- *Describe the relationship between volts, amps, and resistance.*
- *List the major components of electrical wiring used in the home.*
- *Demonstrate how an electrical circuit is protected from overloading.*

The electrical system in a home is designed by the architect or engineer during thc planning phase of the construction. The home owner and architect determine the number and location of lights, electrical plugs, switches, and other equipment. The plans are drawn up. This allows the electrical contractor to make a bid on the job. Standard symbols are used in the electrical business, Fig. 26-1. The symbols describe all aspects of the electrical system for the structure.

To understand the electrical systems, you must first understand some basics about electricity. All of these principles are used in the design of a project's electrical systems.

ELECTRICAL ENERGY

Electrical energy can be generated by forcing electrons to move in a path or circuit. This flow of electrons through a circuit is called electrical **current.** Current is measured in **amperes.**

Electrical current is governed by two factors. The first factor is the amount of **electromotive force** applied to the electrons, or the **voltage.** The second factor is the **resistance** of the circuit to current flowing through it. Resistance is measured in **ohms.**

The voltage depends on the input from the generating device, Fig. 26-2. Resistance, on the other hand, depends on certain factors and elements of the circuit itself.

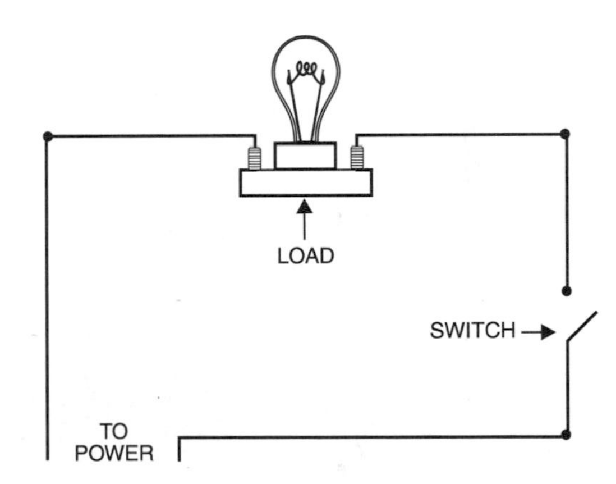

Fig. 26-1. The symbol to the right in the diagram represents a switch. It is the standard symbol for an electrical switch.

One factor that contributes to the resistance of a circuit is the size of the conducting wire. Smaller wire diameters and greater wire lengths increase the resistance. Another key resistance factor is the material used to make the conductor. Each material has a different resistance to the flow of electricity. Aluminum, for example, has a greater resistance than copper. Resistance may also increase or decrease as the temperature of the conducting material changes. These changes vary depending on the conducting material.

The power consuming device in the circuit is called the **load.** The load is still another element of resistance in the circuit. Some circuits have only one load, others will have many loads. The final element of circuit resistance is the switch or circuit controlling device.

OHM'S LAW

There is a direct relationship between electromotive force, electric current, and resistance. This relationship is expressed by **Ohm's law** as follows:

$E = IR$ or $\frac{E}{I} = R$ or $\frac{E}{R} = I$ where:

A B

Fig. 26-2. Voltage depends upon the source. A–C cell battery that produces 1.5 volts. B–Three engine generators that produce 1.5 megawatts of power each. The large silver cylinders above the generators are mufflers for the generators.

E = electromotive force, measured in volts.

R = circuit resistance, measured in ohms.

I = electric current, measured in amperes.

E = IR means that when the electromotive force stays the same, an increase or decrease in the amount of resistance will change the amount of current. For example, if the resistance is increased, the current decreases; if the resistance is decreased, the current increases.

ELECTRIC POWER

Power is another important electrical concept. The rate at which electricity does work is called electric **power.** It is measured in **watts.** However, one thousand watts adds up to one **kilowatt,** which is a more convenient unit of reference for power consumed. If one thousand watts of energy is expended during a period of one hour, one kilowatt-hour of power has been used.

Electric power, in watts, is the product of electromotive force and current. The following formula can be used to calculate power:

$$P = EI \text{ or } P = I^2R \text{ or } P = \frac{E^2}{R}$$

P = power, measured in watts.

One watt of power is consumed when one volt of electromotive force is applied to move one ampere of current through a conductor.

For example, an electric lamp is operating on 120 volts. It is drawing 2 amperes of current and has a resistance of 60 ohms. This lamp will consume 240 watts of power.

$P = EI \quad P = 120 \times 2 = 240$ watts

POWER DISTRIBUTION

Electrical power distribution can mean two different things. It can mean the distribution of electrical power by transformation from the energy source. It can also mean the distribution of electrical power within a building or residence.

Most electric power is produced by mechanical generators. Basically, an electromotive force is generated when wire windings are rotated through a magnetic field within the generator. A generator may be rotated by various means. One method is the engine-driven belt in your automobile. Generators are driven on a larger scale with giant turbines at hydroelectric installations in dams. Coal-powered and nuclear-powered steam turbines generate a large amount of the electrical power consumed by North America.

Electrical power may have to be transported over rough terrain, hundreds of miles from the point of generation to the user, Fig. 26-3. For easier transportation, a **transformer,** Fig. 26-4, may be used to increase voltage and lower the amperage. By raising the voltage and lowering the amperage across power lines, the energy losses in the lines themselves are tremendously reduced. Power output as high as 600,000 volts may be transported over high tension wires.

Near populated areas where the power is to be used, the high voltage lines from the power plant are fed into a substation, Fig. 26-5. The substation is an assembly of transformers designed to lower the voltage to 13,000 volts and convert it to the desired form for local distribution.

Overhead lines or underground cables carry the electrical power throughout a city to transformers or power poles, Fig. 26-6. Overhead transformers further lower the voltage to 240/480 volts or 120/240 volts. Most residences are supplied with 120/240 volts.

Distribution of electrical power throughout a residence starts with power delivery through a **service cable.** This cable usually consists of three conductors twisted inside a heavily insulated covering. The service cable may be either buried or installed overhead.

SERVICE EQUIPMENT

The service entrance conductors enter the residence through a masthead or underground, Fig. 26-7. A minimum of a 100-ampere, three-wire service must be provided for all individual residences according to the **National Elec-**

Fig. 26-3. Electricity is transported over great distances to remote users. Workers are erecting an electrical tower that will carry high voltage lines aloft. (Montana Power)

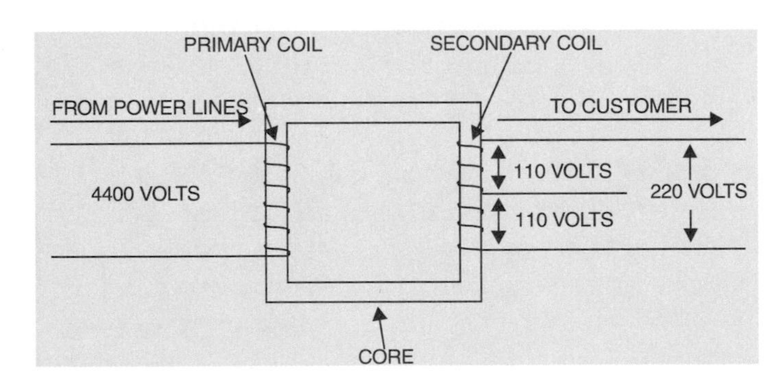

Fig. 26-4. Simple diagram of a transformer. The 4400 volts from the power company is stepped down to a voltage (110 or 220) that your home appliances can use.

tric Code, published by the National Fire Protection Association. For larger residences that are all electric, a 200-ampere, three-wire distribution system is the minimum necessary. The sizes of wire allowed for a given amperage requirement are determined from tables published in the electrical code.

A **kilowatt-hour meter** measures the power consumed by a household. The meter is installed outside the building

Fig. 26-5. Substations such as this one can be found in all cities. They use a series of transformers to reduce the incoming high voltages.

Fig. 26-6. Power lines carry dangerous levels of electricity. Towers keep these lines well away from the general population. (Montana Power)

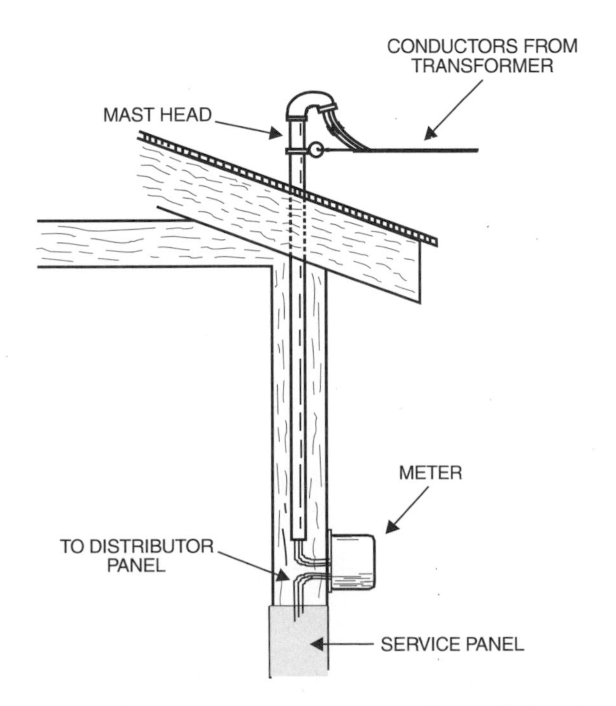

Fig. 26-7. Residential service entrance conductors bring electricity to home owner's service panel.

between the service entrance conductors and the main power distribution panel. The kilowatt-hour meter is the property of the power company. The meter is the end of the line for the power company's responsibility for maintaining the distribution system. The meter is located where it is accessible to the owner and the power company. Beyond this meter, the building owner is responsible for the wiring.

The service panel, distribution box, or main switchboard is the heart of the electrical system of a building, Fig. 26-8. The service panel is a large metal box mounted in an accessible yet out-of-the-way place in the residence.

The equipment placed in the service panel includes the main disconnect switch and a set of **circuit breakers** or **fuses.** Each hot (current carrying) wire of a circuit is protected by a fuse or breaker sized for the capacity of that circuit, Fig. 26-9. The neutral or ground wire is never fused.

Branch circuits go out from the service panel to the various outlets, switches, and receptacles in the house. **Branch circuits** are used to divide the main power line into smaller lines for different sections of a residence. The size of the service panel depends on the total demand of all of the branch circuits. For dwellings, the National Electrical Code recommends a power supply of 3 watts per square ft. (32 watts per square meter) for general lighting.

Three types of circuits are used for total power distribution:

- General purpose circuit.
- Small appliance circuits.
- Fixed appliance circuits.

The **general purpose circuit** includes all lighting convenience outlets except those located in the kitchen. The power requirements for a 3000 sq. ft. (280 m^2) residence is:

3 watts per sq. ft. x 3000 sq. ft. = 9000 watts

For **small appliance circuits,** circuits for the vacuum cleaner, toaster, mixer, etc., at least three additional 20 ampere circuits are necessary. These circuits are in addition to general purpose power requirements.

20 amperes x 120 volts x 3 circuits = 7200 watts

Fig. 26-8. Electrical power for a building is typically controlled in one central area. All of the service panels are located there.

CURRENT-CARRYING CAPACITY FOR COPPER WIRE	
AMPERES	COPPER AWG
20	12
30	10
40	8
50	6
60	4
70	4
80	3
90	2
100	1
150	2/0
200	4/0

Fig. 26-9. Table shows the current carrying capacity of copper wire of different gauges.

Fixed appliance circuits are circuits for garbage disposals, dishwashers, ranges and ovens, clothes washers and dryers, and heating and cooling units. These power requirements are the sum of the power ratings for all of these appliances. The ratings are based on the load of each of these appliances. These ratings are stated by the manufacturer. See Fig. 26-10.

APPLIANCE	AVERAGE WATTAGE
Blender	300
Carving Knife	92
Coffeemaker	894
Dishwasher	1201
Frying Pan	1196
Mixer	127
Oven, Microwave (only)	1450
Range - Conventional Oven	12,200
Range - Self-cleaning Oven	12,200
Toaster	1146
Trash Compactor	400
Waste Disposal	445
Radio	71
Television	145
Hair Dryer	381
Heat Lamp (infrared)	250
Electric Razor	15
Sun Lamp	279
Tooth Brush	1.1
Clothes Dryer	4856
Iron	1100
Washing Machine	512

Fig. 26-10. There is a tremendous range in the power usage by household appliances. Typical values of many items you use daily are listed.

Here, the fixed appliance load is 20,000 watts. However, since it is unlikely that all major appliances will be operating at the same time, the load may be reduced by 25 percent. The reduced fixed appliance load is then 15,000 watts.

The total load requirement of this residence would be:

General purpose circuits	9000 watts
Small appliance circuits	7200 watts
Fixed appliance circuits	15,000 watts
	31,200 watts

Therefore, based on the formula P = EI, we have:

$$\frac{P}{E} = I \quad \frac{31,200}{230} = 135 \text{ amperes}$$

The total current requirement is 135 amperes. With this amount of amperage draw, it would be wise to provide a service of at least 150 amperes. This would allow for an added circuit for future expansion or the addition of another appliance at a later date.

Conductors

Copper wire is used for conducting electrical current. Some older homes may still have aluminum wiring, but most of these have been updated. Wire sizes are designated by a gauge number in American wire gauge sizes. The **gauge numbers** are based on the diameter of the wire.

The current carrying capacity of wire depends on three factors. These factors are the cross sectional area, the type of metal, and the temperature of the conductor. See Fig. 26-11.

Insulation is applied to the surface of a building's wiring to protect the conductor from short circuits. Rubber and various thermoplastic coatings provide insulating qualities. The type and thickness of the insulating coating depends on:

- Maximum voltage and the nature of exposure.
- Whether the conductor will be installed in a dry or wet location.

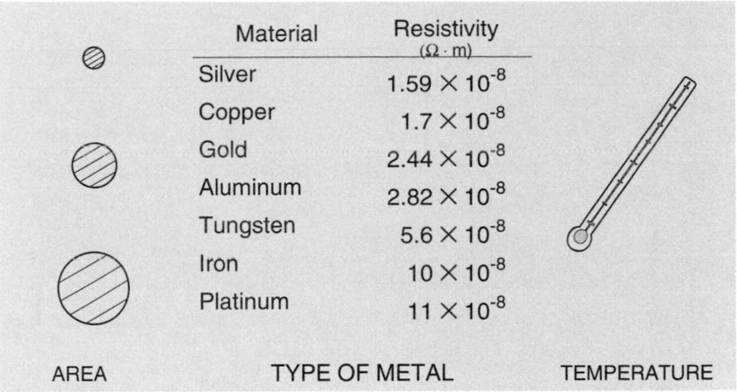

Material	Resistivity ($\Omega \cdot m$)
Silver	1.59×10^{-8}
Copper	1.7×10^{-8}
Gold	2.44×10^{-8}
Aluminum	2.82×10^{-8}
Tungsten	5.6×10^{-8}
Iron	10×10^{-8}
Platinum	11×10^{-8}

AREA — TYPE OF METAL — TEMPERATURE

Fig. 26-11. Three factors affect the current carrying capacity of a metal wire: cross sectional area, type of material, and the temperature of the material.

- Whether the conductor is to be buried in the ground.
- Whether the conductor is encased in concrete.
- Whether the conductor is subjected to corrosive atmospheres.

Receptacles

A convenience **receptacle,** slotted to receive the prongs of plugs, offers a safe means of connecting portable electrical units and appliances into the power circuit. Receptacles for general purpose circuits may be rated at either 15 or 20 amperes. These units are designed for flush mounting. Special receptacles are available for 220 volt appliances. These receptacles are slotted so that the standard plugs for 120 volt appliances will *not* fit. Fig. 26-12 shows a variety of receptacles.

The National Electrical Code currently specifies a three-prong grounded receptacle for all 120 volt convenience outlets. Molded duplex receptacles having two sets of slots are commonly found throughout older residences.

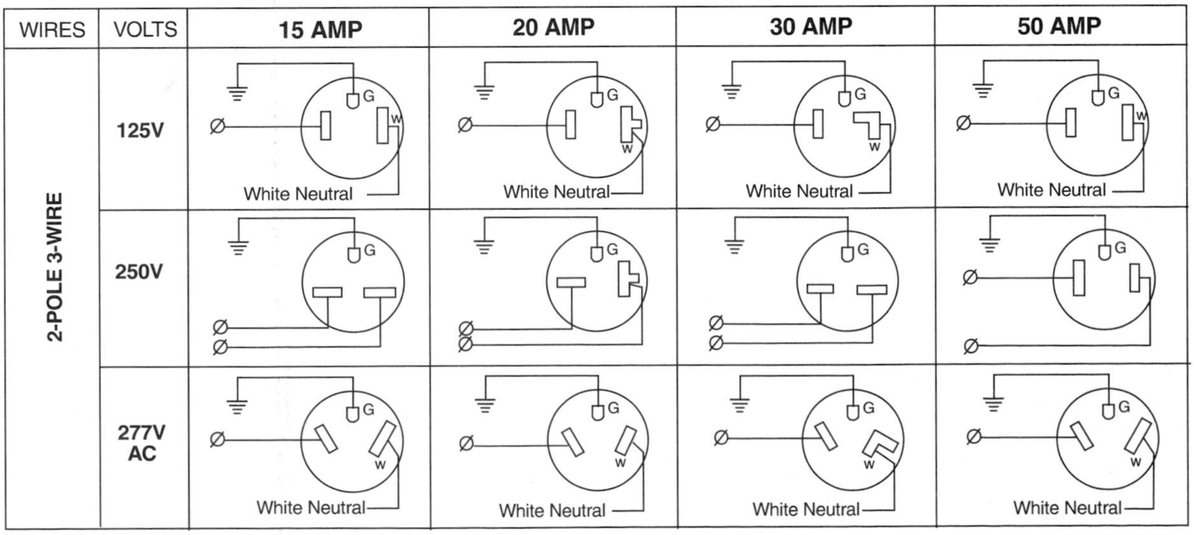

WIRES	VOLTS	15 AMP	20 AMP	30 AMP	50 AMP
2-POLE 3-WIRE	125V	White Neutral	White Neutral	White Neutral	White Neutral
	250V				
	277V AC	White Neutral	White Neutral	White Neutral	White Neutral

Fig. 26-12. Common receptacles and plugs used for different voltages and amperages. All receptacles are now produced with a ground wire for safety.

Switches

A **switch** is the mechanism used to open and close an electric circuit, Fig. 26-13. Switches may be single-throw or double-throw. A **single-throw switch** opens or closes a single wire in a circuit. A **double-throw switch** can be placed in any one of three positions:

- To open a circuit.
- To connect the single wire to the other wire in the box.
- To connect a single wire to either of two other wires in the box.

Switches controlling general lighting or appliance circuits are connected in the hot line of a circuit; the ground circuit is never opened. The switch may be single-throw single-pole type if only one switch is to control one or more fixtures or outlets. If one or more fixtures are to be controlled from two points, two three-way switches are needed. Fig. 26-14 shows the wiring of three types of switches.

Dimmer switches vary the voltage supplied to a fixture. The voltage may be lowered with a variable resistor, transformer, or a transistor, Fig. 26-15. Most dimmer switches used for fluorescent or incandescent lamps are of the transistor variety. Usually, they include an on/off switch to complete the circuit.

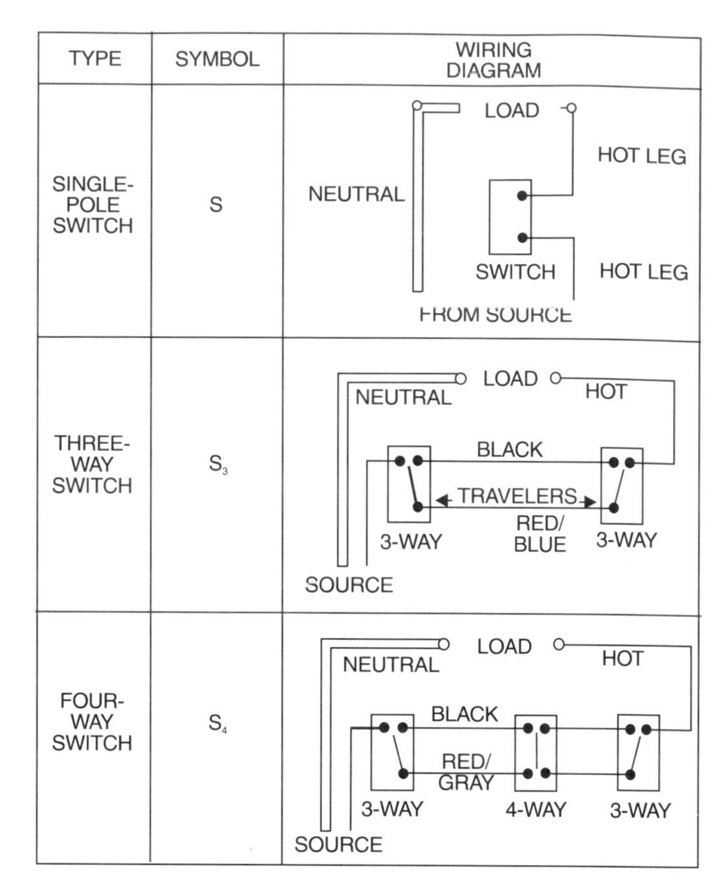

TYPE	SYMBOL	WIRING DIAGRAM
SINGLE-POLE SWITCH	S	LOAD, HOT LEG, NEUTRAL, SWITCH, HOT LEG, FROM SOURCE
THREE-WAY SWITCH	S_3	LOAD, HOT, NEUTRAL, BLACK, TRAVELERS, RED/BLUE, 3-WAY, 3-WAY, SOURCE
FOUR-WAY SWITCH	S_4	LOAD, HOT, NEUTRAL, BLACK, RED/GRAY, 3-WAY, 4-WAY, 3-WAY, SOURCE

Fig. 26-14. Single pole, three-way, and four-way switches used in construction.

COMMON HOUSEHOLD ELECTRICAL SYSTEMS

There are many electrical systems common to all households, lighting systems and the telephone system are two. Systems such as cable television and cable stereo are becoming more common. In addition, in some houses currently under construction, all of these individual electrical systems are being tied into one central electrical system.

LIGHTING

Lighting systems are designed for the purpose of adequately lighting an area. Different lighting tasks require different amounts of illumination for ease and efficiency of performance.

Lamps are available to meet all types of lighting requirements. However, proper lighting entails *more* than just providing the correct amount of light in given areas. The human eye does not adjust readily to extreme contrast of light and darkness. Therefore, general lighting of the area surrounding the work surface must also be considered. Glare from any source in the visual field can cause discomfort. Reflected light may be produced by horizontal surfaces, walls, or unshaded windows.

Fig. 26-13. Ordinary switches for a light.

Fig. 26-15. In addition to turning a light on and off, dimmer switches can vary the amount of electricity that reaches the bulb.

PHONE SYSTEM

People would have a hard time without the ease of communication they now have in their homes. A telephone has almost become a necessity. During the construction of a new home, or during the remodeling of an existing structure, adequate phone lines need to be installed.

A phone system is very inexpensive to install during construction. With new phone communication, computer, and shopping capabilities opening everyday, it is a mistake not to install adequate phone lines, Fig. 26-16.

CABLE SYSTEMS

Cable lines may also be built into homes. Television cable is very common and cable stereo lines are becoming more so. Fig. 26-17 shows a worker installing cable TV and stereo systems in a new home. This technology is now becoming **integrated.** This means that each of the different systems can communicate with each other. This integration allows people to control these systems from any location within the home or even by phone. Many video recorders that are sold today can be programmed over the phone.

Fig. 26-16. It takes well-trained technicians to work their way through the spaghetti-like maze of wire for telephone upgrades and repairs.

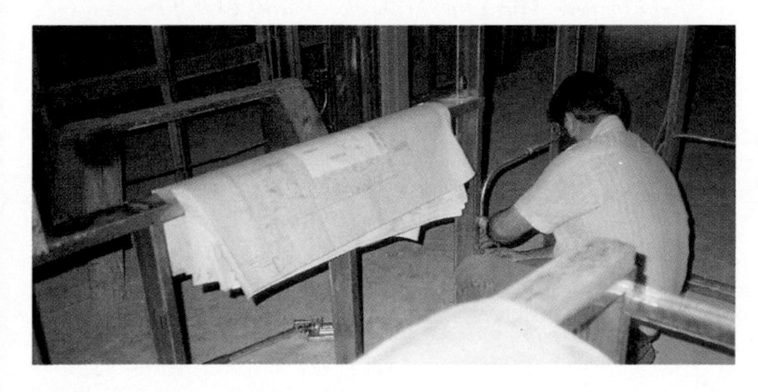

Fig. 26-17. In homes under construction, cable systems are built in to the design. They are often run alongside of the power lines through metal conduit.

SMART HOUSE

A large number of companies have joined together to develop a **Smart House®.** The Smart House concept uses a multi-functional electrical cable. This cable replaces conventional home wiring. The cable contains phone lines and cable television in addition to alternating current. Smart Houses are being tested with conveniences that were unheard of a few short years ago, Fig. 26-18.

Smart Houses have many convenient built-in features. One of the features of the Smart House is a touch screen computer. This computer allows the home owner to control lights, appliances, and the heating and cooling system. Another feature allows the owner to monitor the temperature, lighting, and humidity of any room in the house.

Energy control and savings is important in the Smart House design. The energy efficient features of a Smart House include such things as a central home automation system that allows home owners to set controls for the appliances, heating and cooling systems, lighting and security systems. They also have an energy control system

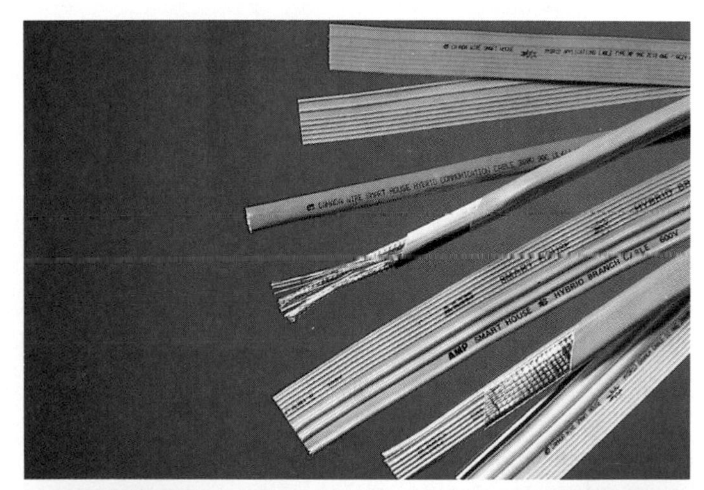

Fig. 26-18. Smart Houses use special wiring systems. Power and communications systems are run together throughout the home. All of the systems may be controlled by computer. (Smart House)

that allows owners to decide how and when to use electricity most efficiently and at the lowest cost. Smart Houses also contain **thermal envelope,** which provides a highly energy efficient building shell. If researchers are correct in their projections, the Smart House could save home owners 50 percent on their energy needs.

ELECTRICAL SYSTEMS IN CIVIL CONSTRUCTION

Often we do not think of the electrical systems that provide convenience and protection to us while traveling. Stoplights are a good example. The stoplight is used throughout the world to control traffic in a safe manner, Fig. 26-19. Stoplights are electrical systems that need to be set up and maintained. Some stoplights are set up to function on an electrical time clock. Others respond to the presence of an automobile. These lights sense the metal in the car through a buried cable in the street. This setup allows traffic in the busy lanes to go though the stoplight for a greater period of time than if the light changed at preset times. With computers that are connected to visual sensors, the light can be controlled by the amount of traffic coming from one lane during different times of the day.

Another electrical system is the railroad signal, Fig. 26-20. Electronic sensors detect the approach of a train and signal the auto traffic to stop. Once the train has passed, the signals stop flashing, and traffic is allowed to resume. The institution of these signals has greatly cut down on the number of automobile versus train accidents.

On the interstate highway system, most interchanges are lighted. The same is true of most streets in cities. These lights provide safety for both pedestrians and traffic. Power lines for these lights may be run in ducts underground or strung overhead from light to light. The lights are controlled by the use of **photocells.** These cells turn on the lights when it becomes dark. Thus, the lights are only on when they are needed. Lights with photocells have no need of an adjustable timer. They also shed light on very overcast days.

Fig. 26-19. Nobody likes to get stopped by a red light, but stoplights provide a much needed safety service.

Fig. 26-20. An automobile is no match for a train. Railroad signals save lives.

UNDERGROUND UTILITY LINES

Most of the time you see overhead electrical power lines stretched from pole to pole. These lines are separated from each other by several feet and need not be insulated, Fig. 26-21. Large transmission towers support the main electrical network throughout the nation.

In recent years, developments in electrical insulation have allowed power companies to bury some of their local electrical lines. Underground electrical lines have several advantages. First, there is a much smaller chance that someone may come in contact with the line and be injured. Secondly, underground power lines are protected from the weather, Fig. 26-22. Underground power lines lead to much less down time for electrical service. A third advantage of underground lines is that they do not disrupt our view of the surrounding landscape.

Of course, there are some disadvantages to underground lines. Underground lines are more expensive to install than overhead lines. The lines are also more expensive to repair if they do break. Additionally, if underground lines are not properly marked, they could be damaged by earthmoving equipment.

Construction workers now install most phone lines underground. The phone companies have gone to underground primarily because there is less maintenance. Within recent years the phone companies have started to bury fiber-optic cables throughout the United States and the world, Fig. 26-23. Fiber-optic cables can carry thousands of times more information than the conventional copper lines allowing extensive visual information to be transferred along with sound. The potential for fiber-optics is tremendous and will have a major impact on our lives.

Fig. 26-21. Lines strung on electrical transmission towers are not covered with insulation. The large air space around them serves to insulate them from each other.

Fig. 26-22. Cold weather and storms pose a constant threat to above ground lines.

Fig. 26-23. Fiber-optic lines are reaching more areas across the country. This machine is burying lines along a stretch of highway.

COMMERCIAL ELECTRICAL UTILITY SYSTEMS

The electrical system in commercial construction is basically the same as that of a home. However, there may be differences in the voltage that the electrical lights operate on and the quality of the electrical equipment. Commercial equipment is relied on to function properly for the safety of the workers and will often be undergoing much more extensive and strenuous use. Therefore, higher quality materials are commonly used. The standard 120 volt outlet that we find in our homes is the same in commercial installations.

There are some commercial electrical systems with unique requirements. These systems require a specialized electrical service, Fig. 26-24. In systems that have a high demand for electrical current, a transformer is placed in the building or several transformers are used throughout the structure.

Fig. 26-24. Generator control room in a large commercial structure.

Fig. 26-25. Industrial electrical circuitry has the same features as residential circuitry, though on a much grander scale.

This circuitry, although more complicated, operates just as the circuits in a home. All of the circuit branches are protected with circuit breakers. The electrical wiring is all copper, though it is of a larger size. All of the electrical motors also operate in the same manner as those in the home, only they are often larger, Fig. 26-25.

SUMMARY

Electrical energy is a great servant that provides us with energy for light and heat. Without an electrical system, few construction projects could be built.

An electrical circuit contains an electromotive force, (voltage), a flow of electrons, (current), and the resistance to the flow of the electrons. The relationship between these three qualities is expressed in Ohm's law. The complete understanding of these forces and their relationship with each other is necessary for someone working on electrical utility systems.

Electrical energy is produced in a generating plant or hydroelectric dam and distributed to customers through high voltage transmission lines. At or near the point of use, the high voltage is reduced to a lower, usable voltage with the use of a transformer.

Electrical systems used in the home are the phone system, heating systems, stereo and entertainment systems. The Smart House is the latest development in quality home living. This system connects the homes utility system to a computer. The computer can monitor all of the functions of the modern home and make adjustments to match our living style and desires.

Electrical systems are used to control traffic. This is done with structures such as stoplights and railroad crossing signals. One of the latest developments in the electrical utility system is the fiber-optic cable, which makes communications cleaner, more reliable, and less expensive.

KEY WORDS AND TERMS

All of the following key words and terms have been used in this chapter. Do you know their meaning?

Ampere
Branch circuit
Circuit breaker
Current
Dimmer switch
Double-throw switch
Electrical power distribution
Electromotive force
Fixed appliance circuit
Fuse
Gauge number
General purpose circuit

Insulation
Integrated
Kilowatt
Kilowatt-hour meter
Load
National Electric Code
Ohm
Ohm's law
Photocell
Power
Receptacle
Resistance
Service cable
Single-throw switch
Small appliance circuit
Smart House®
Switch
Thermal envelope
Transformer
Voltage
Watt

TEST YOUR KNOWLEDGE

Do not write in this book. Please write your answers on a separate sheet of paper.

1. List the three properties present in electrical circuits.
2. Electrical power is measured in ______.
3. What is the equation that represents Ohm's law?
4. The resistance in a conducting wire depends on what three factors?
5. Electrical power distribution can refer to two things. What are they?
6. List three examples each of electrical utility systems used in residential projects.
7. Explain what is meant by a Smart House.
8. List two advantages and two disadvantages of underground electrical utility lines.
9. List two civil electrical systems.

APPLYING YOUR KNOWLEDGE

1. Using a small battery, a switch, some small light bulbs, and wire, construct an electrical circuit to switch the lights on and off.
2. Arrange a field trip to the local electrical company or invite a representative to class. Have the representative explain how the electrical utility company operates in your community.
3. Invite the local telephone company representative to class. Have them explain the importance of fiber optics to the communication field.
4. Make a display of the common components of the home electrical system.
5. Construct a bulletin board showing the importance of the electrical system to residential, commercial, and civil construction.

Portable electric power is important for a new construction site.

27

CHAPTER

HEATING, VENTILATING, AND AIR CONDITIONING UTILITIES

The information provided in this chapter will enable you to:
- *Explain how heat energy is transferred from one place to another.*
- *Describe the need for adequate insulation in buildings.*
- *List three different types of home heating systems.*
- *Analyze the need for energy efficiency in homes and commercial structures.*

Heating, ventilation, and air conditioning systems control the environment in the spaces where you live and work. If these systems do not supply an ample amount of properly conditioned air, you will become uncomfortable. It becomes harder to enjoy the tasks you are trying to do. It is the job of the **heating, ventilation, and air conditioning (HVAC)** contractor and engineer to ensure that the system works properly in a variety of outside weather conditions. Unlike the electrical and plumbing systems, the HVAC system relies on *both* electrical and plumbing lines to operate. Therefore, the HVAC contractor must work closely with these contractors.

HEAT TRANSFER

To work in the HVAC area, you need some knowledge of the process of heat transfer. The design of the heating or cooling systems in a structure relies on these principles. There are three basic mechanisms of heat transfer.
- Conduction.
- Convection.
- Radiation.

CONDUCTION

Conduction is the transfer of heat energy through solid materials, Fig. 27-1, such as metal, wood, or glass. Conduction is accomplished by heat transfer from one molecule to the next. Heat energy always moves from warmer areas to colder areas. If you heat one end of a metal rod, the other end will gradually get warmer until there is no difference in temperature.

This process works in building walls. If you have a building wall and the interior of the wall is warmer than the exterior, the heat will travel from the inside to the outside. The rate of heat transfer is different in each material. Generally, the more dense the material the greater the heat transfer through conduction.

CONVECTION

Convection is the transfer of heat energy by the movement of gases or liquids, Fig. 27-2. There are two types of convection, forced and natural.

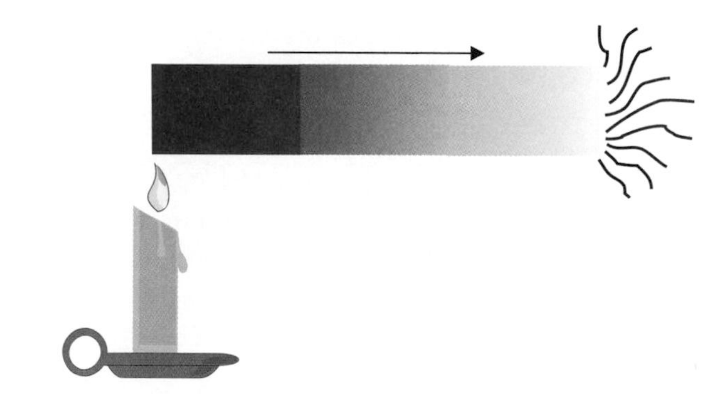

Fig. 27-1. Heat travels from the left to the right side of the bar through conduction.

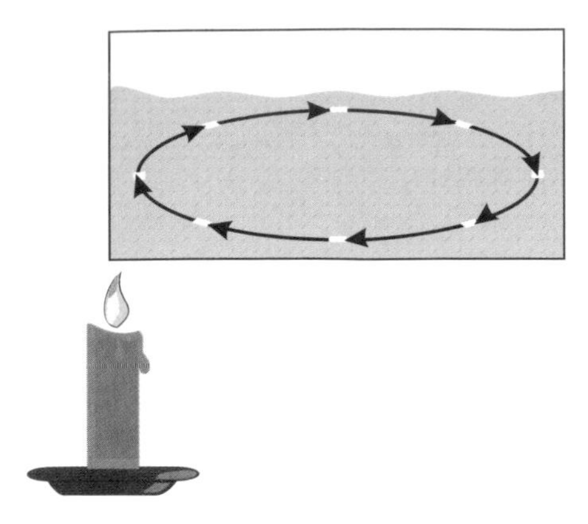

Fig. 27-2. Gases and liquids circulate when heated. This method of heat transfer is convection.

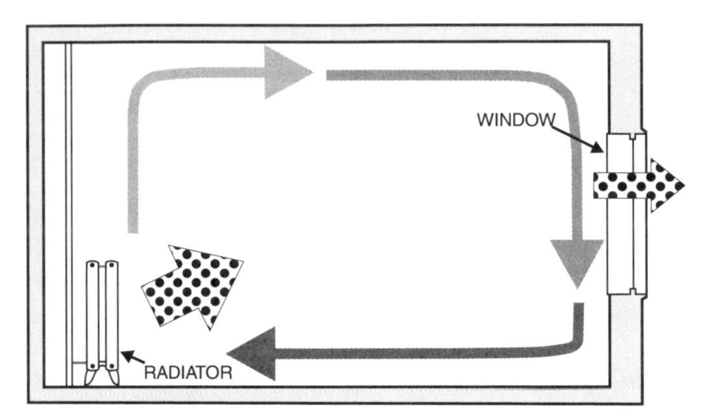

Fig. 27-3. Natural convection in a room. The air near the warm radiator rises pulling in cold air underneath. The warm air is pushed across the room. The air cools by the window and sinks back to the floor.

Natural convection relies on the principle that warm fluids are less dense than cold fluids. Thus, warm air rises and cold air sinks. The warm air in a room transfers its heat to walls and other objects and cools in doing so. This cooled air then sinks to the floor because it is heavier, Fig. 27-3. Smoke stacks use natural convection to move the hot gases up the stack and away from the area, Fig. 27-4.

Forced convection relies on the use of a mechanical device such as a fan to move the air. The fan on an automobile forces the air through the radiator, which cools the engine. Forced air furnaces are another example of forced convection.

RADIATION

Radiation is the transfer of heat energy through electromagnetic waves, Fig. 27-5. The earth receives its heat energy from the sun through radiation. When you are standing next to a hot stove, you are warmed by radiant heat. If the walls of a room are warm, then the people in that room feel warmer than if the walls were cool, even though the temperature is the same. This is a phenomenon of radiation.

Fig. 27-4. Smokestacks are an example of technology taking advantage of natural convection. (Montana Power)

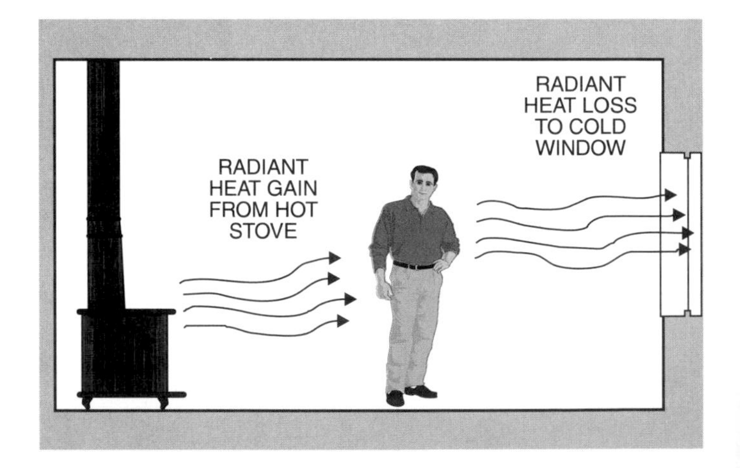

Fig. 27-5. Heat transfer through space by radiation. Radiation heats solid objects without heating the air around them.

TYPES OF HEATING AND COOLING SYSTEMS

Interior spaces are heated or cooled using combinations of the above heat transfer principles. In a **forced air furnace,** fuel is burned in a combustion chamber with a large surface area. Cool air is forced over this surface by a fan. The fan moves the heated air through a system of large sheet metal pipes to the rooms to be heated. The hot air transfers its heat to the room. **Cold air returns,** located at floor level, draw in the cold air and return it to the furnace, Fig. 27-6.

A **hot water boiler** system uses the same method as the forced air furnace, except the fluid used is water instead of air. Once the water has radiated its heat, it is returned to the boiler to be reheated. Cooling devices such as **air conditioners** work using the same process, only the air is forced over cooling coils. Cooling coils are cooled with the use of a refrigeration system.

Electrical heating systems use the electrical energy to heat electric elements. These elements may be in baseboard heaters or the entire ceiling, Fig. 27-7. **Baseboard systems** make use of natural convection to transfer the heat to the room. The air near the floor is heated. The heated air rises while the cooler air settles to the floor. There the cooler air is heated and the cycle repeats. **Radiant heating systems** transfer their heat to the room through radiant energy. Radiant systems send out electromagnetic waves that make objects in the room warmer than the surrounding air.

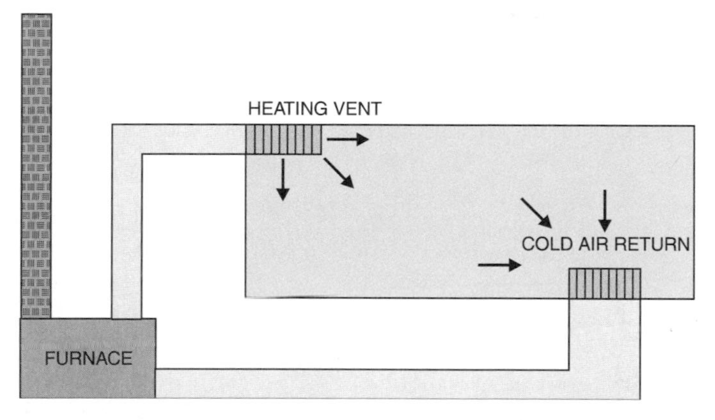

Fig. 27-6. Forced air systems use convection to heat a home or office. Both natural and forced convection aid in the even distribution of heat.

Fig. 27-7. Electric heaters have just been installed across the ceiling in this home under construction. These heaters use electromagnetic waves to keep the occupants warm.

If the climate is not too severe, a heat pump may be used, Fig. 27-8. A **heat pump** can run in both directions. It can cool or heat a building. However, heat pumps need help from a furnace or air conditioning unit in more extreme temperature conditions.

SOLAR HEAT

Solar energy is a plentiful and important energy resource. To face the potential shortage and lower the energy cost for heating our homes and offices, builders have begun to take further advantage of solar energy, Fig. 27-9.

Fig. 27-8. If the climate allows, heat pumps are an efficient method of heating or cooling a structure. (Tom Wood)

Fig. 27-9. Architects design many office buildings to take advantage of the free solar energy. Large energy savings make the buildings attractive to small businesses and large corporations.

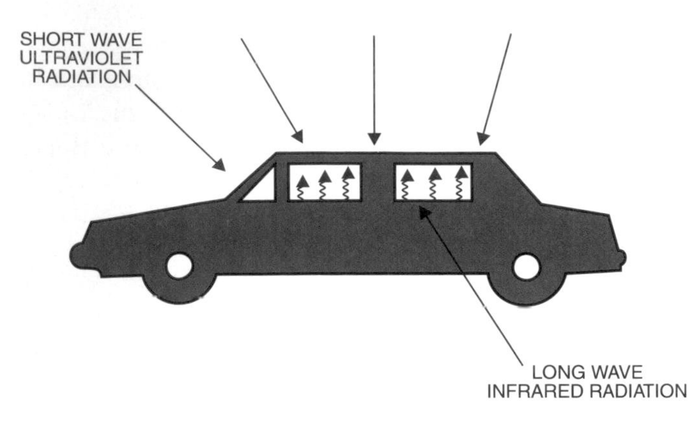

Fig. 27-10. The short waves from the sun easily penetrate the car windows. The longer waves reflected out of the car have much more difficulty getting through the glass. As a result, the energy is trapped in the car.

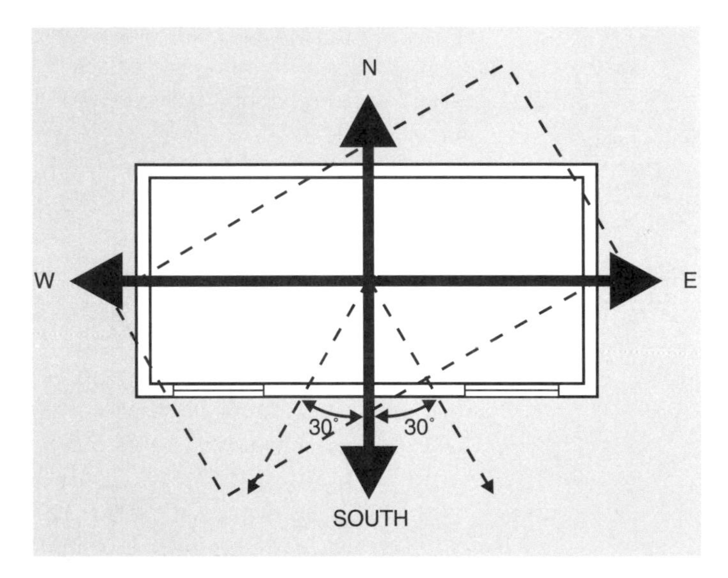

Fig. 27-11. To make the best use of solar radiation, the side of the house with the most glass should face south. A variance of 30 degrees east or west from true south is allowable for the best solar gain.

You have all felt the heat the sun radiates to earth when sitting in a car on a cool but sunny day. This heating is produced by the **greenhouse effect,** Fig. 27-10. The effect happens because the glass of the car windows allows the short electromagnetic radiation into the car. But when this radiant energy strikes the interior surface, the energy is changed to longer wavelengths. These wavelengths do not pass through the glass as easily, so the energy is trapped inside. This heats the interior of the car. This effect is used to supply heat for greenhouses, hence the name.

Any building can make use of this solar heat if it is properly designed. First, a building needs to be facing the proper direction, Fig. 27-11. The side with the most glass should be facing south. Next, some calculations must be made. The amount of glass in the building, the quantity of insulation in the structure, and some solar energy storage

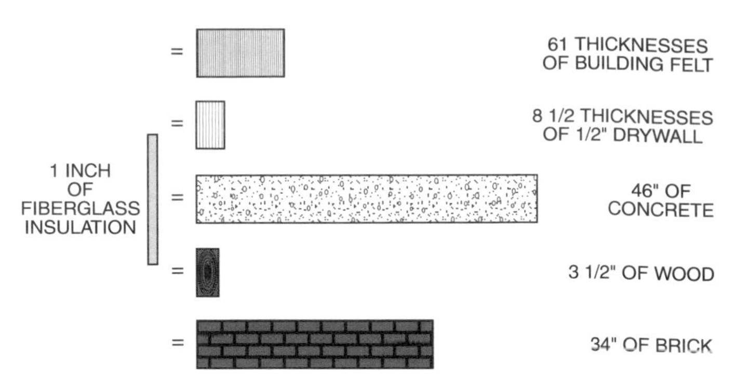

Fig. 27-12. A little fiberglass goes a long way toward energy savings.

system has to be worked out. Many homes, even in some of the northernmost climates, take advantage of solar heat and need little additional energy to supplement in heating.

HEATING AND COOLING BUILDINGS

Heating and cooling a building involves more than installing the mechanical equipment to do the job. The design and construction of the building is very important. The size of the building and materials used in construction bear on the heating and cooling task. In addition, efforts such as adding insulation and reducing air infiltration help reduce the workload of heating and cooling systems.

HEATING AND COOLING LOAD

When a building is in its planning stage, the designers must determine the amount of **heating or cooling load** the building will have. This is determined by calculating the heat loss or gain of the building. Two important factors in determining this load are the climate in which the structure is located and the heat loss through the walls, ceiling, and floor of the structure. Other factors include the amount of glass in the structure and which direction it is facing, heat given off by mechanical equipment in the structure, and the number of people who will be in the structure. Large numbers of people help keep office buildings warmer in the winter.

With all of these factors entered into the calculations, the total heating or cooling load of the building can be determined. This figure allows the designer to properly size the HVAC equipment so the proper climate may be provided.

INSULATION

The heating load of a building can be greatly reduced with the use of proper insulation in the floors, walls, and ceilings, Fig. 27-12. The rate at which materials transfer heat has been carefully studied. Standardized units for measuring this heat transfer have been established.

The most commonly used of these units is the resistance a building material has to heat transfer. This factor is resistivity or R. Review the information on resistivity in Chapter 15. R provides a means to measure a material's ability to resist the flow of heat. Fig. 27-13 shows the R-value for common building materials. The higher the R-value the better the material insulates. You will notice that by doubling the thickness of the material it approximately doubles the R-value.

Types of insulation

Insulation can be made from a variety of materials. Cellulose (made from plant cellulose), glass and mineral fibers, and chemical foams are the most common. These materials insulate well because they break up the space into many tiny cells. These cells then retard the transfer of heat from one side of the material to the other. **Foam insulation** is sprayed in place, Fig. 27-14. It is especially effective because the foam also seals out any air movement between the insulation and the framing.

AIR INFILTRATION

The heating and cooling engineer must also consider how well the structure is built. An estimate is made of how much free air is allowed to pass through the structure. This is called **air infiltration.** Excessive air infiltration adds a major load to the heating and cooling system of the building. Allowing a great deal of air in from the outside defeats the efforts of a heating or cooling system.

Fig. 27-15 shows workers placing a special material on the exterior of a home. This material stops the air from entering the walls of the home and yet allows any excess water vapor to escape. This is the most cost-effective method to ensure that the home is tight and reduce the total heating and cooling requirements.

Air-to-air exchanger

Largely due to new construction materials, construction practices are becoming very refined. Homes can be built so tightly that they may not be healthy to the occupants. In these homes, a mechanical system needs to be installed that forces the stale inside air out of the home and brings in fresh air from the outside. In normal construction this air exchange occurs naturally through the various openings and joints in the construction. This is healthy but very energy inefficient. The warm air leaves through these openings taking the household heat with it. This increases the heating costs.

Super insulated homes are designed to be very energy efficient. Escaping air and heat is not acceptable in their design. To prevent the loss of heat in this type of home an **air-to-air heat exchanger** is installed. This device exchanges the inside air with the outside air but retains approximately 80 percent of the heat from the air. Fig. 27-16 shows the internal workings of such an exchanger. Properly installed and operated, they provide excellent ventilation and retain a great deal of heat that would be lost in normal construction. The heat loss in a super-insulated home is very small.

1/2" Drywall (Gypsum Board)	.45
3 1/2" Fiberglass Insulation	11.00
6" Fiberglass Insulation	19.00
1/2" Plywood	.63
3 1/2" of Wood (Wall Stud)	3.15

Fig. 27-13. R-values for commonly used building materials.

Fig. 27-14. Foam insulation is great for reducing heat loss but bad for your lungs. Workers wear protective clothing, goggles, and face masks while spraying the foam.

Fig. 27-15. Excess air infiltration is one of the largest sources of heat loss in a home. The wrap these workers are installing seals the outside air out of the structure.

COMMERCIAL CONSTRUCTION SYSTEMS

Commercial HVAC systems use special cases of the same systems we have previously discussed. The heating and cooling loads in commercial construction are far greater than those encountered in homes. Similar calculations are completed to design a system that provides the needed indoor climate. Often, all major parts of the mechanical systems of high-rise buildings are placed in one location. The systems may be on the roof or in the basement, Fig. 27-17. Sometimes they are located on specified floors.

Factories provide particular problems when it comes to the movement and treatment of air. Here the air system may have to filter out toxic substances, such as smoke and chemicals, before the air can reenter the environment. In recent years, state and federal regulations have required factories and power plants to follow strict rules on air pollution. Factories have to install stack **scrubbers** to remove nearly all of the pollutants before the gases are allowed through the stack. Fig. 27-18 shows some of this special air handling equipment in a modern steel mill.

Government and company officials continually monitor the exhaust gases to ensure that they are well within the standards. If gases exceed these standards, the offending businesses face heavy fines and may be shut down until the problem is corrected.

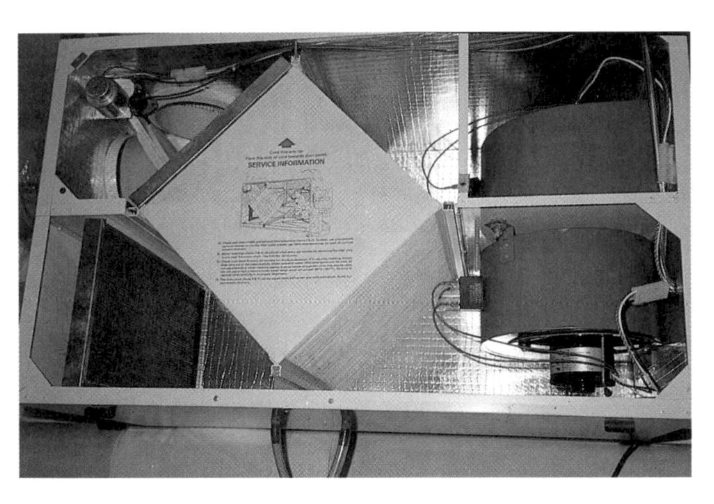

Fig. 27-16. Interior of an air-to-air heat exchanger. Heat exchangers are built into the design of newer super insulated homes.

Fig. 27-17. Large HVAC equipment is often placed on top of factories. This 700 ton cooling tower (condenser) was relegated to the roof.

Fig. 27-18. Modern steel mills have very complex air handling equipment. Tremendous changes have been made in the past few decades.

CIVIL CONSTRUCTION SYSTEMS

Many civil engineering projects have no need for HVAC systems. Open roads and bridges are examples. Nature takes care of their air quality. The tunnel, however, is another matter. In a long tunnel that carries a large volume of auto or train traffic, ventilation is a necessity. Carbon monoxide produced by internal combustion engines is very harmful to our health. It must be removed from tunnels. To accomplish this a large ventilation system must be made part of a tunnel design.

Most often it is a forced air system that brings in fresh air and expels the exhaust gases. In some tunnels, these air circulation fans are over ten feet tall. They turn at high speeds.

Mining shafts and subways are other examples of civil construction projects that rely on extensive ventilation systems to ensure safe working conditions.

SUMMARY

Heating, ventilating, and air conditioning systems provide us with fresh air at the proper temperature to keep us comfortable. Heat is transferred from one place to another through three methods: conduction, convection, and radiation. All heating and cooling systems use one or more of these methods.

To install the proper sized heating and cooling system, heating and cooling loads must be calculated. To calculate the load, the direction the structure faces in relation to the sun will have to be considered. Depending on the structure, heat loss or gain may be controlled with the use of insulation materials.

Ventilation in civil structures such as tunnels and subways must be engineered to ensure proper amounts of fresh air for the people using the system. Heavy industry, where large quantities of potentially toxic gases may be produced, are required to provide a ventilation system that removes the toxic substances from the air before it is returned to the environment.

A properly designed HVAC system, whether it is part of a home, an office building, or a railroad tunnel, allows people to live and work in a healthy environment.

KEY WORDS AND TERMS

All of the following key words and terms have been used in this chapter. Do you know their meaning?

Air conditioner
Air infiltration
Air-to-air heat exchanger
Baseboard system
Cold air return
Conduction
Convection
Foam insulation
Forced air furnace
Forced convection
Greenhouse effect
Heat pump
Heating, ventilating, and air conditioning (HVAC)
Heating or cooling load
Hot water boiler
Natural convection
Radiant heating system
Radiation
Scrubber
Super insulated home

TEST YOUR KNOWLEDGE

Do not write in this book. Please write your answers on a separate sheet of paper.

1. List and give a brief description of the three methods of heat transfer.
2. List three different types of heating systems used in homes and commercial buildings.
3. To receive the maximum amount of solar energy, which direction should a home have its windows face?
4. Name three materials used in home insulation.
5. What does industry use to reduce the toxic emissions leaving its stacks?

APPLYING YOUR KNOWLEDGE

1. Design a simple experiment that will demonstrate one of the methods of heat transfer.
2. Using rigid foam, black paint, and clear plastic, construct a solar collector. Test its effectiveness at different angles to the sun.
3. Collect different samples of insulation and study their insulating qualities.
4. Collect articles about industrial air pollution and display them on the bulletin board. Discuss what is being done by industry to reduce these emissions.

Heating and cooling ducting are affixed to the ceiling, out of the way of the work area of a building.

28
CHAPTER

EXTERIOR FINISHING

The information provided in this chapter will enable you to:

- *List the completion order of exterior finish work to make a building weather tight.*
- *Discuss reasons why exterior finish work is important to the construction project.*
- *Evaluate different building construction exterior sidings.*
- *Identify different materials used for roofing.*

Once the main structural portions of a construction project are complete, the next phase is to complete the exterior finishing work. **Exterior finishing work** includes those items that enable the project to withstand weather elements. This finishing work also includes placing a finish surface on all exterior surfaces. A finish surface is a protective and appearance-enhancing surface.

Almost all construction projects require some exterior finish work. On a road construction project, the exterior finish work includes placing the final black topping or layer of concrete on the road, Fig. 28-1. On a large dam there is the final grout work to smooth the surface of the structure. On buildings, exterior finish work involves many activities. Shingles are installed, doors and windows must be hung, and the exterior siding is attached, Fig. 28-2. This work provides the interior of the structure protection from the weather. The materials used are designed to withstand wind, rain, snow, and extreme temperatures. The materials should also be easy to maintain.

Fig. 28-1. A construction crew is laying asphalt paving on top of a new roadway. The top layer prevents water from seeping into the foundation. It also provides a smooth driving surface.

FINISHING WORK ON CIVIL CONSTRUCTION PROJECTS

The first step in finishing construction projects such as roads and bridges is to make sure that the final leveling of the dirt work is done well. Fig. 28-3 shows a worker using a grader to smooth the 3/4 inch gravel to its final grade.

Fig. 28-2. Aluminum and vinyl siding materials are popular because they require very little maintenance and last a long time.

Fig. 28-3. Before paving a road, the surface must be made relatively smooth. Special tractors, such as this one, level roads to their final grade.

When the roadbed reaches its final level, the paving crew begins. On many roads a layer of asphalt is placed on top. The asphalt is waterproof and prevents water from seeping into the soil below. The soil must be kept dry so it will be able to support the heavy traffic loads. Without the asphalt finish coat, thc roadbcd would get wet, The road would not remain smooth.

On construction projects that use concrete, the finish work is completed as the concrete is placed. Concrete not only provides structural strength, but it is finished smooth to provide a weatherproof surface, Fig. 28-4. This is an advantage of using concrete. Concrete can serve as a supporting structure as well as the finish surface. A sidewalk is an excellent example.

On projects such as dams, the concrete surface is further finished. Dams need extra protection against water. A wash coat of grout is used to seal the dam's surface from water penetration.

Concrete can be finished in a variety of ways for design purposes. The finishing will vary depending on what the architect has in mind. Fig. 28-5 shows a few of the finishes that can be applied to concrete.

On buried pipelines, Fig. 28-6, the finish work consists of backfilling the ditch and compacting the soil. The soil is compacted so that it will not settle and leave a depression over the pipeline.

EXTERIOR FINISHING WORK ON BUILDING CONSTRUCTION

There is a great deal of exterior finishing work done on homes, stores, and other commercial buildings. Some exterior finishing work is done for weather protection. Some work is done for safety. Some work is done simply for appearances.

Fig. 28-4. Structural tasks and finishing tasks blend together when applying concrete to a highway. The concrete provides a strong floor for car and truck travel. The concrete is also smoothed to a finish.

ROOFING

Exterior finish work is easy to identify on building construction projects of both the residential and commercial variety. One of the most critical areas of exterior finish is the roof. A good roof is required to protect the building and its contents. Builders today have a very large variety of roofing materials to select from. The choice of roofing material depends on several factors. These factors are:

- The slope of the roof.
- The cost of materials.
- The service life of the materials.
- The weather conditions under which the roof must function.

Fig. 28-5. Concrete finishes vary with their purpose. A–Brushed finish on a sidewalk. B–Spiral from the slip form can be seen on this concrete base of a street light. C–A rough pocketed finish was used on this interior stairway.

Fig. 28-6. After a pipeline is constructed and planted, the land must be returned to its original condition. (Montana Power)

The unit of measure for roofing materials is the **square.** A square of roofing material covers an area that is 10 ft. by 10 ft., or 100 square ft. All roofing materials are sold by this measure. The calculation of the weight of the roof is determined in the **pounds per square.**

Slope and pitch of roofs

Proper drainage is essential to the life of a roof. Even flat roofs have a slight slope built into them that allows the water to drain off the roof into a drainage system. In cold climates, the roofing material will quickly deteriorate due to frost action if the roof does not have the proper slope.

The **slope** of a roof refers to the number of inches the roof rises for each unit of run. Twelve inches is one unit of run. A unit of pitch is 24 in., Fig. 28-7. For example, a roof with a 4/12 slope is the same as a roof with a 4/24 or 1/6 pitch. The slope communicates to the contractor how the roof should be built. On building plans, the slope is shown by the use of a small triangular symbol placed near the profile of the roof, Fig. 28-8. The rise is indicated for each 12 in. of run. Review Chapter 24.

Flat roofs

Flat roofs are those roofs that have very little slope. Slopes of less than two inches of rise to each foot of run

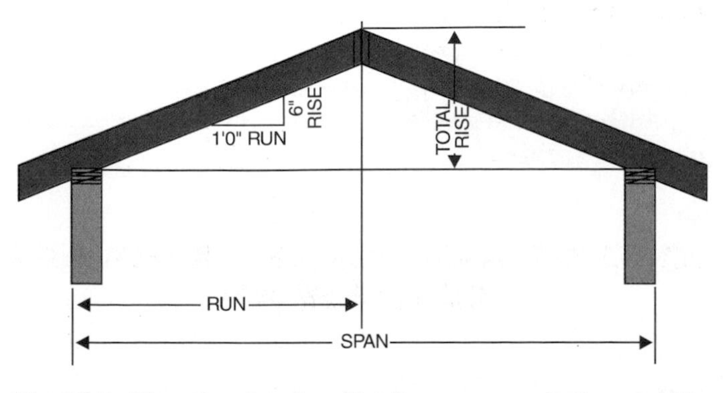

Fig. 28-7. Slope is a function of unit run - one unit of run = 12 in. Pitch is a function of unit span - one unit of span = 24 in.

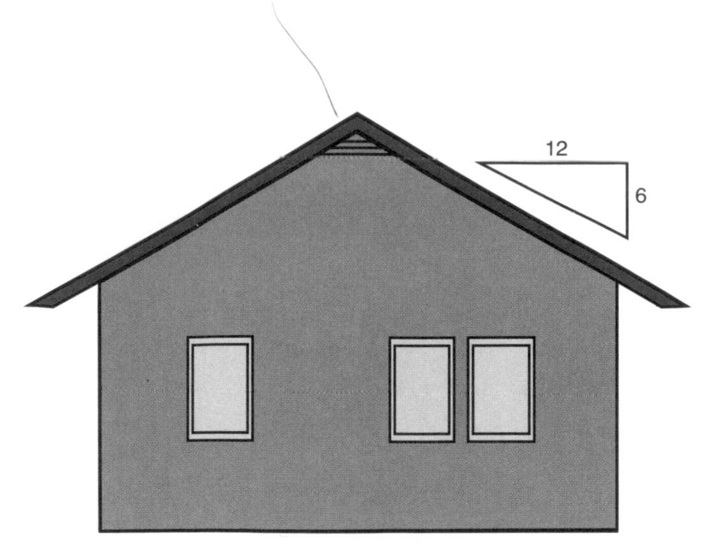

Fig. 28-8. The slope of a roof is shown on the elevations of the plans.

are considered flat. Water and snow will drain off these roofs more slowly. They must use a roofing material that does not allow moisture to work under the flat surface. Roofs of this type use a built-up roofing system or a membrane roofing material, Fig. 28-9.

The **built-up roof** is made from succeeding layers of felt flooded with hot bituminous liquid asphalt or coal tar pitch. These materials seal the layers together. This makes a rigid roof covering. This roof is then covered with small gravel to provide further weather protection.

Membrane roofing material consists of a rubber-like substance that forms a continuous membrane over the entire roof area. This membrane is attached to the structure only at the edges of the building. It is free to adjust to the movements of the building caused by changes in temperature. The membrane is held in place on the roof by a thin layer of gravel. The advantage of the membrane roof is its ability to adjust for temperature fluctuation. The roof will not buckle under extreme temperatures.

Sloped roofs

Sloped roofs have a slope of more than 2 in. of rise in 12 in. of run. These roofs use shingles or metal roofing materials. **Asphalt shingles** are made from materials that are saturated with asphalt and then coated with high heat melting flexible asphalt. Mineral granules are pressed onto the asphalt coating providing color and a fire resistant surface. Fig. 28-10 shows common styles of asphalt shingles.

Shingles may also be made from wood, concrete, or clay. Wood shingles are made from decay-resistant woods. If the wood pieces are sawed to shape, they are called shingles. If they are split to shape, they are called shakes. Wood

A

B

Fig. 28-9. Water eventually works its way into flat roofs. A–Workers are removing the old roofing material. B–New roofing materials are applied. Today's materials resist water and frost action better than those in the past.

Fig. 28-10. Asphalt shingles come in many styles and colors. (Lowden, Lowden and Co., Velux-America Inc.)

shakes and shingles produce a very attractive roof, Fig. 28-11.

Shingles made from clay or concrete have a very long life and are fireproof, Fig. 28-12. The supporting structure of the building, however, must be designed to support this heavy roofing material.

Metal roofing materials are factory finished in a variety of colors. They are becoming very popular in many regions of the country. These materials are easy to install due to the large size of each piece. They are also fireproof.

The selection of the roofing material depends on the slope of the roof as well as the exterior appearance the owner and architect wish to achieve. The availability and cost of the roofing material also play important parts in the final selection of a roofing material.

Fig. 28-11. Roofers are finishing the shingles on one side of a new home. Wood shingles are more expensive than their asphalt counterparts, but they produce a very attractive roof. (Lowden, Lowden and Co.)

Fig. 28-12. Clay shingles are common in the southern and western United States.

INSTALLING DOORS AND WINDOWS

The next phase in exterior finishing consists of installing doors and windows, Fig. 28-13. Rough openings for the doors and windows are made slightly larger than the actual size of the door or window unit. This allows for small adjustments when installing these units. Doors and windows should be installed perfectly level and plumb. If this is not done correctly, any settling of the building or slight error in construction would cause the doors and windows to function improperly. Fig. 28-14 shows a window that has just been installed. Notice the space between the window frame and the wall framing members.

Window units

Windows originally were used for admitting daylight and later for ventilation. Today, we have a great variety of window types, Fig. 28-15. Some of these window types are:

- Single hung.
- Double hung.
- Casement.
- Awning.
- Sliding.
- Hopper.
- Fixed.

Fig. 28-13. Proper installation of a door is important. Improperly installed doors are one of the biggest sources of energy losses. A properly installed door also presents a desirable front for a home or office. (Therma-Tru, Division of LST Corp.)

Fig. 28-14. A properly installed window. (Lowden, Lowden and Co.)

Except for the fixed type, all refer to the methods by which they open for ventilation purposes.

Generally, all windows are glass panels secured in metal, wood, or plastic frames. Each frame is referred to as a **unit.** They come in standard sizes from each window manufacturer. Fig. 28-16 shows sizes of a particular manufacturer's windows. Note that the sizes are provided for overall dimension of the unit, the rough opening, the glass size, and the sash opening. Each of these figures is important to the designer. The figures show the designer if the window meets various codes. Glass size, free opening size, and how the unit will appear in the completed building may all be specified in building codes.

Window units are manufactured complete with hardware, **weatherstripping,** and operating mechanism. They are ready to be installed into the building. Also, each unit may be ordered with different types of glass. Options range from **single glazing,** where only one pane of glass is in each unit, to **triple glazing,** where each unit has three panes of glass. The triple-glazed window is installed in buildings where energy conservation is of prime importance. Window units may also be ordered with special light and heat reflecting properties. These properties also make them more energy efficient. On window units that open, removable screens are common to keep insects out of the building.

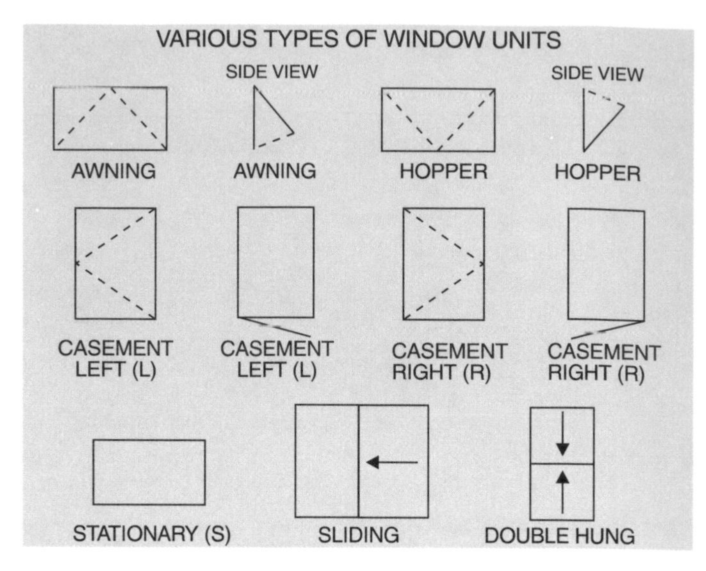

Fig. 28-15. Common windows used in residential construction.

Window selection is an important consideration when considering the overall appearance of the structure. Building designers use windows to give a building an attractive appearance, Fig. 28-17. Windows such as skylights have gained popularity. **Skylights** are window units that are installed in the roof and provide natural lighting for interior spaces. Well-placed skylights provide approximately five times the natural light that a wall unit brings into a room. These units are generally double glazed to reduce heat loss. They must be carefully installed to ensure they will not leak.

Exterior door units

Residential doors are manufactured from wood or metal. They may be insulated to reduce heat loss. Commercial doors are generally heavier than residential doors. In addition to wood and metal, they may also be made from thick glass set in a metal frame. Doors may be ordered as a complete unit with the door set in a **door jamb** ready to be placed into the rough opening. Doors and jambs may also be purchased separately. If the door is purchased separately, the carpenter must set the door into the frame and install all of the necessary hardware.

Exterior doors come in a variety of sizes. Though 3 ft. by 6 ft. 8 in. is becoming the standard for homes, commercial buildings often use wider doors. Quality doors provide both protection and privacy. Like windows, doors aid in the style of the structure.

Doors may be made to swing to the right or left and inside or outside, Fig. 28-18. Exterior doors on homes generally swing into the living area. Doors may also open by sliding to the side, Fig. 28-19. In the case of garage doors, they may open by moving overhead on a track. On commercial buildings, such as schools, building codes require that the doors swing out. This provides a faster escape for the occupants in case of fire.

8.16/An

Andersen® WINDOWALLS®

PRIMED CASEMENT WINDOWS

1′-10 1/2″

UNIT DIM	2-3 3/4	4-4	4-7 1/2	6-4 1/4	8-4 1/2	10-4 3/4
RGH. OPG	2-2	4-2 1/4	4-5 3/4	6-2 1/2	8-2 3/4	10-3
SASH OPG	1-10 1/2	3-10 3/4	4-2 1/4	5-11	7-11 1/4	9-11 1/2
GLASS*	20 1/4					

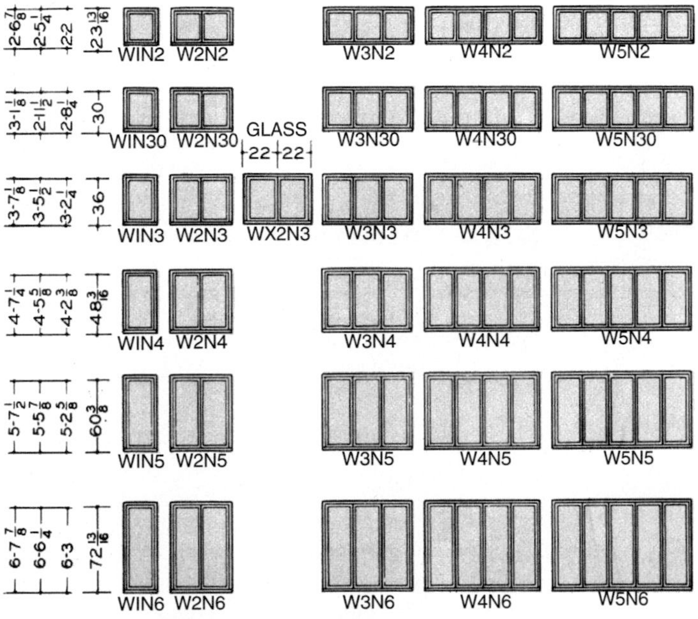

2-6 7/8
2-5 1/4
2-2
2-3 13/16
WIN2 W2N2 W3N2 W4N2 W5N2
3-1 1/8
2-11 1/2
2-8 1/4
30
WIN30 W2N30 W3N30 W4N30 W5N30
GLASS
22 22
3-7 1/8
3-5 1/2
3-2 1/4
36
WIN3 W2N3 WX2N3 W3N3 W4N3 W5N3
4-7 1/4
4-5 5/8
4-2 3/8
48 3/16
WIN4 W2N4 W3N4 W4N4 W5N4
5-7 1/2
5-5 7/8
5-2 5/8
60 3/8
WIN5 W2N5 W3N5 W4N5 W5N5
6-7 7/8
6-6 1/4
6-3
72 13/16
WIN6 W2N6 W3N6 W4N6 W5N6

CASEMENT PICTURE WINDOWS MULTIPLES OF 1′-10 1/2″

UNIT DIM	4-4	6-4 1/4	8-4 1/2	8-4 1/2
RGH. OPG.	4-2 1/4	6-2 1/2	8-2 3/4	8-2 3/4
SASH OPG.	3-10 3/4	5-11	7-11 1/4	7-11 1/4
GLASS*	43 7/16	67 11/16	91 15/16	43 7/16

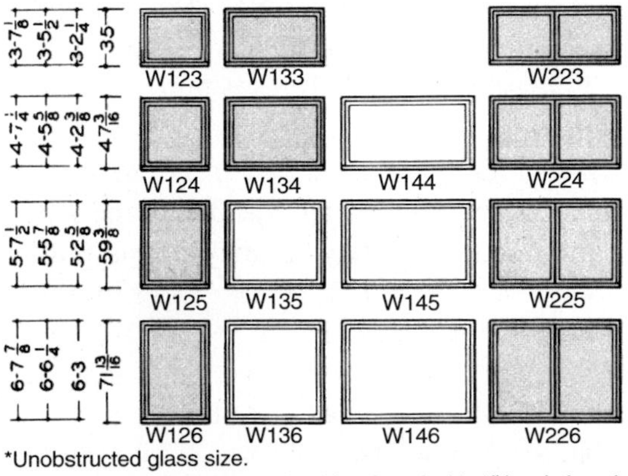

3-7 1/8
3-5 1/2
3-2 1/4
35
W123 W133 W223
4-7 1/4
4-5 5/8
4-2 3/8
47 3/16
W124 W134 W144 W224
5-7 1/2
5-5 7/8
5-2 5/8
59 3/8
W125 W135 W145 W225
6-7 7/8
6-6 1/4
6-3
71 13/16
W126 W136 W146 W226

*Unobstructed glass size.

Picture window sizes shown in white glazed with 1″ insulating glass.

CAUTION: Unless specifically ordered, Andersen windows are not provided with safety glass, and, if broken, the glass could fragment causing injury. Many laws and building codes require safety glass in locations adjacent to or near doors. Andersen windows are available in safety glass which may reduce the likelihood of injury when broken. Information on safety glass is available from your local Andersen supplier.

NOTE: For information on egress specifications refer to page 48.

TYPICAL INSTALLATION DETAILS
Scale: 1 1/2″ = 1′0″

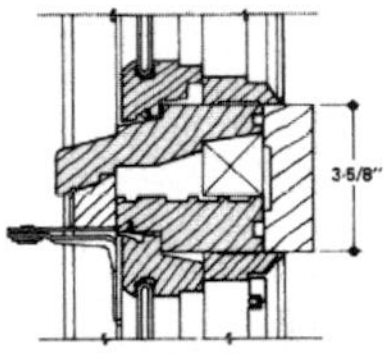

3-5/8"

TRANSOM

Detail showing casement units stacked to form transom. Interior and exterior trim not furnished. Standard exterior head casing is removed from lower unit.

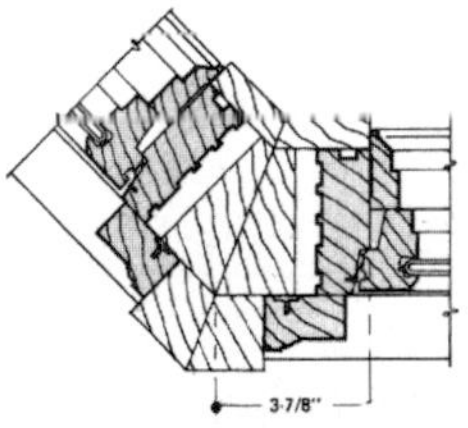

3-7/8"

45° ANGLE BAY SUPPORT MULLION

Mullion detail of 45° angle bay installation using standard units. Filler members between units furnished by others.

CAUTION: In masonry construction allow adequate clearance at sill for caulking and dimensional change of framework.

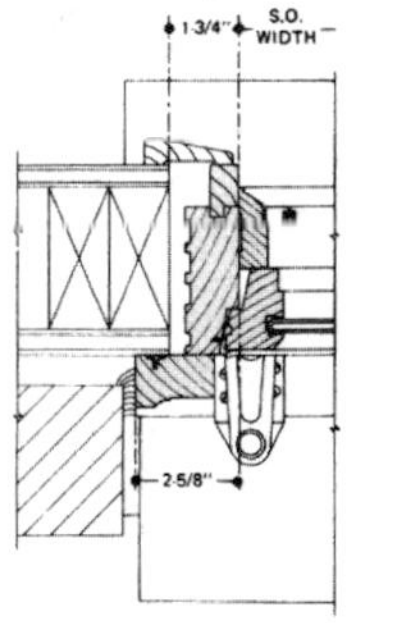

1-3/4"
S.O. WIDTH
2-5/8"

BRICK VENEER

Brick veneer installation with 1/2″ sheating and interior finish.

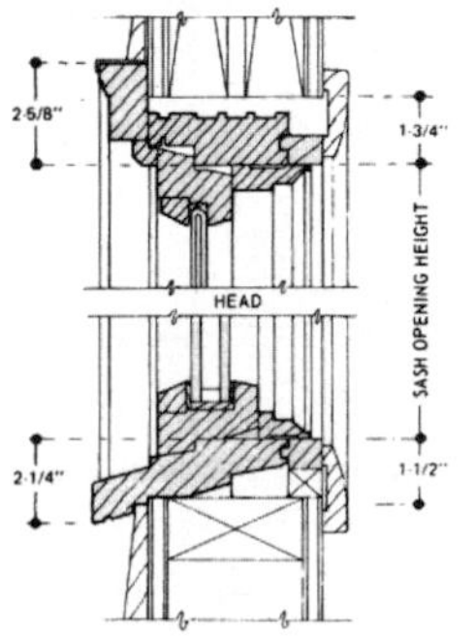

2-5/8"
1-3/4"
HEAD
SASH OPENING HEIGHT
2-1/4"
1-1/2"

PICTURE WINDOW VERTICAL DETAIL

Casement unit installed in frame wall construction. Head section shows sash glazed with double-pane insulating glass. Sill section shows sash glazed with 1″ insulating glass.

Fig. 28-16. Page from a window manufacturer's booklet. The booklet shows the dimensions of all of the manufacturer's window units. (Andersen Windowalls)

EXTERIOR WALL COVERINGS

When all of the exterior doors and windows have been installed, the exterior wall coverings are put in place.

Fig. 28-17. Windows have a dramatic effect on the appearance of a room. (Velux-America Inc.)

Exterior wall coverings are materials used to make the building look better while protecting the structural parts from weather damage. Wall coverings are made from a variety of materials. Wood, brick, stone, vinyl, and prefinished metal are common choices.

Wood siding

Wood siding is made from species of wood that have a natural resistance to decay such as western red cedar or redwood, Fig. 28-20. Siding products are also manufactured from reconstituted wood materials such as particleboard and hardboard. With these products, the siding is generally preprimed to ensure that the finish coat of paint has good weathering characteristics. Plywood products are also commonly used as exterior siding materials and come in a variety of patterns and finishes. Wood siding may be installed horizontally or vertically, depending on the architectural style.

Aluminum siding

Aluminum siding consists of sheets of aluminum with a color finish of enamel baked on. Aluminum siding is popular

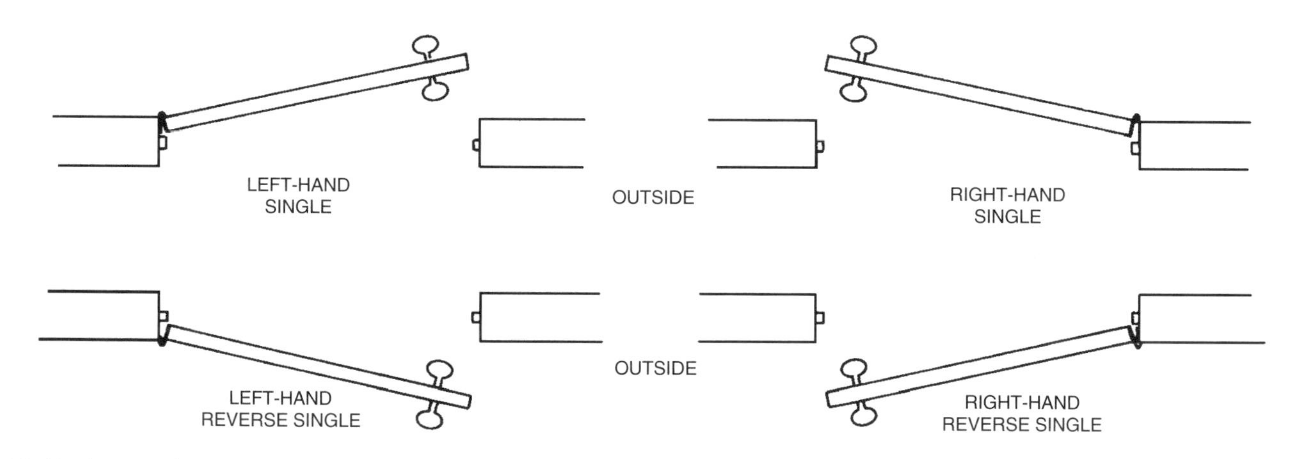

Fig. 28-18. Door swings are identified by location of hinges and direction of swing.

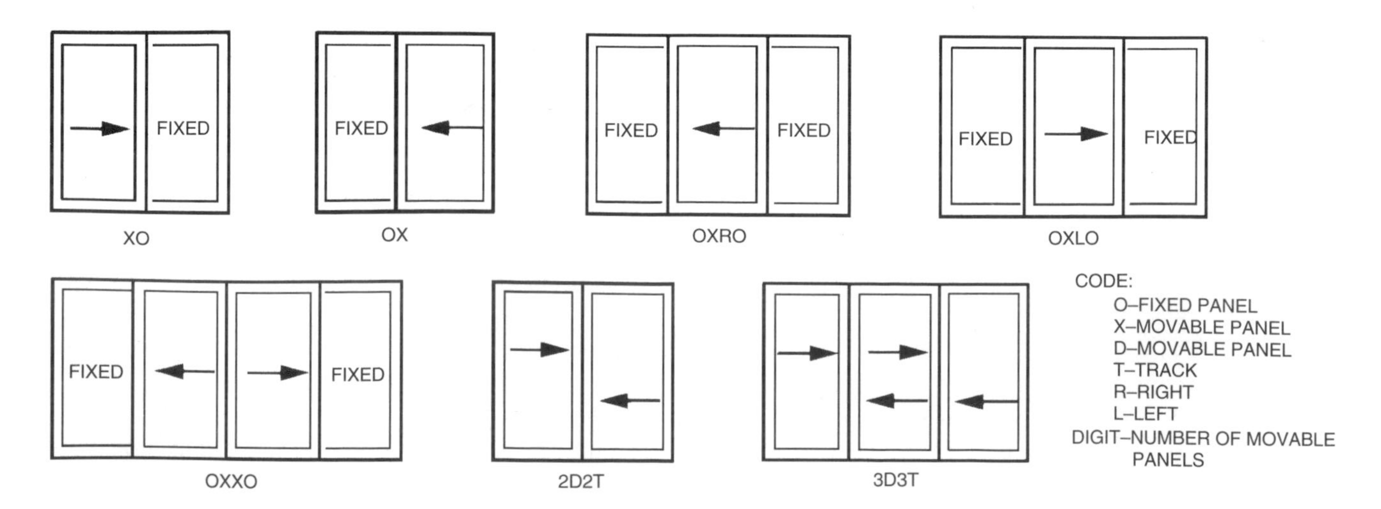

Fig. 28-19. Descriptive code for fixed and movable sliding glass doors can be matched with sketches of door panels.

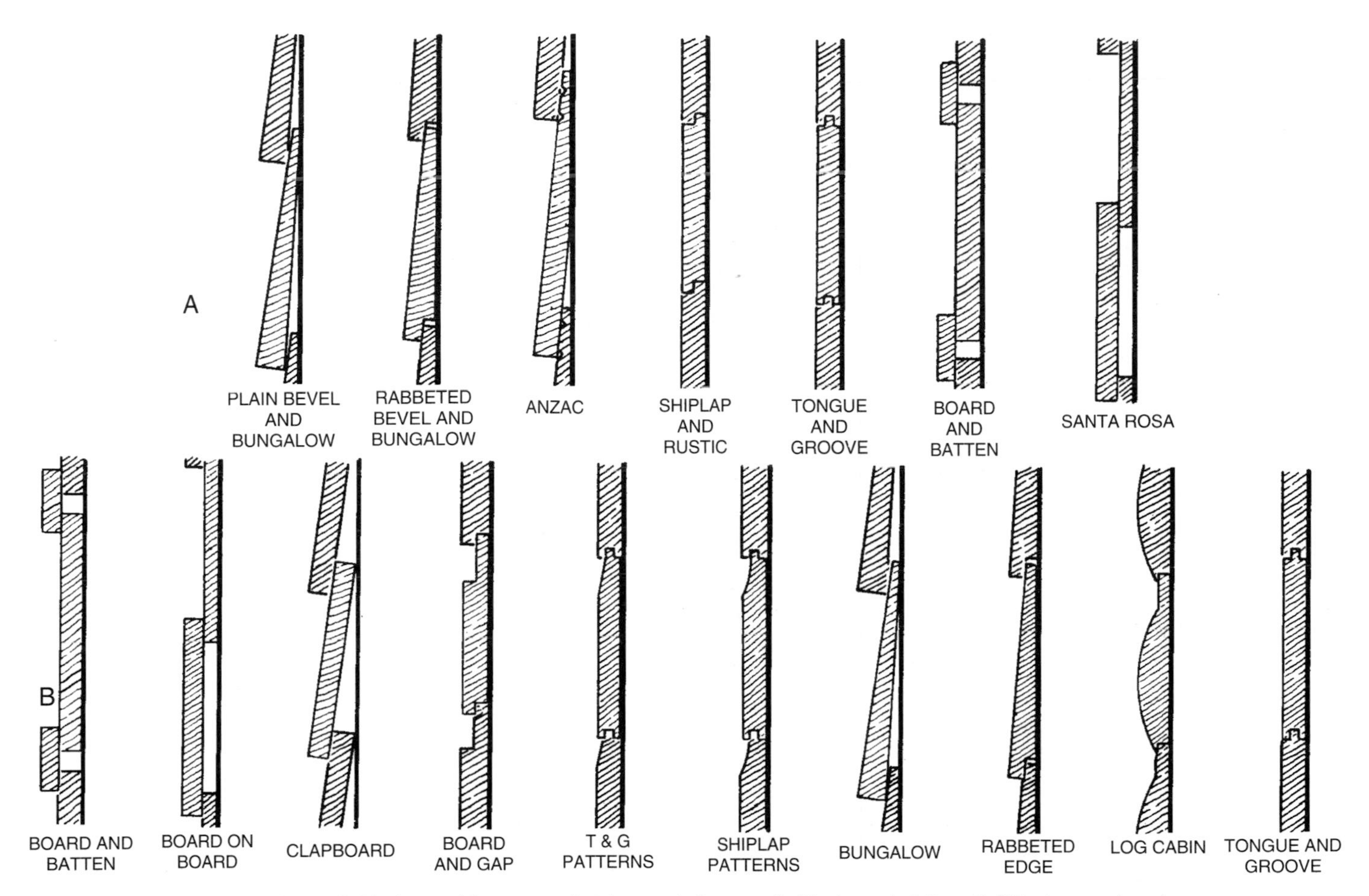

Fig. 28-20. Wood siding is available in a wide range of styles and shapes. A–Redwood siding. B–Western red cedar.

because it is a long lasting and low maintenance finish. Accumulations of dirt may be simply hosed off. Aluminum siding can be installed over wood, concrete block, and other surfaces.

A variety of colors and textures are available for aluminum siding. Some panels are textured with a wood grain to give the appearance of painted wood siding. The aluminum panels come prepunched for easy assembly.

Aluminum siding should be electrically grounded. If wiring were to come in contact with the panels, an electrical hazard could be created. Grounding with a No. 8 wire or larger is recommended by the Aluminum Siding Association.

Vinyl siding

Vinyl siding is formed with sheets of very strong vinyl. Vinyl siding is another popular alternative to wood siding. Like aluminum siding, the vinyl is long lasting and easy to clean. It is produced in many colors and textures.

Vinyl siding also has an advantage over aluminum siding. Aluminum siding must be produced in strips that have been precut to fit a structure. This adds to the cost. Vinyl siding is easier to cut and affix. Thus, vinyl siding is mass produced in standard lengths. Workers cut the vinyl strips to fit the structure at the site.

Brick and stone

Siding for offices and factories is generally made of a material that requires little or no maintenance. Commercial buildings often use masonry products like brick or stone, Fig. 28-21. These materials provide not only the durability but also the fire protection required in commercial construction.

Stucco is a masonry product that makes an excellent exterior covering for buildings. Fig. 28-22 shows workers installing stucco to the exterior of a building. Once the siding is installed, the structure is weather tight and the interior finish and finishing details can be completed.

SUMMARY

Exterior finish work is that portion of the construction project that enables the finished structure to withstand the forces of nature. Without the exterior finish work, construction projects would deteriorate quite rapidly from wind, water, sunlight, and freezing temperatures.

Finishing a civil construction project may consist of placing a layer of asphalt or concrete over the substructure of the road, or the final grouting on a canal or dam. For these projects, the finish work is designed to keep the

Fig. 28-21. Worker cuts stone on the face of a building. Stone provides a stately appearance with little or no maintenance.

Fig. 28-22. This worker is applying the base coat for a stucco wall.

structure serviceable for a long period of time. Less consideration is given to the appearance.

In commercial construction, the exterior finish work must be such that it can stand up to the forces of nature as well as large volumes of human traffic using the structure. Thus, very durable materials are often used for the finish work on these buildings.

In residential construction, the exterior finish is important for both its function and its looks. Since there are not as many people using the structure, the materials do not have to be as durable as in commercial construction.

Roofing materials, whether for residential or commercial construction, are measured by the square. A square of material covers 100 square ft. Today there are a large number of different roofing materials. Each has advantages and disadvantages.

Windows and doors are commonly installed in buildings as complete units. The window and door units are manufactured in factories and delivered to the construction site ready to install into the structure. Once these components are installed, the building becomes more weather tight.

The last exterior finish feature to be installed on buildings is the siding. Siding materials can be made of wood, composites, metals, or masonry products. The primary functions of siding are to protect the building and make the exterior look pleasant.

KEY WORDS AND TERMS

All of the following key words and terms have been used in this chapter. Do you know their meanings?

Aluminum siding
Asphalt shingles
Built-up roof
Door jamb
Exterior finishing work
Exterior wall covering
Flat roof
Membrane roofing material
Pounds per square
Single glazing
Skylight
Slope
Sloped roof
Square
Stucco
Triple glazing
Unit
Vinyl siding
Weatherstripping
Wood siding

TEST YOUR KNOWLEDGE

Do not write in this book. Please write your answers on a separate sheet of paper.

1. List the proper order for installing exterior finish work on a building.

2. List one reason why highways are covered with a hard surfacing material such as concrete or asphalt.
3. What is the unit of measure for roofing materials?
4. What four things are important to consider when choosing a roofing material?
5. Name three different kinds of window types that are used in construction.
6. The standard size of an exterior residential door is ________ x ________.
7. List four different kinds of exterior wall coverings used on homes.

APPLYING YOUR KNOWLEDGE

1. Carefully observe various wall coverings in your area and report on their condition. Which types seem to be holding up the best?
2. Take photos of several civil construction projects in your community such as roads, bridges, dams, and canals. Pay close attention to the exterior finish used in each, and take note of their condition.
3. Interview a highway or street engineer. Discuss the difference in the cost and service life of an asphalt road surface and a concrete road surface. Report your findings to the class.
4. Make a poster using cutouts from magazines that show the great variety available for exterior sidings. Make another cutout showing the variety of window and door units available.
5. Calculate the number of squares needed to roof your home.

Heavy equipment was necessary to complete this exterior finishing task.

29

CHAPTER

INTERIOR FINISHING

The information provided in this chapter will enable you to:

- *Describe the major steps of interior finish work.*
- *State three reasons why drywall is used in interior finish work.*
- *List four types of finish flooring.*
- *Discuss the advantages of suspended ceilings.*
- *Identify standard dimensions for kitchen cabinets.*

One of the final phases of a construction project is the completion of the interior of the structure. Interior finishing is done only to construction projects that enclose an interior space. Projects such as roads, bridges, and transmission towers do not have an enclosed interior space. Consequently, there is no interior finishing work to complete.

Interior finishing work consists of placing final coverings on the ceilings, walls, and floors, as well as installing the cabinetry and millwork. Also, the utility systems roughed in during earlier phases are completed during this phase of construction. This includes finishing the electrical wiring, installing the plumbing fixtures, and completing HVAC equipment.

INTERIOR CEILING AND WALL COVERINGS

There are many different materials and methods used in creating interior walls and ceilings. Drywall, plaster, and various masonry products all find their way into homes and commercial structures.

DRYWALL OR GYPSUM BOARD

The most common material used to cover walls and ceilings in residential and commercial construction is **gypsum board**, or **drywall,** Fig. 29-1. This material is used because it is easy to install, inexpensive to purchase, and provides fire protection to the structure. Few other materials have *all* of these advantages.

Drywall is made from **gypsum,** a naturally occurring mineral. Gypsum is mined, crushed, and then wrapped with a covering of heavy paper. Gypsum board comes in sections that are four ft. in width and up to 16 ft. in length. Most gypsum boards are either 1/2 in. or 5/8 in. thick.

Drywall is installed using special nails or screws over wood or metal framing members, Fig. 29-2. After the material is installed, the joints are finished with a special joint compound and tape, Fig. 29-3. Gypsum is relatively soft, so the outside corners are reinforced with metal angles, Fig. 29-4.

In the past, many structures used **plaster.** The plaster was troweled on by hand over wood lath. This is an excellent interior wall system, but it is more labor intensive than drywall. Consequently, plaster is out of the price range for many construction projects.

Installing drywall

When drywalling the interior of a building, workers first install the ceiling panels. After all of the ceiling panels have been nailed or screwed in place, the walls are installed. Drywall is placed horizontally across the walls. The top sheet is installed first, Fig. 29-5. This allows for the fewest number of joints on the job. It also places the single horizontal joint in an easy position to tape.

After all of the drywall panels are fastened in place, the joints are taped. To **tape** a joint, a coat of **taping compound** is applied to the joint. The taping compound is applied along shallow channels formed by the tapered edges where the boards are butted against each other and along both sides of inside corners. Then, a specially made paper tape is embedded into the coat of taping compound. The tape is pressed smoothly into the wet taping compound. After allowing the first coat to dry, the compound

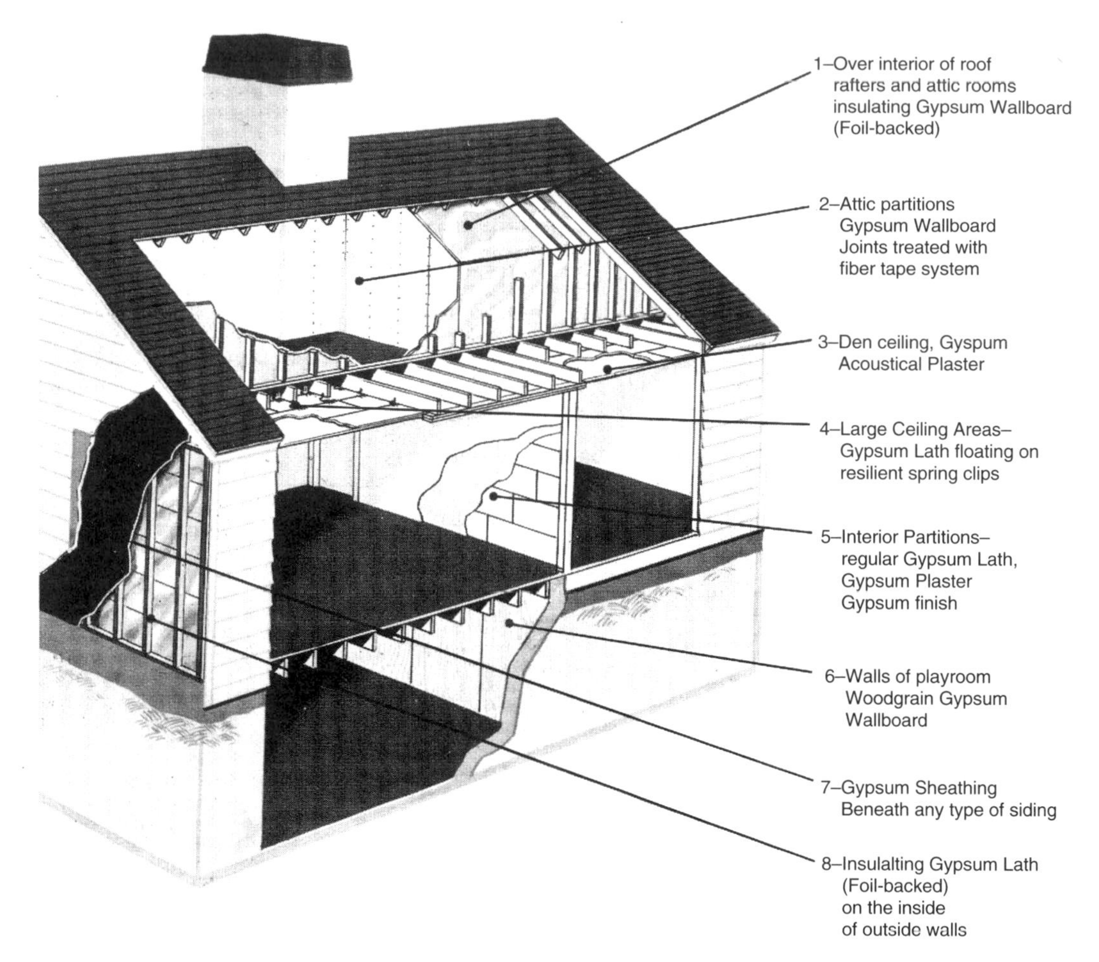

Fig. 29-1. Diagram shows many uses of gypsum material in residential construction.

Fig. 29-2. These workers are attaching drywall to the ceiling framing members with special nails.

is lightly sanded. Second and third coats of the taping compound are applied, Fig. 29-6. These coats are also sanded when dry. When the taping process is completed, the joints should be unnoticeable. The entire wall becomes one smooth surface.

COVERING DRYWALL

One of the reasons that drywall is so popular is its versatility. There are many different ways to cover drywall to give a room an attractive and interesting look. Some of the materials used to cover drywall are:

- Paint.
- Wood paneling.
- Wood trim.
- Wallpaper.

Paints

After the drywall has been installed it still needs a finish applied. Paint is the most commonly used surface finish. Paint provides a serviceable, sanitary, clean, and colorful finish. Paints come in a large variety of types, textures, and colors. Paint is easy to apply and, with modern water base paints, the clean up of tools has become easy as well.

Paint consists of finely ground solids suspended in a liquid. When the paint is applied to a clean, dry surface, the solids form a continuous durable film. The finely ground solid particles are known as **pigments.** The liquid

Fig. 29-3. After the joint compound dries, it will be sanded smooth.

Fig. 29-4. Gypsum board crumbles easily. Metal angles are used to protect the vulnerable corners.

Fig. 29-5. The top sheets of drywall are installed first. Sections are measured and cut out of the drywall so that it will fit around a building's mechanical systems.

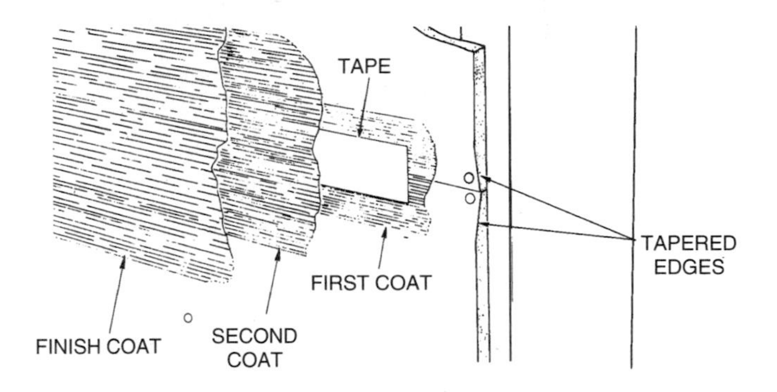

Fig. 29-6. Reinforcing joints with tape prevents cracks from appearing at filled drywall joints. The joint fill and first coat can be joint compound or all-purpose compound. The second and third coats should be finishing compound or all-purpose compound.

portion of the paint is called the **vehicle.** When paint is applied, the vehicle portion of the paint evaporates leaving the pigment to form the paint film on the surface. Fig. 29-7 shows a label from a can of paint. The label describes the percentages of solids and vehicle in the paint.

The two most common types of paints are oil-base paints and water-base paints. As their names imply, the **oil-base paints** use oils that are refined from plants and animals for the vehicle. **Water-base paints** use water as the main vehicle. Pigments give both paints their body and color.

Besides the vehicle and the pigment, modern paints include many other substances to give a paint desired characteristics. Some additives make the paint last longer. Other additives make the paint dry faster. Some other additives make the paint penetrate the surface deeper. Still other ingredients give paint a gloss, or **satin**, finish.

Some paints are designed especially for certain surfaces. Certain paints work better on metallic surfaces, while others are better for porous wooden surfaces. There are

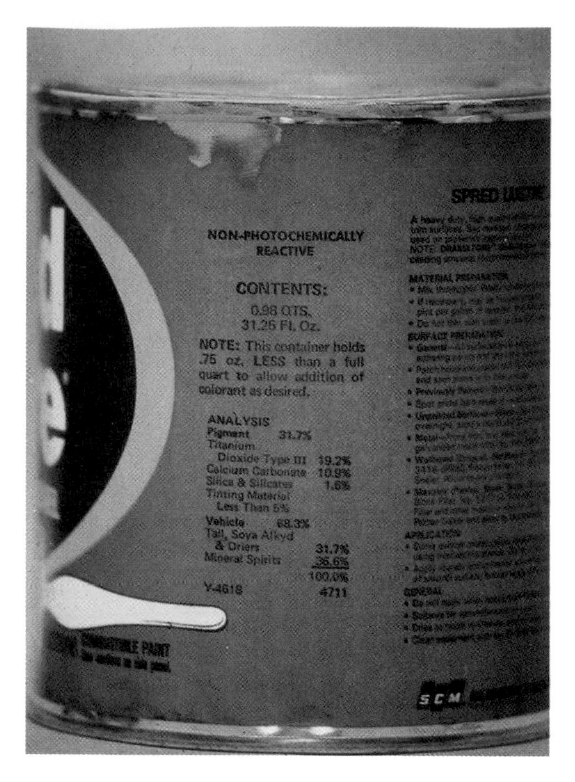

Fig. 29-7. When a can of paint sits for an extended period of time, the vehicle and pigments tend to separate. The can of paint should be shaken vigorously before being used.

paints designed to be used for areas that have already been painted. This saves a great deal of time and effort when trying to repaint a room. The selection of modern paints is almost endless.

Fig. 29-8. Drywall provides a fire resistant backing for wood paneling. (Georgia-Pacific Corp.)

Fig. 29-9. Wood wall coverings can be used to create a variety of effects. (Lindal Cedar Homes, Inc.)

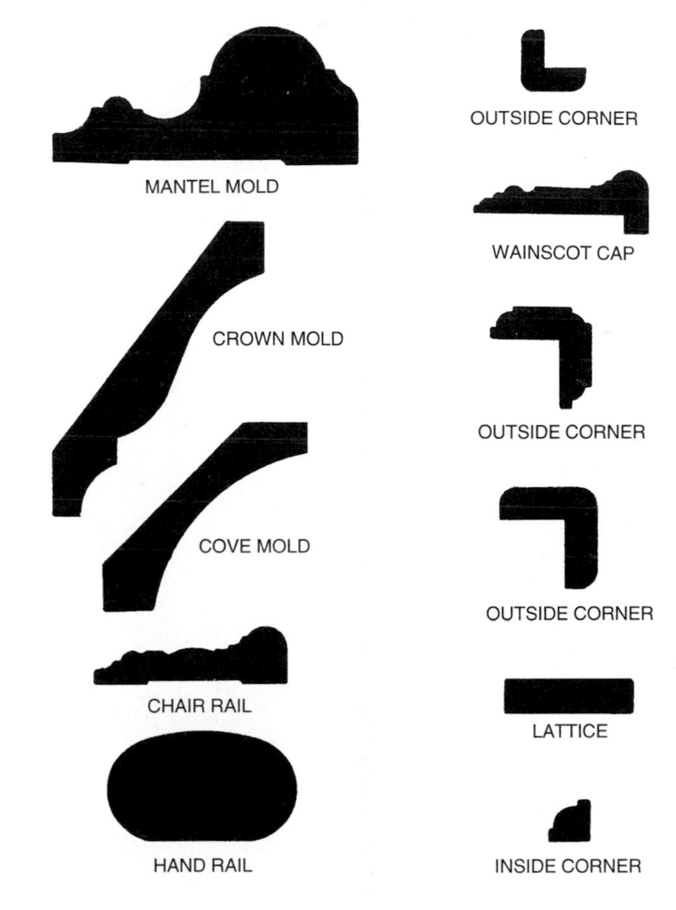

Fig. 29-10. Shapes of stock moldings. Many stock moldings are made from fir and pine.

Wood materials

Wood is commonly used in homes and offices to provide an attractive, high-quality finish. Generally, wood wall covering products are installed over drywall, Fig. 29-8. This provides fire resistance to the structure. Fig. 29-9 shows several types of wood interior systems. Wood paneling may be installed in sheets or may be installed as individual pieces. Lumberyards carry a large variety of wood paneling materials.

Wood trim, Fig. 29-10, is also extensively used in modern interior finish work. The interior trim may be made from common soft or hard woods, or can be made from very exotic woods. These trims are used in the highest quality homes and offices. Interior wood trim is available in a large number of standard patterns. It also can be custom-made to meet the desires of the owner or architect.

Wood can be given a variety of different colors using a **stain.** To protect the wood's surface, transparent finishes have been developed. These **varnishes** and **lacquers** protect while allowing the natural beauty of the wood to be seen. These finishes are transparent liquids that dry to a clear and tough finish.

Wallpaper

Drywall may also be covered with **wallpaper,** Fig. 29-11. These wall coverings are made from paper, cloth, plastic, or foil. Wallpaper may be selected from thousands of patterns and designs. Special skills and tools are helpful to do a professional job of installing wall coverings. However, many home owners choose to hang their own wallpaper. An advantage of wallpaper is that it can be easily replaced. This renews the appearance of a room at a relatively low cost.

MASONRY MATERIALS

Interior walls may also be covered with masonry materials. These materials include stone, brick, and ceramic tile, Fig. 29-12. If masonry materials are to be used for the

Fig. 29-11. These are just a few of the thousands of designs available with wallpaper. (Therma-Tru, Division of LST Corp.)

Fig. 29-12. Tile is carefully installed in a shower.

interior of the structure, careful planning should be done prior to construction. The structure must be designed and built to support their significant weight, Fig. 29-13.

Commercial and industrial buildings that are constructed from concrete may not have any special interior wall treatments. The concrete itself is simply cleaned. In some commercial buildings the surface may be sandblasted. This reveals the stone in the concrete for an attractive appearance.

Ceramic products, Fig. 29-14, are often used for interior wall coverings especially in areas around water. Bathrooms and swimming pools are commonly surrounded by ceramic tile. Ceramic tiles come in a very large variety of sizes, colors, textures, and patterns. When applying the tile, the first step is to install concrete board on the walls. Then

29-14. Ceramic tile is water resistant and ideal for the area around a tub. However, tile is very slippery. Tubs are now often built with safety features. This tub has an additional step to enter as well as metal handles inside the bath.

Fig. 29-13. Stone makes an exciting interior, but architects must create a design to support the considerable extra weight.

the tile can be attached to this board using adhesives or cement grout.

SUSPENDED CEILINGS

In buildings where it is important to have access to the utilities installed overhead, a suspended ceiling is often used, Fig. 29-15. With a **suspended ceiling,** the wall materials are installed first. Then, wall angles are attached to the interior walls, Fig. 29-16. Next, main and support channels are installed. These channels are supported by wires attached to the framing members above. After this entire framework is installed, the individual ceiling tiles are set into position. This type of ceiling construction is attractive, and it allows easy access to an overhead utility system. This system is commonly used in commercial buildings and the basements of homes.

Fig. 29-15. This suspended ceiling has a series of panels that are easily removed.

Fig. 29-16. After installing the wall materials, the first step in placing a hanging ceiling is to attach wall angles.

FLOORS

The finish flooring system varies with the type of construction. If the structure has a concrete floor, the finish floor may be as simple as cleaning and placing a finish on the existing concrete. This finish is created by cleaning the floor with chemicals (mild etching) and then polishing the floor with a rotary action. This is common in some areas of factories and schools. A concrete floor may also serve as the base for ceramic floor tile, asphalt tile, seamless vinyl, linoleum, and carpeting.

If the floor system is wood framed, then the plywood subfloor is put down first. This subfloor is covered with a second layer of plywood or particleboard underlayment, Fig. 29-17, upon which the finish floor may be laid. The choices for the finish floor are: hardwood flooring, asphalt tile, linoleum, seamless vinyl, and carpeting.

Generally, the finish flooring is one of the last items to be installed in a structure. This reduces the chance that workers from other trades will soil the flooring while completing their own finish work. The owner has a large variety of specialized finish flooring systems from which to choose. Fig. 29-18 shows one home owner finishing his floor with tile.

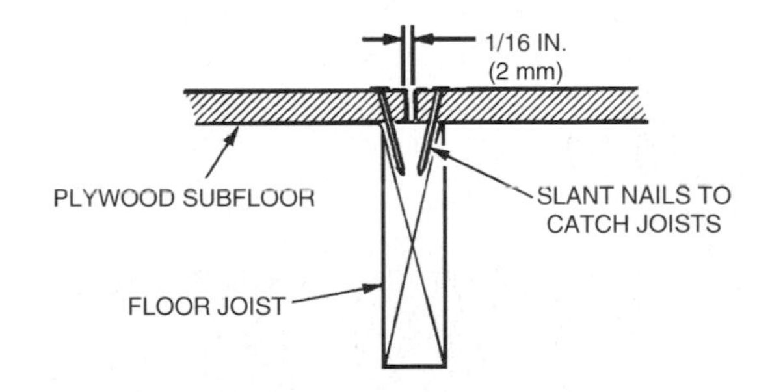

Fig. 29-17. Plywood subflooring must have clearance between sheets to allow for expansion. Space also helps to eliminate squeaking in the subfloor.

Fig. 29-18. Small floor tiles come in sheets, making installation easy.

CABINETRY

Cabinets furnish specialized storage for many areas of a building. Kitchen cabinets help the homemaker keep cooking utensils orderly. They also provide countertop work area, Fig. 29-19. Cabinets are used throughout the home in the dressing area, dining, laundry, serving area, and in a variety of minor storage areas, Fig. 29-20.

Cabinets must be carefully planned and well built. Functional kitchen cabinets must be adaptable to changing conditions. They must take care of different kinds of appliances, Fig. 29-21. Cabinet interiors should be suitable for storing many sizes and shapes of dishes, silverware, packaged goods, and canned goods.

Fig. 29-19. A selection of kitchen cabinets. The style of a kitchen can be created by choice of cabinets alone. (Timberlake Cabinet Co., Merillat Industries Inc.)

Fig. 29-20. You will find cabinetry of some sort in almost every room in a home. (Timberlake Cabinet Co.)

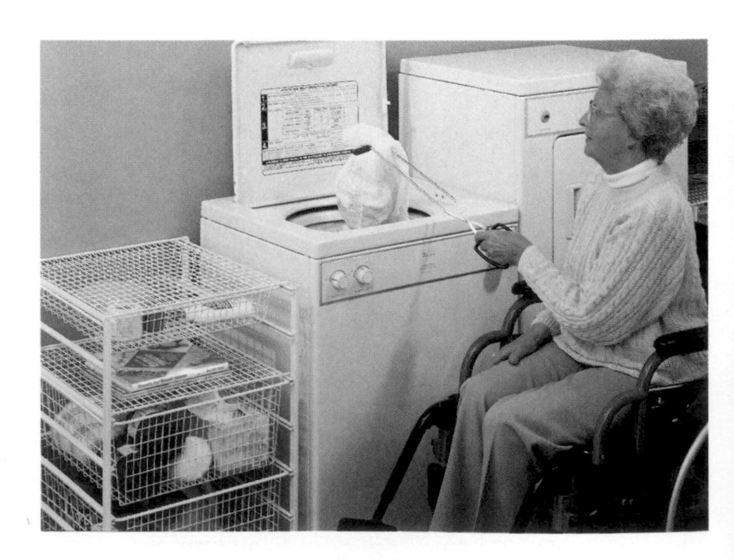

Fig. 29-21. Cabinetry is not always the best design solution. Here, the easy access of wire baskets proves handy. (Whirlpool)

Cabinetry of commercial buildings may be designed for display purposes of merchandise as in Fig. 29-22. These cabinets use different methods and materials for attaching the doors.

There are three basic methods of producing cabinets.

- Some are built on the construction site.
- Some are built at a mill and supplied in a knocked-down condition.
- Some are mass-produced in a factory and delivered assembled.

Fig. 29-22. Display cabinetry is generally made of glass on all sides facing the consumer.

STANDARD CABINET SIZES

Architectural drawings include the location and design of cabinets for the building. Specific details concerning cabinetwork are sometimes included in the drawings. There may be elevation drawings as in Fig. 29-23. Dimensions are placed on the drawings. Written specifications describe the type of jointery as well as the kind and quality of material to be used.

Overall heights and general features of built-in cabinets are standard, Fig. 29-24. Base cabinets for the kitchen usually are 36 in. (91 cm) high and 24 in. (61 cm) deep, including the countertop. For most people, this is a comfortable working height while standing. Customized cabinetry is built for people with different needs.

Most upper cabinets are 30 in. (76 cm) high and 12 in. (30 cm) deep. The distance between the base cabinet and the upper cabinet is usually 18 in. (46 cm). Writing desks should be 30 in. (76 cm) above the floor for convenient height. A bathroom vanity will be about the same height.

In most kitchens, cabinets do not go all the way to the ceiling. The space above often is closed with a drop ceiling. If there is no drop ceiling, it is better to have two sets of cabinets. Small ones are placed toward the top. Larger ones below are for the storage of most-used items.

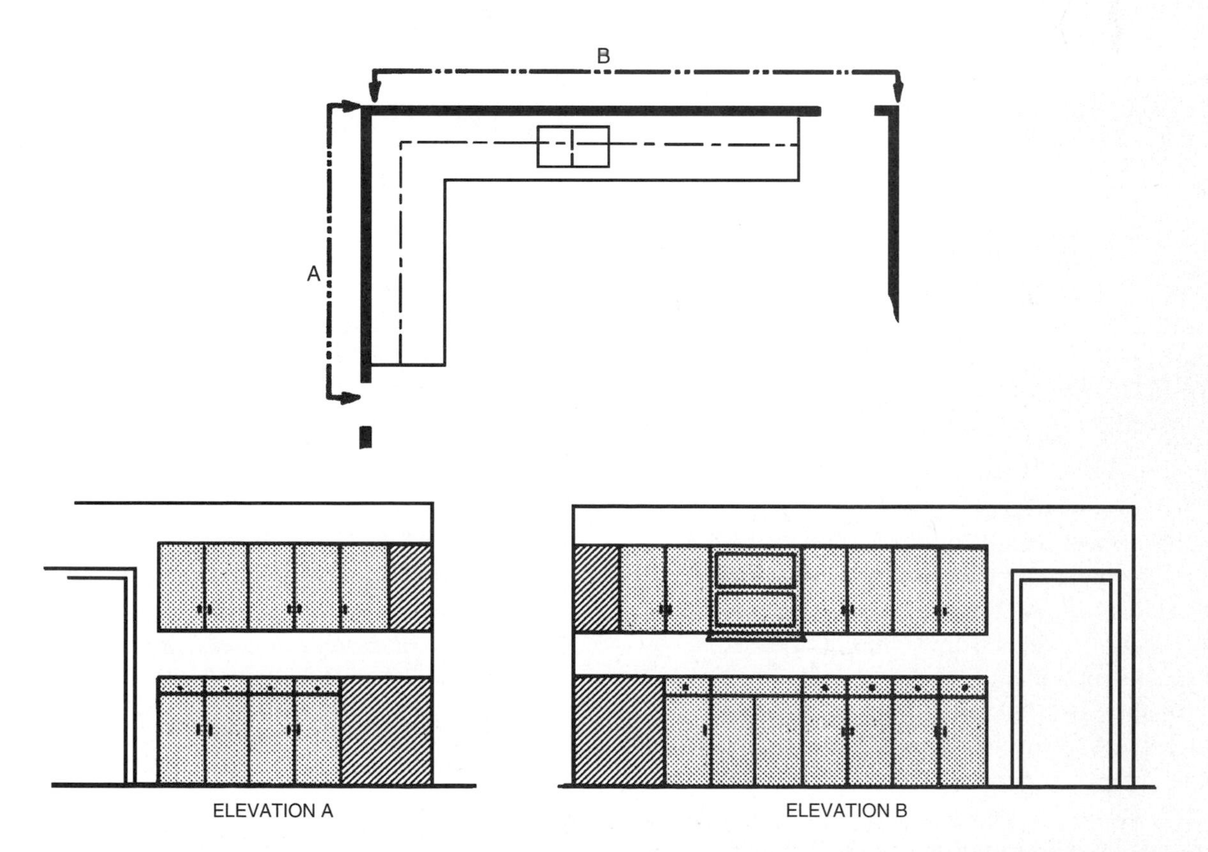

Fig. 29-23. Elevation details of cabinets found on the architectural drawing. Though not shown here, dimensions would be included.

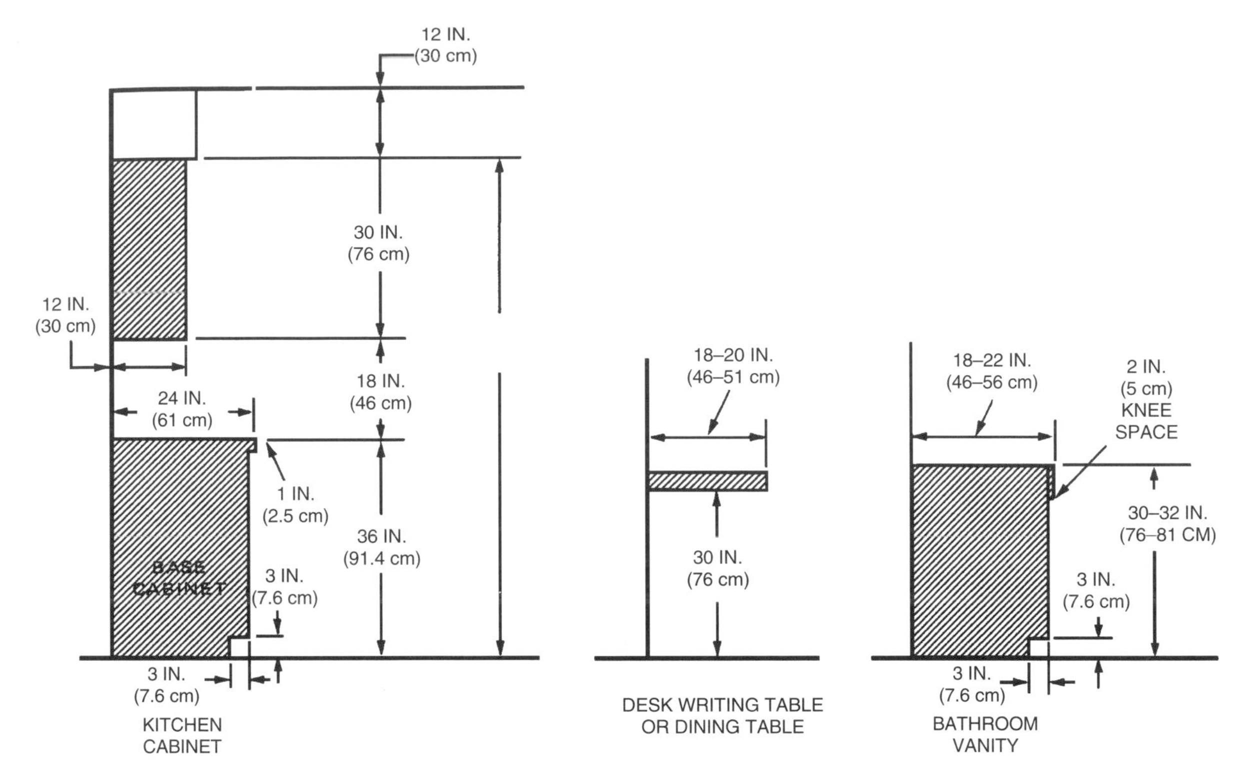

Fig. 29-24. Kitchen cabinets and other cabinetry are built to basic standard measurements.

All base cabinets are built with a **toe strip** or **toe space** to leave room at the floor. Wall units are constructed in much the same way as base units.

DOOR CONSTRUCTION

Doors for cabinets must be attractive as well as useful. The main function of doors is to close off storage space. However, doors should also add interest to the outside of the cabinet. Glass or metal grillwork is sometimes used to show off fancy china, glassware, or silverware, Fig. 29-25.

Fig. 29-25. Cabinetry may be used as a display case in addition to storage space. (Drexel Heritage Furnishings, Inc.)

DRAWER CONSTRUCTION

Drawers must be soundly built. In kitchen cabinets, drawers are used for storage of silverware and cooking utensils. Drawers make excellent storage space for three reasons.

- It is easy to arrange the contents through use of dividers.
- Drawers are relatively clean and keep contents dust free.
- Drawers conceal items from sight until needed.

Drawer construction is often a good indication of overall furniture quality. If the drawer joints are well made and if the drawer slides easily when pulled by the corners, the furniture is usually of good quality.

Drawers, like doors, are classified in two general types: flush-mounted drawers and lip, or offset, drawers. **Flush drawers** close even with the cabinet. These drawers must be carefully fitted and are commonly found in furniture construction. Fig. 29-26 shows flush drawers in kitchen cabinets. **Lip drawers** or **offset drawers** fronts have a 3/8 x 3/8 in. (10 x 10 mm) rabbet along their edges that overlap the opening. The offset drawer allows a looser fit.

SHELVES

Shelves are used widely in cabinetwork. They are often found in wall units where drawers are not used. In some

Fig. 29-26. Flush drawers have less room for inaccuracy in fitting. For flush drawers to look right, time and care must be spent.

designs, it may be necessary to fit and glue the shelves into dadoes cut into the side pieces as in Fig. 29-27. The **dado joint** provides greater structural strength to the unit.

Whenever possible, it is a good idea to make a shelf adjustable. Such shelves can be adjusted for different sizes. See Fig. 29-28.

COUNTERTOPS

The work area of the cabinet is the countertop. This surface should be rugged, beautiful, and waterproof while resisting stain and heat. See Fig. 29-29. The countertop is often in a color that accents basic room color. Counter material should also be easy to clean.

Countertops are made up in many kinds of material. The **plastic laminates** are the most popular. Plastic laminates are rugged, waterproof, and economical as well. Laminate comes in many colors and decorator patterns.

CABINET HARDWARE

Cabinet hardware includes knobs, pulls, hinges, catches, and other metal fittings that are put on the cabinet. See Fig. 29-30. The hardware is prefitted before the finish is applied.

Drawer pulls usually look best when they are located slightly above the center line of the drawer front. However, they are centered horizontally. **Door pulls** or knobs on the base unit doors are located somewhere in the top one-third of the door. For doors in the upper units, the pulls are located in the lower one-third of the door.

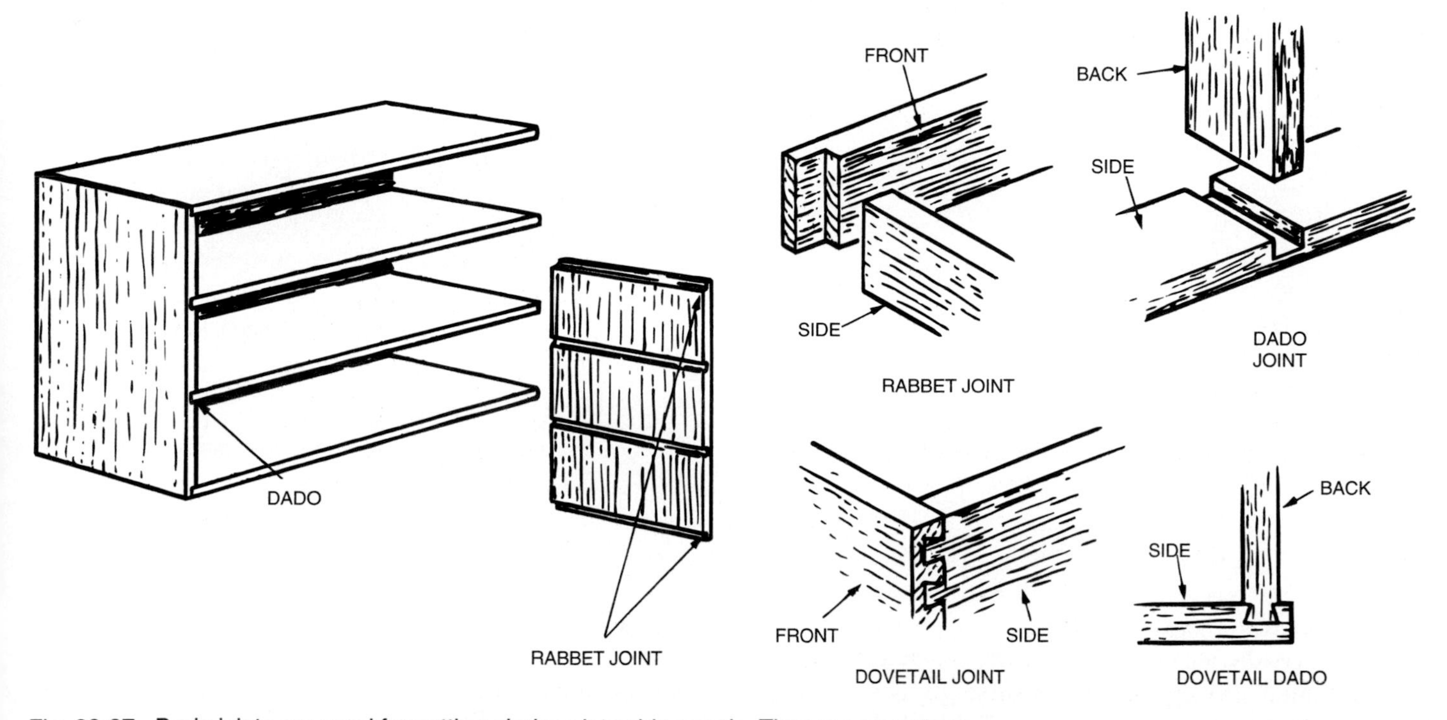

Fig. 29-27. Dado joints are used for setting shelves into side panels. They are very strong.

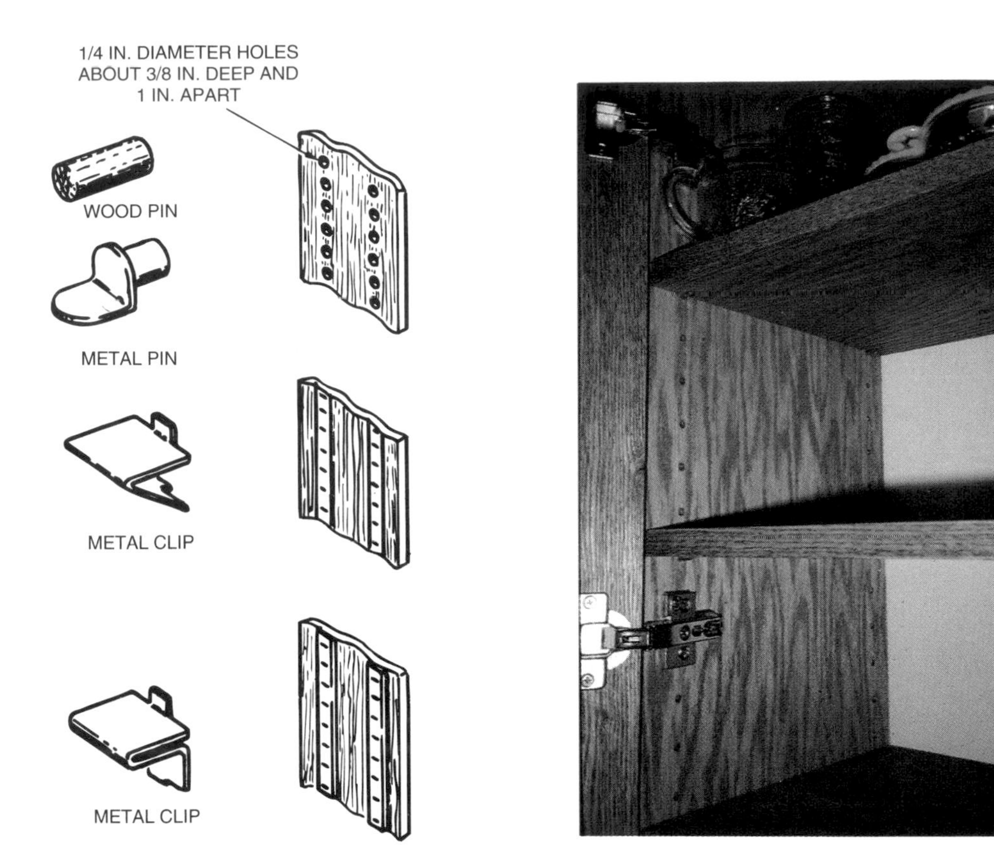

Fig. 29-28. Three methods of supporting adjustable shelves. These simple forms are easy and economical to install. Metal strips, metal insert pins, and clips are available from hardware dealers.

Fig. 29-29. Spills, hot pots and pans, and sharp knives are daily ordeals for kitchen counters. Countertops must be made out of rugged materials. (DuPont)

The most common material used for wall and ceiling coverings is drywall. This material provides fire resistance to the structure and is also economical to install and finish. Its surface can be painted, wallpapered, or it can serve as a base for thin paneling.

Once the drywall is installed, the finish trim work can be completed. Wood materials are most common for trim materials, although metal and plastics are also used. In special areas, ceramic or masonry materials are used to provide an attractive and easily maintained surface.

Floors can be finished in a variety of ways. If they are concrete, they may simply be cleaned and perhaps sealed. They may be covered with carpeting, linoleum, tile, or wood. The owner and the building designer must consider the conditions the floor will be used under to choose the best type of flooring material.

Cabinetry consists of all the built-in cabinets that are installed in a structure. Cabinets are designed in many different forms to perform a variety of tasks. A set of quality cabinets can enhance the beauty of most rooms.

SUMMARY

Structures with interior space such as homes, offices, and factories generally require the interior space to be finished. The interior finish allows the structure to be cleaned easier, provides more pleasant surroundings, and provides for better safety.

KEY WORDS AND TERMS

All of the following key words and terms have been used in the chapter. Do you know their meaning?

Cabinet
Cabinet hardware

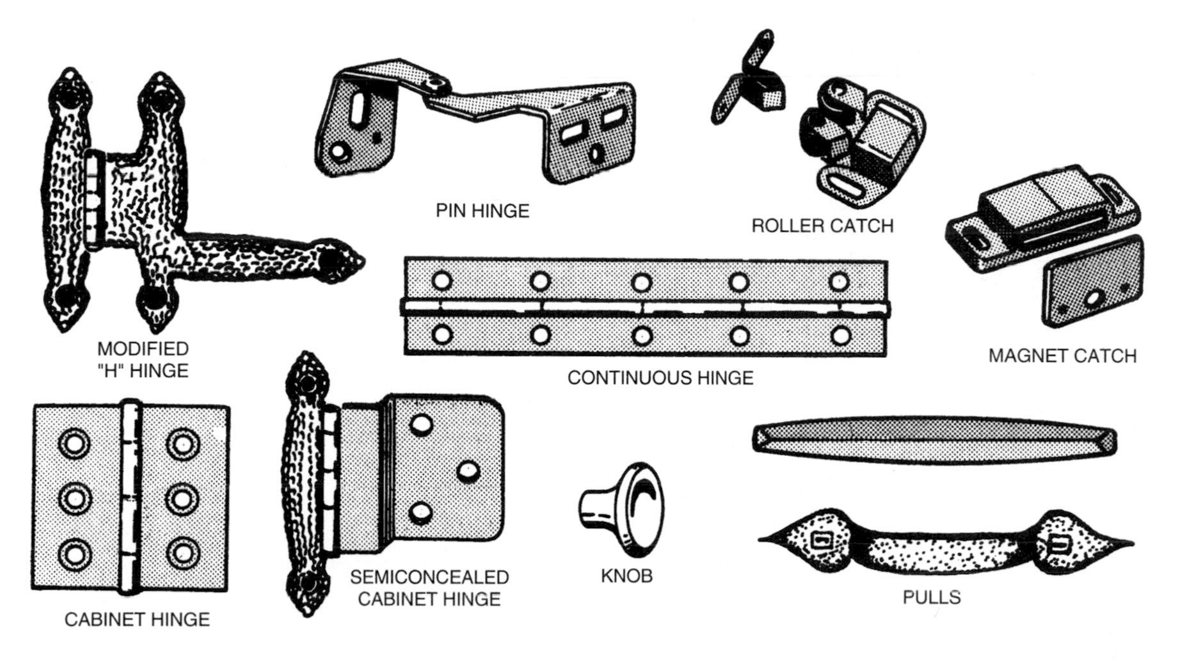

Fig. 29-30. Types of hardware for cabinet doors and drawers.

Dado joint
Door pull
Drawer pull
Drywall
Flush drawer
Gypsum
Gypsum board
Interior finishing work
Lacquer
Lip drawer
Offset drawer
Oil-base paint
Paint
Pigment
Plaster
Plastic laminate
Satin
Stain
Suspended ceiling
Tape
Taping compound
Toe space
Toe strip
Varnish
Vehicle
Wallpaper
Water-base paint
Wood trim

TEST YOUR KNOWLEDGE

Do not write in this book. Please write your answers on a separate sheet of paper.

1. What are the major components of interior finish work?
2. The most common type of material used to cover the interior of a building is _________.
3. Paint consists of what two major components?
4. List three kinds of masonry or ceramic materials used for interior finish work.
5. What is a major advantage of a suspended ceiling?
6. The standard height of a kitchen counter is _____ in.

APPLYING YOUR KNOWLEDGE

1. Collect various pieces of drywall. Test the material for hardness, strength, and screw holding ability. Complete a report on the advantages and disadvantages of using drywall material as an interior covering.
2. Build a small wall section a few feet high and install drywall on the studs. Finish the joints in your wall section. How much skill do you feel is required to make a living at this trade?
3. Obtain several kinds and brands of interior paint, and paint several pieces of drywall. After the paint has had the proper time to dry, design and conduct a testing procedure to determine which paint has the best ability to withstand repeated washings.
4. Have your instructor remove several ceiling panels of a suspended ceiling in your school. Notice how the main and secondary rails are supported as well as the utilities that are hidden by the suspended ceiling.
5. Conduct a field trip within your school and keep track of the various flooring types used in different areas. List the reasons why each floor covering was used in each area.

ELLIOTT CORP.

Attractive office interiors improve employee moral.

FINISHING DETAILS

> *The information provided in this chapter will enable you to:*
> - *Create a landscape design for a small construction project.*
> - *Develop a short punch list for a construction project.*
> - *Explain the advantages of a warranty to both the contractor and the project owner.*

The final step in the construction of any project is to complete all of the details on both the inside and outside of the structure. This may include some painting, landscaping, and checking that the entire project operates and functions as it was designed. If necessary, final adjustments are made so it will operate properly.

A final inspection is conducted before any construction project is turned over to the owner, Fig. 30-1. The final inspection should catch any unfinished details. A list of the unfinished details is created. This list is called a **punch list.** If the quality of workmanship of the project was good, the punch list will be relatively short. If the contractor tries to cut corners during the construction, problems will show up during the completion of the construction. These problems may be costly to fix and repair. Here, the finishing details phase may take a long time.

ELECTRICAL FINAL CHECKLIST

1. Outside receptacle installed and on ground-fault circuit-interrupter protection (GFCI) 210-8A-3.
2. Bathroom and garage receptacles on (GFCI) 210-8A-1, 210-8A-2, 210-52G.
3. Receptacle outlet installed behind gas range 210-50B.
4. Appliance branch circuits in kitchen and laundry 220-4B-C.
5. Receptacles within 6 feet of laundry 210-50C.
6. Location of bell transformer not inside of panels 100-3B.
7. Switched receptacles or lighting outlets 210-70A.
8. Proper polarity of fixtures and receptacles 200-10C, 410-23, 410-47.
9. Completed panel board circuit directory 110-22, 384-13.
10. Switches and receptacles installed properly 110-12.
11. Grounding electrode conductor connected to a ground rod with an approved ground clamp 250-81A, 250-115.
12. Connection to ground clamp accessible 250-112.
13. Location of HVAC equipment disconnecting means 440-14.
14. Quantity of circuit breakers, CTL type in panel board 384-15.
15. Smoke detector completed and working, O.P.F.D. Ordinance 69.
16. Telephone outlet at water meter Ordinance 2084.

Fig. 30-1. Example of a final electrical checklist for the construction of a new home. The numbers at the end of each item relate to relevant sections of the National Electrical Code.

LANDSCAPING THE PROJECT

No construction project is complete until the surrounding area is graded and landscaped. Preparing a landscape plan is the first step.

LANDSCAPE PLAN

If a project is large, Fig. 30-2, a **landscape architect** prepares the **landscape plan.** If the construction work is a small project, the architect or owner prepares the landscape plan. In a project, such as a shopping mall, there is also *interior* landscaping to consider. A landscape plan may be for something as small as an apartment garden or as large as a park, highway, golf course, or complete town.

The finished landscape should take three things into consideration.

- The need the project is to serve.
- The surrounding environment.
- How people will interact with the project.

The landscape architect works from a set of plans that shows the locations of all human-made and natural features in the project. The planner visits the site many times and consults with the owners, Fig. 30-3. The planner verifies

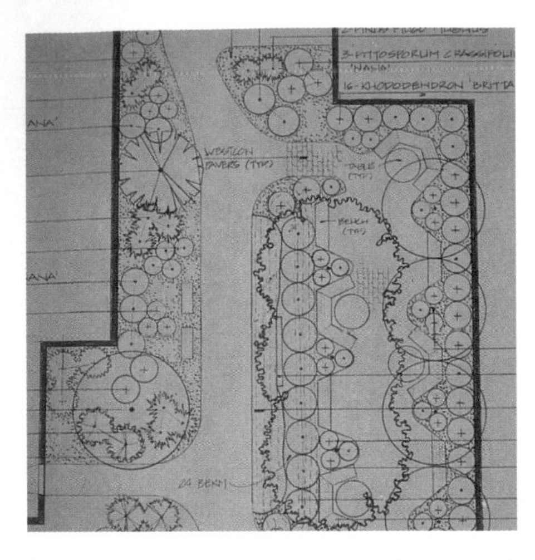

Fig. 30-2. Shown here is a segment of a large landscape plan. The location and type of plants, shrubs, and trees are listed. (Skip Pass, Skagit Valley Community College)

Fig. 30-3. Landscape planners present the owners with a computer copy of the landscape. This lets the planner and owners work together to make changes.

the locations for particular plantings and design features. The complete plan and sometimes a model are then presented to the owner for approval.

For smaller projects, the owner can visit with local **nurseries,** Fig. 30-4, to obtain their advice. This service is provided by nurseries that wish to sell plants, trees, and shrubs to the owner of the project. Since landscape plantings take several years to reach their full beauty, a landscape design may take place in several stages. Plants will change from stage to stage. These changes allow the

Fig. 30-4. Nurseries are good places to go for landscaping advice as well as landscaping materials.

grounds to look pleasant while some of the plants are reaching their full growth. If at all possible, the architect may work around trees that are already on the property.

LANDSCAPING

The actual landscaping begins while construction on the project is in its final stage. All human-built features are in place and the exterior of the project completed, Fig. 30-5. At this time, the final cleanup of the property can take place. The landscape has been left at a rough grade up until this point. Now, the contour specified in the design is created. Major trees and shrubs are hauled into place and planted, Fig. 30-6.

Lighter work is done next. All of the heavy traffic is off the area. If grass is to be planted, the soil can be prepared for the lawn. Lawns may be achieved by two methods–seeding the lawn or by purchasing sod. **Seeding** involves loosening the soil and spreading out grass seed. When placing **sod,** you are purchasing squares of grass that have already matured. The squares of sod are then staked to the ground.

Both systems require prior removal of rocks and debris from the soil. The advantage of the sod is that it takes a very short time to achieve a quality lawn. Applying sod, however, is much more costly than seeding the lawn. Fig. 30-7 shows the completed project.

Landscaping on highway projects usually follows the same procedure as landscaping a building except the final grading is done with large machines. The area is seeded with a hardy grass seed that is sprayed on the soil surface. A **mulch** is sprayed along with the seed that will help establish the grass. This is a fast and cost effective means of establishing good ground cover. Good ground cover helps prevent soil erosion.

FINAL INSPECTION

Before you purchase any item, you look the item over carefully to see that it meets your standards. When buying

A

B

Fig. 30-5. A botanical garden taking shape. A All of the major structural work on the building has been completed. No more heavy construction machinery needs to be run over the land. B–One month later, the first trees and shrubs have been installed. (Skip Pass, Skagit Valley Community College)

Fig. 30-6. Four months after beginning the landscaping, most of the major landscaping has been accomplished. Large trees have been planted and large rocks have been moved into place. (Skip Pass, Skagit Valley Community College)

Fig. 30-7. The completed landscape. The smaller plants have been placed and sod has been put down.

a pair of shoes, for example, you are careful to see that the toe is not scratched or that the stitching is of good quality. If the quality is poor, you select another pair or ask for a reduced price on the damaged merchandise. It is no different for a construction project. When the construction project is completed and all of the final finishing is done, it is time for the **final inspection.**

FINAL INSPECTION TEAM

On large construction projects, the final inspection is completed by a team, Fig. 30-8. The team consists of the contractor, the architect or engineer, and the owner (or an individual who represents the owner). This team carefully reviews the project. They determine if all systems are in place and working properly. As a method to ensure that the project is turned over in proper working order, part of the payment is usually withheld. The money is turned over when the project meets the agreed upon specifications. This ensures that the work is completed in a timely manner and that it is of high quality.

The inspection team makes notes on what needs to be corrected and creates their punch list, Fig. 30-9. The punch list includes all items that need to be repaired or fixed before the final payment is given to the contractor. After the corrections are made, another inspection is conducted. During this inspection, all of the former trouble spots are checked. If all corrections are satisfactory, the final payment is made. On the other hand, if more corrections need to be made, additional inspections are conducted. Final payment is withheld until all corrections are done properly.

Fig. 30-8. After the completion of this remodeling project, the contractor, owner, and architect perform a final inspection.

Fig. 30-9. During the final inspection, a punch list is created. Here, the owner and architect are writing down a brief list of items that the contractor (right) needs to fix.

SELLING THE JOB

Once all corrections are complete, the project is ready to be turned over. On large road construction projects this is often referred to as **selling the job.** This means that the portion of the road that is completed is paid for. In most cases, the party paying is the state, county, or city.

In construction projects such as homes, the home owner asks the contractor to sign a lien waver. A **lien waver** is a legal document stating that if the contractor has any outstanding debt on materials or labor used in the project, the suppliers and laborers cannot hold the owner of the home responsible for the debt. When the lien waver is signed, the owner gives the contractor payment in full. The project is turned over to the customer.

WARRANTIES

Quality contractors will provide the owner of the newly constructed project with a promise that the materials and workmanship are of good quality, Fig. 30-10. This is a **warranty.** The warranty is for a specified period of time, usually one year. If during this time, problems arise with either the materials or the workmanship, the contractor agrees to fix the problem at no cost to the owner.

An advantage of this type of agreement is that it maintains the good reputation of the contractor. In a project as complicated as a home, large building, highway, or bridge, unforeseen problems may arise in a short period of time. The quality contractor recognizes this possibility and is willing to stand behind his or her work. A warranty also protects the owner from some unforeseen defect in the construction materials or poor workmanship.

SUMMARY

The finishing details are the final phase of most construction projects. If the project is very large, some of the finishing work on one part of the project may be completed while other parts of the project are still in earlier phases.

Landscaping is one of the last construction activities completed on a job. First, the site is cleaned up and all construction debris is removed. Then, the landscaping begins using the plans that were designed before construction started. Outdoor landscaping includes planting grass, trees, and shrubs. Quality landscaping puts the finishing touches to a job and enhances the entire project.

Before the job is turned over to the owner, a final inspection is conducted. During this inspection, the owners or their representatives review the quality of the work. They make sure that the finished job matches the design. During this inspection a punch list is made. The punch list contains those items that need further finishing or adjustment. After the contractor has reworked all of the items on the punch list, the final payment is made to the contractor.

Quality contractors ensure the quality of their work by issuing a warranty for one year after the job is completed. During this time, if something needs repair, and it is determined that the problem lies with the contractor, the contractor will repair the problem without cost to the owner. This agreement provides protection for the owner and, at the same time, maintains the reputation of the contractor.

Fig. 30-10. With all construction projects, there is the possibility that something will go wrong. There are many complex factors that affect the construction of this bridge, house, and large building. Warranties protect the new owners of all of these projects.

KEY WORDS AND TERMS

All of the following words and terms have been used in this chapter. Do you know their meaning?

Final inspection
Landscape architect
Landscape plan
Lien waiver
Mulch
Nursery
Punch list
Seeding
Selling the job
Sod
Warranty

TEST YOUR KNOWLEDGE

Do not write in this book. Please write your answers on a separate sheet of paper.

1. True or False. All landscaping is done outside of a project.
2. What three things should a landscape plan take into account?
3. What people make up the final inspection team?
4. When does the contractor receive final payment for a construction project?
5. What is the standard length of time on a contractor's warranty?
6. What is the advantage of the warranty to the project owner?
7. What is the value of a warranty to the contractor?

APPLYING YOUR KNOWLEDGE

1. Make a drawing of your school yard, and then plan a new landscape design for the school. Make a model of your proposed design.
2. Visit a nursery and talk with the owner. Try to gain a better understanding of landscape design and the growing of ornamental plants.
3. Carefully inspect your technology education laboratory. Make a detailed list of building items that need to be repaired.
4. Request a sample punch list of a construction project that is being completed in your area. Review the list and note the detail that goes into it.
5. Write up a short warranty on what you expect from the workmanship on a small construction project. Discuss in class why you listed the items you have in your warranty. Discuss why it is important to have a warranty from the contractor on construction projects.

31
CHAPTER

SERVICING CONSTRUCTION PROJECTS

The information provided in this chapter will enable you to:
- *State the difference between maintenance, repair, and remodeling.*
- *Explain why it is important to maintain construction projects.*
- *Discuss and give examples where preventive maintenance can be used in the construction industry.*

With any construction project, there is normal wear and tear caused by routine activities as well as some gradual **deterioration** due to weathering. After the period of time that the project is under warranty expires, it becomes the responsibility of the owner to provide for the maintenance and repair of the property. In some cases, this maintenance is very minor and easy to accomplish. On the other hand, if the property has been allowed to deteriorate through years of neglect, the maintenance may be extensive and very costly to complete. In extreme cases, it is more cost effective to demolish the structure.

MAINTENANCE OF CONSTRUCTION PROJECTS

Maintenance are those activities that take place to keep items in proper condition. This includes activities such as replacing lightbulbs, cleaning the carpet, painting the exterior surface, repainting the traffic lines on the road, and resurfacing a highway, Fig. 31-1. Maintenance can be conducted on a regular time schedule, or it can be initiated after some problem arises.

Fig. 31-1. Roadways have a limited lifespan. Resurfacing must be done from time to time. (Athey Products Corp.)

PREVENTIVE MAINTENANCE

If maintenance activities are done on a routine schedule, it is called **preventive maintenance.** This is done on those areas and items where a failure can cause a safety hazard or extensive repair costs. An example of preventive maintenance is the replacing of bulbs in traffic lights. The traffic office knows the expected life of the lightbulbs and makes sure to replace the lights before they burn out. A burnt out traffic light could cause a serious situation. This preventive maintenance, Fig. 31-2, provides us with safe traffic signals.

On the other hand, the light in a room is not replaced on a regular schedule. When the bulb burns out it is replaced. This type of maintenance is completed when the need arises. The failure of the bulb does not pose a danger. Since the lights in a home will eventually burn out, most households keep a supply of replacement bulbs on hand to quickly replace the light.

Fig. 31-2. A worker is involved in routine maintenance. The lightbulbs are replaced on a schedule.

Fig. 31-4. Time takes its toll on all structures. Scaffolding has been erected to correct the accumulation of damage caused by the weather.

Painting a bridge is also maintenance. The bridge is painted to resist rusting, which weakens the structure. When the windows of a home are painted, the paint resists the penetration of water into the wood and stops the wood from decaying. This is preventive maintenance. Fig. 31-3 shows a road crew resurfacing a section of road. **Resurfacing** seals the road to prevent moisture from entering the roadbed and softening the substructure. This is also a form of preventive maintenance.

REPAIR

Repair, Fig. 31-4, deals with more extensive work than the repainting of a bridge or the resurfacing of a road. **Repair** jobs consist of rebuilding parts of a construction project. Fixing a wall that was damaged is repair work. Parts of the wall have to be reconstructed. This activity is far more extensive than applying a coat of paint to the surface of the wall.

Fig. 31-3. Resurfacing a road helps prevent damage to the roadbed. A damaged roadbed would require much more extensive and expensive repairs.

A section of road might have to be replaced due to poor design or the forces of nature, such as a flood or a hurricane. The individual who does the repairs must know the specific building practices that were used to construct that portion of the project in the first place.

REMODELING

Another form of servicing construction projects is **remodeling.** As time passes, a construction project may no longer meet needs, desires, functions, or space requirements. Offices must be expanded for new employees. Homes are expanded for new family members. Roads are widened for more traffic. The world is constantly changing. In order to meet new needs and demands that arise, construction projects must be remodeled.

The remodeling industry is a very rapidly growing industry. Often, it is far more economical to remodel a structure to meet the current needs than to build a new one, Fig. 31-5. This is true for most roads, dams, and buildings. Remodeling includes many things. Major parts of a structure may be added or removed. Walls within a structure can be rearranged. Also, the utilities and mechanical equipment may be **modernized** during remodeling to take advantage of the latest technology. Remodeling a highway may consist of widening a road. Two lane roads are turned into four lane roads to carry more traffic safely.

When remodeling, building contractors must study the original plans carefully. This lets them know what problems may arise. If a wall is to be removed, calculations must be done to determine how to support the structure

Fig. 31-5. The upper floors of this structure are being remodeled. Chutes carry the garbage quickly and safely to the bins below.

the wall. If the wall contains utilities, such as electricity and plumbing, these utilities have to be rerouted. This can cause major changes in the utility system.

During any remodeling of a structure, new building codes become applicable for the entire structure. Building codes keep changing. Buildings do not always have to update with each new change. However, when remodeling a structure, the new codes are applied to the whole building, not just the section being remodeled. Major changes may be necessary over the entire structure in order to meet all the requirements. All of these items should be reviewed carefully before the decision to remodel is made. The property owner must determine if remodeling is cost effective.

DEMOLITION

After careful study, the decision might be reached that remodeling is not cost effective. It may be possible that the constructed project was seriously damaged by a flood, an earthquake, or simply through old age. The safety of the structure is in question. In these cases, the most cost-effective option may be to **demolish** the structure and reclaim the land for some other purpose. Modern explosives allow the demolition of structures without damage to nearby structures. Large modern equipment is often used to remove small structures. Fig. 31-6 shows a structure being demolished.

Before the actual demolition of a building begins, a **salvaging** operation takes place. Many times there are valuable construction components that can be saved for use in other construction projects. Examples are the doors, windows, electrical equipment, and plumbing fixtures. Scrap steel from buildings brings in a respectable price. Copper pipe and wiring are valuable recyclables. Aluminum siding and fixtures are also recycled. Even large trees can be removed from the site to be used elsewhere.

Another form of salvaging is the removal of entire buildings from a site. Fig. 31-7 shows a home that is being moved to a new location. Although it is costly and difficult to move entire structures, sometimes it is cost effective if the building is in good repair and can be moved without major problems.

With proper maintenance and repair, construction projects can last many years. The Statue of Liberty, Empire State Building, Hoover Dam, Brooklyn Bridge, and the lock and dam system on the Mississippi River are all historic structures in the United States that remain in good condition after many years of service. With modern building practices and scheduled maintenance and repair procedures, construction projects can be expected to last longer than they did in the past.

SUMMARY

After a construction project is completed, daily use makes maintenance a necessity. Maintenance may be very simple such as cleaning, painting, and keeping a structure functioning properly. If maintenance is neglected, major renovation may be necessary to maintain the structure. If a leaky roof is not repaired when it is first noticed, damage from the leak over the years may lead to the need for extensive repair over the entire structure.

Fig. 31-6. A backhoe is used here to tear down the walls of a building. A bulldozer waits in the background.

Fig. 31-7. Moving an entire house is no easy task. The route chosen must be closed to traffic, and the time of many workers is necessary. Nevertheless, the benefits can outweigh the costs. (Larry Mayer)

Maintenance consists of relatively minor repairs and upkeep of items. Repair is more extensive and usually requires specialized skills and equipment to return the construction project to its original condition.

Remodeling is the changing of a constructed project due to changes in the demands of the owner. This may be true for homes, office buildings, roads, and bridges. Careful analysis should be used to determine if it is economical to remodel a structure or demolish the original construction and begin anew.

KEY WORDS AND TERMS

All of the following words and terms have been used in this chapter. Do you know their meanings?

Demolish
Deterioration
Maintenance
Modernize
Preventive maintenance
Remodel
Repair
Resurface
Salvage

TEST YOUR KNOWLEDGE

Do not write in this book. Please write your answers on a separate sheet of paper.

1. Give three examples of normal maintenance procedures that are done to constructed projects.
2. How is preventive maintenance different from general maintenance?
3. Why are stoplights maintained on a regularly scheduled basis?
4. What is the major difference between maintenance and repair?
5. What is the difference between repair and remodeling in construction?
6. When remodeling, new building codes are applicable to how much of the new building?
7. List five items that may be salvaged from a building before it is demolished.

APPLYING YOUR KNOWLEDGE

1. Make a list of several items in your school that need maintenance.
2. List three maintenance activities you do at home. Discuss your list with those of your classmates.
3. Ask the custodian of the school to come to class and discuss what maintenance procedures are used in the school.
4. Set up a maintenance schedule for various items. Use your home, your room, or some entertainment equipment as your focus.
5. Make a bulletin board display of the various types of maintenance and repair items done to keep constructed projects safe and efficient. Break the items down into lists of those items that must be taken care of on a daily, monthly, and yearly schedule.

USEFUL INFORMATION

The 1960 General Conference on Weights and Measures formally gave the revised metric system its new title "System International de Unite." The name has been shortened everywhere to "SI."

SI UNITS OF MEASURE

SI is by far the best system of measurement and calculation known. It is extremely convenient to use because of its base of 10. SI provides greater speed in use, and it is widely used by the scientific world. It is also used by most of the world's population. A total measurement system, SI metric consists of the following:

1. Seven base units.
2. Two supplemental units.
3. Derived units that have special names and any number of combinations.
4. Non-SI units.

SEVEN BASE UNITS

QUANTITY	SI UNIT	SI SYMBOL
Length	meter	m
Mass (weight)	kilogram	kg
Time	second	s
Temperature	kelvin	K
Electric current	ampere	A
Luminous intensity	candela	cd
Amount of substance	mole	mol

Base units are standards based on a natural phenomenon. The standards can be reproduced and without variance. Length, or one meter, is equal to the distance traveled by light in a vacuum during 1/299,792,458th of a second.

DERIVED UNITS

The SI derived units are formed by a simple mathematical multiplication and/or division of two or more SI base units. See Fig. 32-1 Therefore, the advantage of the SI system lies in the fact that the derived units are found by dividing or multiplying by base units. This is called a coherent system. Derived units are shown in Fig. 32-2.

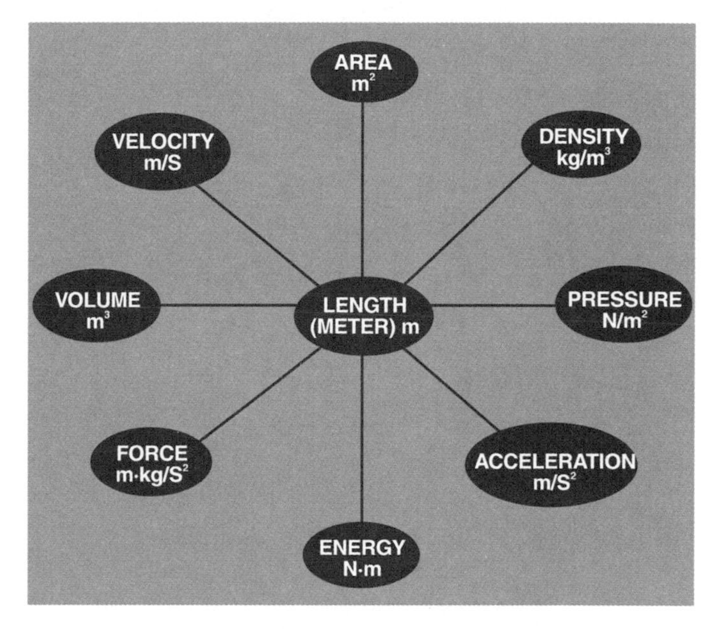

Fig. 32-1. The key to SI metrics is the meter.

SUPPLEMENTAL UNITS

There are other units that are used so often it has become convenient to give them names and symbols of their own. They are called supplemental units.

The supplemental unit is so-called because the General Conference on Weights and Measures has not said whether they are base units or derived units. Supplemental units include:

QUANTITY	SI NAME	SI SYMBOL
Plane angle	Radian	rad
Solid angle	Steradian	sr

NON-SI UNITS

The General Conference on Weights and Measures has recognized that certain units are not part of the SI system. But these units are so widely used and accepted that they cannot be ignored or replaced. In most cases, the non-SI unit is:

1. An extension of a base unit.
2. A multiple or submultiple of a base unit.
3. A unit used with the SI in a specialized field.

There are 11 non-SI units shown in Fig. 32-3.

QUANTITY	SI UNIT	SI SYMBOL
Energy (heat, work)	Joule	J
Force	newton	N
Power	watt	W
Electrical charge	coulomb	C
Electrical potential (voltage)	volt	V
Electrical resistance	ohm	Ω
Electrical conductance	siemens	S
Electrical capacitance	farad	F
Electrical inductance	henry	H
Frequency	hertz	Hz
Magnetic flux	weber	Wb
Magnetic flux density	tesla	T
Illumination	lux	Lx
Pressure	pascal	Pa
Luminous flux	lumen	Lm

Fig. 32-2 SI derived units.

PREFIXES

The multiples and sub-multiples of all SI units are formed by means of SI prefixes, Fig. 32-4. SI prefixes are based on the power of ten. Prefixes eliminate insignificant digits and decimals by the indicated order of magnitude. The powers of ten are, however, used in making calculations.

QUANTITY	SI NAME	SI UNIT	BASE UNIT
Liquid volume	liter	l	meter
Area	square meter	m^2	meter
Solid volume	cubic meter	m^3	meter
Temperature	centrigrade	°C	kelvin
Angle	degree	. . . °	radian
	minute	. . . ′	radian
	second	. . . ″	radian
Time	minute	min	second
	hour	h	second
	day	d	second
Mass	tonne	t	kilogram

Fig. 32-3. Non-SI units.

The prefix multiples are derived from Greek terms while the prefixes of sub-multiples are derived from Latin terms. All the SI prefixes may be used. The choice of prefix is governed only by its appropriateness in a particular circumstance. It is recommended that only one prefix be used in forming the value. For example:

Preferred:

Mg (megagram) or (metric ton) = 1,000,000 grams

Not:

kkg (1000 kilograms) = 1,000,000 grams

Prefix	Symbol Base Unit	Exponential Power	Decimal Value
tera	T	10^{12}	1 000 000 000 000
giga	G	10^9	1 000 000 000
mega	M	10^6	1 000 000
<u>kilo</u>	k	10^3	1 000
hecto	h	10^2	100
deka	da	10	10
SI base unit (no prefix)			1
deci	d	10^{-1}	0.1
<u>centi</u>	c	10^{-2}	0.01
<u>milli</u>	m	10^{-3}	0.001
micro	μ	10^{-6}	0.000 001
nano	n	10^{-9}	0.000 000 001
pico	p	10^{-12}	0.000 000 000 001
femto	f	10^{-15}	0.000 000 000 000 001
atto	a	10^{-18}	0.000 000 000 000 000 001

Note: The three prefixes underlined are the most commonly used.

Fig. 32-4. SI prefixes.

USING PREFIXES

Numbers are added, subtracted, multiplied, and divided in exactly the same manner as any decimal number. When using the SI metric system, a step-by-step procedure will reduce the possibility of error when changing the prefixes. See the example shown in Fig. 32-5.

DRAWING TO SCALE

All construction drawings are drawn to scale except for schematics and tables. Drawing to scale refers to a drawing that has been reduced proportionally from actual size. Reduced size allows the total drawing to be placed on the drawing sheet. In some cases the drawing may be enlarged proportionally, especially a detail for clarity purposes.

0.452	Prefix of kilo	kilometer
4.52	Prefix of hecto	hectometer
45.2	Prefix of deka	dekameter
452.0	Base unit of	meter
4520.0	Prefix of deci	decimeter
45200.0	Prefix of centi	centimeter
452000.0	Prefix of milli	millimeter

Fig. 32-5 Examples of using prefixes.

RECOMMENDED SI SCALES FOR CONSTRUCTION AND MAP DRAWINGS

Recommended SI scales for construction and map drawings are listed in the table in Fig. 32-6.

TYPE OF DRAWING	SI SCALE	NEAREST ENGLISH EQUIVALENTS
Engineering Drawing	1:1	Full size
	1:2	6 in. to 1 ft.
	1:5	3 in. to 1 ft.
	1:10	1 in. to 1 ft.
	1:20	1/2 in. to 1 ft.
Architectural and Working Drawings	1:50	1/4 in. to 1 ft.
	1:100	1/8 in. to 1 ft.
	1:200	116 in. to 1 ft.
Site Plans	1:500	1/32 in. to 1 ft.
	1:1000	1 in. to 100 ft.
	1:2000	1 in. to 200 ft.
Surveys	1:5000	1 ft. to 1 mile
	1:10,000	6 in. to 1 mile
	1:20,000	3 in. to 1 mile
	1:50,000	1 in. to 1 mile
Maps	1:100,000	1/2 in. to 1 mile
	1:200,000	1/4 in. to 1 mile
	1:500,000	1/8 in. to 1 mile
	1:1,000,000	1/16 in. to 1 mile

Fig. 32-6 Recommended SI scales for construction and map drawing.

COMPARING THE COMMON MEASURING UNITS

To convert the many individual customary measurement units that we use each day, simply multiply the unit and a common numerical factor. The resultant figure is the SI equivalent base unit and prefix. See Fig. 32-7.

USEFUL CONVERSIONS

The construction industry operates on standard sizes; quality control demands it. The conversion of American construction will, no doubt, bring about revision of these standard sizes. See Fig. 32-8. For example, the standard size of a sheet of plywood is 48 in. by 96 in. Direct conversion would be 1219.2 millimeters by 2438.4 millimeters. The American National Metric Council has recommended that when standards are revised they be changed to read to the nearest whole metric value. The above plywood size may be established at 1200 mm by 2400 mm.

Think of the many other changes this one standard revision will make. In residential construction, a sheet size change will affect the stud spacing, joist spacing, and ceiling heights.

ENGLISH CUSTOMARY UNIT	MULTIPLIED BY	(equals)	METRIC EQUIVALENT
Length:			
inches	25.4		millimeters
feet	30.48		centimeters
yards	0.9		meters
miles	1.6		kilometers
Area:			
square inches	6.5		square centimeters
square feet	0.09		square meters
square yards	0.8		square meters
square miles	2.6		square kilometers
acres	0.4		square hectometers (hectares)
Mass:			
ounces	28.0		grams
pounds	0.45		kilograms
tons (short ton, 2000 lb.)	0.9		megagrams (metric ton)
Liquid Volume:			
ounces	30.0		millimeters
pints	0.47		liters
quarts	0.95		liters
gallons	3.8		liters

Fig. 32-7. Approximate conversion of customary to metric units.

Lumber	board feet	to	cubic meter or (board meter measure)
Plywood, particle-board, hardboard, celotex	square feet	to	square meter
Molding and trim	linear foot	to	meter
Steel beams	pounds/foot	to	kilograms/meter
length	feet		meters
height	inches		centimeters
width	inches		centimeters
thickness	inches		millimeters
Sheet metal, thickness	decimal inch	to	millimeter
sheet size	inches	to	centimeters
Nails	pound	to	kilogram
Water, paint, other liquids	gallon	to	liter
Cement (sack)	pound	to	kilogram
Sand	short ton	to	metric ton
Gravel	short ton	to	metric ton
Window units	inches	to	centimeters

Fig. 32-8. Useful conversions to be used in the construction industry.

The universal adoption of the SI metric system is underway. It is essential that we learn the basics of this system of weights and measurements. All of the old standard sizes will eventually be changed or converted to metric sizes.

TECHNICAL TERMS

A

Adhesion: the process of bonding pieces together with a different material.

Adjustable rate mortgage: a mortgage in which the interest rate is allowed to move up or down according to a national interest rate index.

Adobe: a brick made of natural, sun-dried clays and a binder.

A-frame: a roof system in which the roof serves as the walls on two sides. Conventional walls are placed on the other two ends.

Aggregate: sand and larger particles of stone, rock, or other material that make up about 65 to 80 percent of the total volume of concrete.

Agriculture: the process of growing and cultivating plants.

Air chamber: a piece of the supply system that holds a small amount of air to cushion the water and ease the pressure on the pipes when the valve is closed suddenly.

Air conditioner: a device used to control temperature, humidity, cleanliness, and movement of air in conditioned space.

Air infiltration: the leakage of free air into a structure through doors, cracks, windows, and other openings.

Air-supported roof: a roof construction similar to a dome structure. It varies in that the roof is supported by a few pounds per square foot of air pressure.

Air-to-air heat exchanger: a device that exchanges the inside air with the outside air but retains approximately 80 percent of the heat from the air.

Alloy: a combination of two or more pure metals that yields certain qualities. For example, steel is an alloy of iron.

Amortization: the gradual removal of the financial commitment of a mortgage by periodic payments.

Ampere: the unit used to measure current.

Anaerobic: living or functioning without air.

Analysis: the part of the design process where the architect calculates how strong the components need to be to support the design.

Antenna: a communication component that transmits and receives radio waves.

Anticipating: studying the potential problems that might occur when a particular action is taken.

Appropriation: the assignment of large amounts of money, usually by federal, state, and local governments, to finance large construction projects of a public nature.

Arbitration: the intervention of an impartial third party between the labor union and company management to settle a dispute.

Architect: the person who designs the construction project.

Architectural drawing: a sketch that shows many details of the physical form of the project such as elevations, floor plans, foundation plans, sections, and details.

Architectural symbol: a standardized sign representing walls, wall materials, and methods of construction.

Artist's rendering: very real looking sketches that give a client a better idea of what the designs will look like when built.

Asphalt: a bituminous material that is strong, durable, and highly waterproof.

Asphalt shingles: a roofing substance made from materials that are saturated with asphalt and then coated with high heat melting flexible asphalt. Mineral granules are pressed onto the asphalt coating to provide color and a fire resistant surface.

Attractive nuisance: a doctrine of law requiring precautionary steps to be taken to prevent children from being injured on or around a construction site. It is based on the theory that small children, who are unaware of the dangers, are attracted to the construction site by the mounds of dirt, tractors, and equipment.

Automating: the process of completing projects with machines.

B

Bar chart: a schedule that focuses on the time needed to complete each task and subtask.

Baseboard system: a heating system that uses natural convection to transfer the heat to the room.

Batter board: a horizontal board supported by posts or stakes securely anchored to the ground used to accomplish offsetting.

Beam: a smaller horizontal support structure.

Bearing wall: a wall that will support the beams above.

Bedrock: the hard layer of rock under the looser surface material.

Bid: a charge for construction that includes the cost of materials and labor.

Birds mouth: the vertical and horizontal cut made near the bottom of a rafter so it has full bearing on top of the wall.

Bond: a written agreement that is sold to individuals or lending institutions with the understanding that the seller will repurchase each at a later date for an additional amount of money.

Box beam: wood beams that are made in the shape of a box.

Branch circuit: a mechanism used to divide the main power line into smaller lines for different sections of a residence.

Brick: small, masonry blocks of inorganic nonmetallic material hardened by heat or chemical action.

Bridging: the use of framing members to help evenly distribute the load placed on one joist to the joists on each side of it.

British thermal unit (Btu): the most common way engineers use to measure the heating or cooling capacity of equipment. One Btu is equal to the amount of heat needed to raise the temperature of one pound of water one degree Fahrenheit.

Builder's risk insurance: a means of protecting all parties against physical damage to the insured property during the construction period. It provides reimbursement based upon actual loss or damage rather than any legal liability that may be incurred.

Building codes: laws or ordinances that apply to the materials selected and used in building projects. They are based on local standards which govern the quality and characteristic properties of the building materials.

Built-up roof: a rigid roof made from succeeding layers of felts flooded with hot bituminous liquid asphalt or coal tar pitch.

Bull float: a piece of equipment that smooths and further consolidates the surface of the concrete.

C

Cabinet hardware: metal fittings that are put on cabinets, such as knobs, pulls, hinges, and catches.

Cabinets: specialized storage areas of a building.

Cable-supported roof: a roof construction that uses strong steel cables to support the roof.

Cant: a log that has been slabbed (cut) to square it up on two or four sides.

Capital: the financial backing of a project.

Ceramics: a range of materials that have a crystalline structure, are inorganic, and can be either metallic or nonmetallic. Ceramic materials are generally stable and are not greatly affected by heat, weather, or chemicals. They have high melting points and are stiff, brittle, and rigid.

Chip removing tools: tools with which some of the material is lost while performing the separating process.

Chunkrete: a concrete that uses large chips of wood as an aggregate.

Circuit breaker:. a protective device that can open or close an electrical circuit.

Civil construction: structures consisting of highways, bridges, dams, canals, and utility structures.

Clean-out: a capped opening that allows easy access to the insides of the pipeline.

Clear: the removal of undesired trees, large rocks, fences, and even other buildings on a construction site.

Cliff dwellings: structures built into cliffs.

Clinker: stony matter that has specific chemical and physical properties that give the cement its hydraulic characteristics.

Cohesion: the combining of materials where the adhesive is made of the same substance as the materials being bonded together.

Cold air return: a device located at floor level that draws in the cold air and returns it to the furnace.

Collective bargaining: negotiations between an employer and a labor union.

Combining: the process of attaching two or more pieces of material together to form a single piece.

Commercial building: a building in common use by the general public, such as schools, banks, stores, churches, office complexes, sports centers, malls, libraries, and government buildings.

Communications breakdown: the lack of clear and effective communication between all workers.

Completed value concept: an insurance policy based on the assumption that the value of a project increases at a constant rate during the course of construction. The policy is written for the value of the completed project but the premium is based upon a reduced or average value. The coverage provided for the project is over the actual work completed with the standard materials at any given time.

Completion schedule: a provision in a contract that details when construction can begin and when it is to be completed.

Composite: a material composed of two or more substances that are bonded together.

Compression forces: forces on a substance that press inward.

Compressive strength: the ability of a substance to hold up under heavy compression forces.

Concrete: a material made of cement and aggregates.

Concrete blocks: hollow, masonry units of portland cement, sand, and fine gravel aggregates.

Conditioning: a process that changes the characteristics of a material but does not change the physical dimensions of the material.

Conditions of termination: circumstances under which either the owners or the contractors may end the contract without completion of the project.

Conduction: the transfer of heat energy in a solid material from one molecule to the next.

Conductivity (K): the amount of heat that is allowed to pass through a single material when there is a temperature difference of one degree Fahrenheit.

Conservation: the preservation and protection of something such as energy from loss or waste.

Consolidate: to work out air from concrete with the use of a vibrator.

Construction: the technology that deals with the design and building of structures.

Construction scheduler: a person who understands all the important components of a construction project and estimates the length of time each component will take to complete.
Contingencies: items of expense that are left out, not foreseen, or forgotten on the original estimate.
Contour map: A map based on survey data drawn to show the changes in elevation of a piece of land. Lines connect like elevations.
Contract: an agreement between two or more parties that describes in detail the responsibilities of each party.
Contractor: the person who is under contract to erect a structure.
Contractor's license: a permit or authorization issued to contractors to protect the public against the consequences of incompetent workmanship and deception.
Controlling: directing how and when tasks are done by supervising workers, coordinating their work, and assuring a constant supply of materials.
Convection: the transfer of heat energy in a gas or liquid caused by the movement of the gases or liquids.
Coordinates: a set of numbers that leads you to a point on a chart.
Core: the inside of a plywood sheet.
Cost-plus contract: a contract stating that the contractor will receive a set percentage or a fixed amount over and above the total costs of completing the project.
Critical path method (CPM): an open-ended process that assesses the effect of all variations, changes, extra work, or deductions upon the time of completion and upon the cost of the work.
Crown: a bow or arch in a long, wood framing material.
Current: the flow of electrons through a circuit.
Curtain wall: the wall sections of each floor are hung on heavy girders that are already in place.
Custom manufacturing: the process of companies making small numbers of products designed to the customer's specifications.

D

Dado joint: a joint that provides greater structural strength to a unit.
Dam: a structure that contains water in a particular location for storage purposes or sometimes to produce power.
Decking: the floor of a bridge.
Demographics: the study of population and its characteristics.
Demolish: the process of destroying or removing structures.
Demotion: the change of jobs to one with less responsibility or to one that requires less knowledge or skill.
Depreciation: the loss of estimated value of a piece of equipment after it has been bought and used.
Desalinization: the process of removing salt from seawater.
Designing: an organized sequence of steps that leads to a final decision.
Detailed blocks: a concrete block with a patterned face.
Deterioration: the process of materials breaking down.
Diagonal brace: a wood board that adds strength to a wall and ensures that a wall will remain plumb.
Dimensioning: a clear definition of part size requirements.
Dimmer switch: a mechanism used to vary the voltage supplied to a fixture.
Dome roof: a roof construction designed to transfer the roof load to a high strength outer ring.
Doorjamb: the vertical strip that forms the side of a doorway.
Door pull: a knob on a door.
Double plate: the plate to which the tops of all walls are connected.
Double-throw switch: a mechanism used to open a circuit, to connect the single wire to the other wire in the box, or to connect a single wire to either of two other wires in the box.
Drain, waste, and vent system (DWV): the drain and waste pipes that carry away the waste water to the sewer system and the venting system to allow the waste plumbing system to work safely.
Drawer pull: a knob on a drawer.
Drilled shaft foundation: a round hole drilled into the earth and then filled with concrete.
Drywall: the most common material used to cover walls and ceilings in residential and commercial construction.

E

Earth filled dam: a dam that uses very large amounts of earth to build a large bank of earth and stones. This bank holds back the water.
Economical: cost efficient.
Economic rewards: the money and fringe benefits such as life and hospital insurance, social security, workmen's compensation, and paid vacation a worker receives for performing certain tasks.
Electrical power distribution: the distribution of electrical power by transformation from the energy source. It can also mean the distribution of electrical power within a building or residence.
Electromotive force: the electric pressure caused by a flow of electrons from one point to another; voltage.
Elevation drawing: a sketch that establishes the vertical location and gives a general idea of how the finished construction will look.
Elevator: a mechanical device that transports people from one level to another.
Energy: the force that drives a construction project.
Energy farm: a facility that concentrates solar energy to produce heat and electrical energy.
Entrepreneur: a person who strikes out on their own to establish a business.

Environment: the land, vegetation, and animals around us.

Environmental impact study: research on the long-range impacts major construction projects have on the local environment.

Equilibrium: the point at which the moisture content in wood is the same as the surrounding air. At this point, the shrinking and swelling of the wood will be at a minimum.

Estimate: to approximate the length of time it will take to complete a job.

Estimating: the careful determination of probable construction cost of a given project.

Excavate: the process of removing earth and rock from a construction site.

Expansion joints system: a roofing system that allows the roofing material to expand and contract.

Exterior finishing work: items that enable the project to withstand weather elements.

Exterior plywood: plywood that holds its original form, shape, and strength with repeated wetting and drying.

Exterior wall coverings: materials used to make the building look better while protecting the structural parts from weather damage.

F

Fascia board: a board that runs horizontal between the ends of the rafters and often attaches them together.

Fastening and combining tools: equipment used to fasten materials together.

Feedback: a monitoring system that determines if a goal has been achieved and to what degree. Also the information that this system provides.

Ferrous metal: a metal that contains a large percentage of iron.

Field office overhead: all the costs that can be readily charged to a project.

Field recruiting: a representative of the construction firm covers college campuses in search of appropriate graduates to work for the firm.

Final inspections: the owner, contractor, and lending agency tour the construction project to determine if the completed work meets the original specifications.

Financing: the money obtained for a construction project.

Finishing tools: instruments that aid in the completion of the final stage of a construction project.

Fire code: public safety regulations guarding against fire.

Fire sprinkler: a sprinkler head that is designed to turn on when the temperature at the ceiling level reaches 170 degrees. Once activated, the sprinkler sprays water until it is shut off manually by the fire personnel responding to the fire.

Fixed appliance circuit: a circuit for fixed appliances, such as dishwashers, ranges and ovens, clothes washers and dryers, and heating and cooling units.

Fixed contract: a contract stating that the builder agrees to furnish all materials and labor to complete the project for a set sum of money.

Flat roof: a roof with a slope of a few inches per foot or less.

Flat sawed: lumber sawed in slices parallel to one side of the log.

Floor joist: a small floor framing material that supports the floor above.

Flush drawer: a drawer that closes evenly with the cabinet.

Foam insulation: a material sprayed in place to restrict the flow of heat escaping from an area. It also seals out any air movement between the insulation and the framing.

Forced air furnace: a heating system in which fuel is burned in a combustion chamber with a large surface area. Cool air is forced over this surface by a fan. The fan moves the heated air through a system of large sheet metal pipes to the rooms to be heated. The hot air transfers its heat to the room.

Forced convection: movement of fluid by mechanical force such as fans or pumps.

Foreclosure: a legal proceeding in which a lending firm takes possession of the mortgaged property of a debtor who fails to live up to the terms of the contract.

Forming: the process of molding material into a particular shape and size.

Fossil fuel: fuels created from decaying plant and animal matter.

Foundation: the part of construction that ties the construction project to the earth.

Function: the ability of a structure to meet the need it is built to fulfill.

Functions of management: the planning of the project, the organization of the labor, tools, and materials, and the control of the progress and completion of the project.

Fuse: a protective device that opens an electric circuit during a surge of excessive current.

G

Gable: the triangle formed at the end of the structure by the meeting of the two slopes on a gable roof.

Gable roof: a roof which slopes in two directions.

Galvanizing: the formation of a mechanical barrier against moisture and the prevention of oxidation, or rusting, of an iron or steel base.

Gambrel roof: a roof that uses two different slopes on each side of a gable roof. The first slope is quite steep, while the upper slope is more gentle.

Gasohol: a fuel that chemically converts the excess production of grain into alcohol, which is then blended with gasoline. This fuel burns well in automobiles and helps conserve nonrenewable resources such as oil.

Gauge number: a number based on the diameter of the wire that designates wire sizes.

General contractor: an individual who obtains an overall contract for building a residence, a road, a dam, or a commercial building.

General provisions: a document stating the conditions and responsibilities of all people involved in the construction project.

General purpose circuit: any lighting convenience outlet except those located in the kitchen.
General safety concepts: specific safety rules for tools and equipment.
Geodesic dome: a special type of truss system that transfers the load of the roof equally and uniformly to all members of the structure.
Geology: the study of the rocks and soil.
Geothermal: an energy source in which the earth's core produces heat in the form of hot water and steam. This hot water and steam may then be used to generate electricity.
Girder: the main horizontal support structure that sustains vertical loads.
Glass block: two airtight glass shells that are fused together to provide good insulation, low maintenance, and controlled daylight.
Gluelam: beams that are built up by the use of successive layers of lumber.
Goal: the defining of a problem to be solved.
Grade beam construction: the forming of a beam on grade, at ground level.
Graded: lumber classified according to strength, appearance, or usability.
Graduated payment mortgage: a mortgage in which the monthly payments are low at first but become higher at a future date. It is based upon the idea that the buyer's income will increase as the buyer's monthly payments increase.
Gray water system: a water conservation system that stores, filters, and reuses a portion of household water.
Great Wall of China: a defensive wall between China and Mongolia built approximately 2500 years ago.
Greenhouse effect: the effect occurs when the short electromagnetic waves of the sun penetrate a surface (such as the earth's atmosphere or glass) and the longer reflected waves cannot penetrate as effectively. Energy is trapped beneath the surface in the form of heat.
Grievance: a complaint by a worker that he or she has not been treated fairly in some aspect of work or pay.
Ground water: the water below the ground's surface.
Gypsum: a naturally occurring mineral. It is used commonly in construction.
Gypsum block: a hollow unit made of gypsum and a binder of vegetable fiber, mineral fiber, or wood chips.
Gypsum board: the most common material used to cover walls and ceilings in residential and commercial construction projects.

H

Hardboard: a panel made of wood chips that have been exploded, leaving the fibers and lignin (nature's basic building block of wood). These are fused under heat and pressure into a hard, long-lasting board.
Hardware: utility items that are used in putting together the finished building.
Hardwood: wood produced from broadleaf or deciduous trees that is generally used for wood furniture, decorative interior paneling, and as interior trim.
Header: a horizontal framing member placed at the openings in walls for doors or windows to support the structure above.
Heating or cooling load: the amount of heat or cool air that a building will have to provide. It is determined by the heat loss or gain of a building.
Heating, ventilating, and air conditioning (HVAC): the systems that control the environment in the spaces where you live and work. If these systems do not supply an ample amount of properly conditioned air, you will become uncomfortable.
Heat pump: a device that may cool or heat an area.
Heat transfer: the process of transferring energy in the form of heat from one place to another in order to maintain temperatures that are within our comfort range.
Heavy building construction: large buildings that are used for commercial purposes.
Heavy steel framing: large office buildings shaped by steel I beams secured with bolts or by welding.
Hip roof: a roof that slopes in four directions.
Home office overhead: the cost of doing business and the fixed expenses that must be paid by the contractor.
Hot water boiler: a heating system in which water is circulated through heating coils.
Housekeeping: the management of a clean working area in the laboratory and on the construction site.
Hydraulic cement: volcanic ash mixed with limestone, burned, and combined with water and small pieces of rock to form a strong and lasting concrete.
Hydroelectric dam: a barrier that produces electrical power that is used to power our homes, offices, and industries.

I

I beam: an I-shaped steel or iron support column. It is the most common shape of support columns in large buildings.
Igneous rock: stone that is produced by heat and pressure. Such stone is produced naturally through volcanic activity and the pressure exerted by shifting of the earth's surface.
Impact: the effect of construction on the land.
Industrial complex: a building where various types of manufacturing are taking place.
Industrial construction: structures consisting of electrical power plants and various materials processing plants.
Infiltration: the amount of air allowed to enter or escape the home.
Information: the flow of data needed to construct a project.
Input: a component of production consisting of information, materials, tools, labor, capital, time, and energy.
Input phase: the gathering of ideas for possible solutions to a problem.

Inspection: a thorough investigation and evaluation of the technique, materials, and work quality of the contractor.

Insulating glass: a unit of glass consisting of two or more sheets of glass separated by an air space.

Insulation (electrical): a material applied to the surface of a building's wiring to protect the conductor from short circuits.

Insulation (heat): materials that prevent or slow down the heat transfer between buildings and the surrounding area to help achieve energy savings.

Integrated: the ability of technological systems to communicate with each other.

Interdependence: technology clusters that rely on each other to provide information, equipment, structures, or transportation.

Interest: the price paid for borrowing money. It is usually stated as an annual percentage rate of the amount borrowed.

Interest dispute: an argument over the terms and conditions of a collective bargaining agreement.

Interior finishing work: the process of placing final coverings on the ceilings, walls, and floors, installing the cabinetry and millwork, and completing the utility systems.

Interior plywood: plywood bonded together with glue that is not waterproof but will maintain strength when subjected to occasional moisture.

Internal search: the process of filling vacancies in a company by promoting employees.

J

Job analysis: a person who assembles data concerning the type of work newly hired employees would be expected to accomplish.

Job descriptions: statements regarding the general duties and specific tasks a worker must perform.

Jointery: the process of holding materials together with a specialized joint. No mechanical fasteners or adhesives are used.

Judicial system: judges and juries who determine how documents should be interpreted.

Jurisdictional strikes: strikes caused by a disagreement between two unions over the assignment of work or jobs.

K

Kiln: an oven.

Kiln burned: bricks made of clay or shale and burned to harden.

Kilowatt: one thousand watts.

Kilowatt-hour meter: a meter that measures the power consumed by a household.

Kit form: a construction process in which the builder erects the structure from plans provided by the manufacturer.

L

Labor: the human resource needed for construction.

Labor agreement: the total relationship between a union and management. Agreements are of two types, simple and supplementary.

Labor union: a group of organized workers who wish to bargain collectively.

Lacquer: a transparent liquid that protects the wood's surface while allowing the natural beauty of the wood to be seen. It dries to a clear and tough finish.

Land based drilling rig: a piece of oil drilling machinery that can be moved from one location on land to another.

Landfill: areas where large quantities of solid waste are buried between layers of earth.

Land line: a communication link that relies on copper or fiber-optic lines to directly link one location to another.

Landscape architect: a person who works from a set of plans that shows the locations of all human-made and natural features in the project.

Landscape plan: a design or blueprint that shows the locations of all human-made and natural features in the project.

Lateral progression: the transfer of a job to another of the same level and pay.

Launch pad: a platform from which space crafts can be launched.

Layout and measuring tools: equipment used to transfer distances and angles from the specifications to the construction site.

Legal provision: a legal document composed of instructions to the bidders, bond forms, and owner/contractor agreements.

Letter of commitment: a legal confirmation of a loan.

Lien: a legal claim against the property of another for the satisfaction of a debt.

Lien waiver: a legal document stating that if the contractor has any outstanding debt on materials or labor used in the project, the suppliers and laborers cannot hold the owner of the home responsible for the debt.

Lifting and holding tools: equipment used to lift material some distance and temporarily hold large pieces of material in place while applying permanent fastening methods.

Light construction: the construction of homes, small stores, and offices.

Light weight steel framing: a structure that uses beams of steel in place of wood framing members.

Lip drawer: a drawer with a front having a 3/8 x 3/8 in. rabbet along its edge that overlaps the opening. This rabbet allows a looser fit.

Load: power consuming device in a circuit.

Loan: an amount of money that is lent to a company or individual for a period of time. The loan must be paid back with interest.

Local responder: a transmitter tower that picks up a low power signal and relays it to more powerful transmitters. These, in turn, send the signal through standard communications networks.

Locate: the marking off of where a structure is to be established.

Lock: an enclosure used to lift and lower ships.

Long range planning: the company sets goals for the future to keep the company active and prosperous.

Low heat transmission glass: glass treated on one side with a very thin metallic coating that helps reduce heat loss in winter and heat gain in the summer.

Lumber core plywood: plywood that has a core of solid wood.

Lumberyard: an area where wood rests until it is sold.

M

Maintenance: activities that take place to keep items in proper condition.

Mansard roof: a roof similar to the hip roof. It varies in that the slope of the roof is much steeper and the top is flattened off.

Manufacturing: the processes of changing materials into more usable forms in a manufacturing plant or factory.

Masonry: alternate foundation materials, such as concrete block, rock, clay bricks, and large blocks of cut stone.

Masonry wall: wall of clay and concrete blocks as well as native stone that provides a good looking and low maintenance wall system. It can be a support wall for the structure above, or function as a covering for the wall system used to support the structure.

Mass production: the production of a large number of a particular item or product.

Material: the part and physical resource used in a structure.

Mechanical and electrical drawing: a sketch that shows the plumbing, heating/cooling systems, electrical requirements, ventilating systems, and lighting of the project.

Mechanical fastener: a piece of material that attaches materials together using some type of clamping action that makes the pieces behave as one unit.

Mechanic's lien: a claim filed against the real estate involved in the construction when persons furnishing materials or labor are not paid.

Mediation: the intervention of an impartial person between the labor union and company management to help the parties come to an agreement.

Membrane roofing material: a rubber like substance that forms a continuous membrane over the entire roof area.

Mental attitude: the operator's proper frame of mind and respect for tools and equipment.

Metamorphic rock: stone formed by the gradual change in the character and structure of igneous and sedimentary rock.

Mixing tools: equipment used to thoroughly mix materials to produce a quality finished product.

Modernize: the process of updating utilities and mechanical equipment in taste, style, or usage.

Modular construction: the process of constructing two or more sections of a large structure in a factory and then assembling them on the site.

Monolithic slab foundation: a very solid and rigid foundation that floats on top of the ground.

Mortgage: a written contract pledging the property as a loan that a borrower gives to a lender as security for the payment of a debt.

Mulch: a protective covering that is sprayed along with the seed to help establish grass.

N

National Electric Code: a set of rules and regulations to be used by an electrician when installing electric wiring, appliances, and machinery.

Natural convection: movement of a fluid caused only by temperature differences (density changes).

Negative progression: the demotion to a job with less responsibility or to one that requires less knowledge or skill.

Nonferrous metal: a metal that contains little or no iron.

Nonrenewable: anything such as oil that cannot be easily replenished. Once it is used up, it is gone.

Nursery: a place which grows and sells plants, trees, and shrubs.

O

Offset: the temporary establishment of the outside dimensions of the structure.

Offset drawer: a drawer with a front having a 3/8 x 3/8 in. rabbet along its edge that overlaps the opening. This rabbet allows a looser fit.

Offshore oil platform: a piece of oil drilling machinery that is towed by ships from location to location.

Ohm: a unit of electrical resistance.

Ohm's law: an electrical circuit law that states that when the electromotive force stays the same, an increase or decrease in the amount of resistance will cause a direct change in the amount of current. It is often stated E = I R.

Oil base paint: paints containing oils that are refined from plants and animals for the vehicle.

Operator's manual: instruction book for tools and equipment.

Organizational strike: a strike in which employees try to force the employer to deal with the specific union as the collective bargaining representative.

Organizing: the assigning of different tasks and resources to meet the planned goals of a project or company.

Origination fee: a fee added onto a mortgage which is equal to 1 or 2 percent of the mortgage.

Originator: the person who conceives the idea of the construction project.

Output: the completed project.

Output phase: the decision that is based on the best information available at the time.

Overburden: excess material overlying a deposit of coal.

P

Paint: a mixture of finely ground solids suspended in a liquid that forms a continuous durable film when applied to a clean dry surface.

Panalized construction: prefabricated building structures.

Panama Canal: the best known example of the lock system. A waterway for ships to move across the isthmus of Panama.

Particleboard: a panel made of wood fibers that are bonded with glue and pressed into sheets under high pressure and temperatures.

Payment schedule: a provision in a contract that allows contractors to redeem their money invested in materials and labor as early as possible. A percentage of the total amount of the contract is paid to the contractor at specified times.

Penny: the term used to describe the length of a nail.

Permafrost: a layer of permanently frozen earth.

Personal interview: a meeting at which a representative of the company obtains information about an applicant's qualifications for the job.

Photocell: a cell that turns on lights when it becomes dark.

Physical environment: the actual work conditions of a company.

Pier foundation: a bell bottom pier is formed by a shaft that is flared out at the bottom. A bell bottom pier will spread the weight over a greater area.

Pig iron: the product of the blast furnace. It contains many impurities and must be refined further.

Pigment: the finely ground solid particles of paint.

Pile: a column of wood, steel, concrete, or combinations of these materials that is driven into the earth to provide a solid foundation upon which to build.

Pile driver: a device that drives a pile into the soil.

Pipeline: a line of pipe that uses the fluid motion of liquids and gases to transport fluids from one place to another.

Placement agencies: organizations that find prospective employees for companies.

Plain sawed: lumber sawed in slices parallel to one side of the log.

Planning: the process of setting up goals, policies, and steps for an activity.

Plaster: a pasty substance that hardens when dry. Plaster is used for coating walls and ceilings.

Plastic: a synthetic polymer.

Plastic laminate: a rugged, waterproof, and economical counter top material.

Plate: the horizontal framing member to which vertical pieces are nailed.

Plumb: a term used to describe a wall that is absolutely vertical.

Plywood: a construction material made of thin sheets of wood (veneers) bonded together with glue.

Points: typically a 2 or 3 percent charge for any type of long-term mortgage. It compensates the lender for the difference in the agreed rate and the rising interest rates as the mortgage matures.

Polymers: natural and synthetic compounds made of molecules.

Portland cement: a bonding material of limestone and clay that is burned at a higher temperature to produce concrete.

Positioning: the operator's location in relation to the piece of equipment to be operated.

Positive progression: the promotion of employees to positions with more responsibility and generally a larger salary.

Pounds per square: unit of measurement of the weight of the roof.

Power: the rate at which electricity does work.

Power of eminent domain: the government's ability to force individuals to sell their property to the government for a reasonable price.

Prefabricate: to make building components in a factory prior to the construction of a building.

Pressed boards: a construction material composed of any vegetable, mineral, or synthetic fiber mixed with a binder and pressed into a flat sheet.

Prestressed concrete: a precast structure of concrete and reinforced steel.

Preventive maintenance: maintenance activities that are done on a routine schedule.

Primary processing: the gathering of raw materials.

Principle: the amount of a loan.

Process: a purposeful action that assembles various inputs into a desired product.

Processing phase: the careful review of ideas and information that are processed into some usable form.

Profit: the amount of money added to the total estimated cost of the project to compensate a company and its owners for the use of their money, the risks they have taken, and managing the construction job.

Progress chart: a schedule that focuses on the cost of each job and the progress of the contractor.

Public liability insurance: protection against liability for the injury or death of a person.

Pumping station: a station that helps in the movement of the fluid from place to place.

Punch list: a list of finishing details.

Pyramids: massive limestone structures built in Egypt approximately 4500 years ago.

Q

Quality control: to build the best structure for cost; to produce with consistent quality.

Quarter sawed: lumber that is first quartered and then cut at right angles to the exterior of the log.

R

Radiant heating system: a heating system in which warm or hot surfaces transfer their heat through radiant energy into the space to be conditioned.

Radiation: the transfer of heat energy through electromagnetic waves.
Rafters: a series of beams that support a roof.
Rammed earth structure: structures of soil and other binding materials that are mixed and then packed into forms.
Raw material: material that is in the crude form.
Receptacle: a slotted point along an electrical circuit to which a cord plug is attached for the purpose of using the current supplied by that circuit.
Reconditioned: the repairing of tools and equipment.
Recruiting: the process of actively seeking new employees.
Recycle: to reuse resources.
Refinement of ideas: compiled information in support of various solutions that are needed to prove the value of one design over all others.
Refinery: large quantities of equipment that connect into a maze of pipes, fittings, tanks, and pumps.
Reinforced concrete: concrete strengthened with steel.
Remodel: the process of changing a building's structure so that it will meet new needs, desires, functions, or space requirements.
Renegotiable rate mortgage: a mortgage that is automatically renewed every three to five years. Payments could go up or down, depending on whether interest rates are higher or lower.
Renewable: anything, such as fuel from plants, that can be replenished easily.
Renovate: to restore a building to a former state.
Repair: the processes of rebuilding parts of a construction project.
Research and development: the investigation of structural elements and subsystems of the design.
Residential construction: the construction of homes for people to live in.
Resistance: the opposition to current flow. It is measured in ohms.
Resistivity (R): the ability to resist the flow of heat.
Responsibilities of the parties: the specific activities and services for which owners and contractors are liable that are clearly listed in the contract.
Resurface: the process of sealing the road to prevent moisture from entering the roadbed and softening the substructure.
Retaining wall: a wall structure that is used to hold back or support the earth.
Retirement: the withdrawal from the work force.
Ribbed slab: a combination of a concrete girder and a floor slab that is cast into one unit.
Ridge: the location where the two slopes meet on a gable roof.
Ridge board: a framing member that runs horizontal near the center of the structure.
Rights dispute: a difference of opinion over what a collective bargaining agreement means.
Rise: the vertical distance from the top of the wall to the top of the roof.
Roof system: a horizontal structure that connects the wall structure together and provides protection from the elements.
Run: the horizontal component of the width covered by a rafter.

S

Safety constraints: devices that protect a worker's sight, hearing, lungs, and skin.
Safety zone: the safe distance an operator should maintain from the cutting area of a tool.
Salvage: the process of saving valuable construction components that can be used in other construction projects before the demolition of a building begins.
Sand-lime: a brick composed of sand and lime hardened under pressure and heat.
Satellite: an object orbiting in space, which may relay communications signals.
Satin: the glossy finish of paint.
Scale: a drawing that has been reduced proportionately from actual size so that it will fit on a drawing sheet. Small units of measure stand for larger units.
Scheduling: the process of determining the order and timetables of production activities and the arrival of the right materials, people, and equipment at the job site at the proper time.
Scrubber: a device that removes nearly all of the pollutants in a structure.
Seamless pipe: a pipe produced by forcing a steel rod over the pointed nose of a piercing mandrel. The mandrel punches a hole in the center of the rod, while rolls on the outside control the diameter.
Seasoning: the process of drying lumber to the point where it is ready to be used.
Secondary processing: the transformation of raw materials into usable industrial materials.
Sedimentary rock: stone composed of silt or the skeletal remains of marine life that have been deposited by ancient seas.
Seeding: the process of loosening the soil and spreading out grass seed.
Selling the job: the project is ready to be turned over and paid for once all corrections are complete.
Seniority: a privileged status established by the number of years spent with the employer.
Separating (materials): the process of dividing a piece of material into two pieces.
Separating (occupational): the process of discharging, relocating, laying off, or retiring of personnel.
Separating tools: equipment used to remove unwanted material and also to make two or more pieces out of one.
Septic tank: a sewage settling tank in which part of the sewage is converted into gas and sludge before liquid waste is drained off underground to a drain field.
Service cable: a cable that delivers power for the distribution of electrical power throughout a residence.
Setting: the process of chemically combining water and cement to form a new compound.

Settlement: the agreement resulting from a strike by the parties.

Sewage treatment plant: a treatment plant that removes impurities from water to make it safe to be returned to a river or lake.

Shearing: a separating process that does not produce any chips and therefore no material is lost.

Sheathing: the subfloor material that serves as the base for the finish floor that is installed later in construction.

Shed roof: a roof that slopes in only one direction in a single plane.

Shelter: a structure that covers and protects people from the elements.

Shop steward: a union officer who represents a group of workers.

Shut off valve: a valve that allows the repair of the supply pipe's fixture without shutting off the entire supply system.

Sill plate: a beam that anchors the floor joists to the foundation.

Simple agreements: documents that establish a few elementary rights and duties of the employer and employee and prescribe a general method of handling grievances.

Single glazing: a window unit containing only one pane of glass.

Single-throw switch: a mechanism used to open or close a single wire in a circuit.

Site plan: a drawing which contains records such as the soil test boring schedule, excavation limits, paving requirements, river control, or detour construction.

Skylight: a window unit that is installed in the roof and provides natural lighting for interior spaces.

Slip form: the form used in the slip casting. In slip casting, slip (clay suspended in water) is poured into forms. When the wall reaches the correct thickness, the excess slip is removed.

Slope: the number of inches the roof rises for each unit of run.

Slump: the distance the wet concrete will settle under its own weight.

Small appliance circuit: a circuit for the vacuum cleaner, toaster, mixer, etc.

Smart House®: a home that uses a multi-functional electrical cable instead of conventional home wiring.

Smokestack: a pipe or funnel that transports hot gases from one place to another.

Social environment: the actions and interactions of workers with the people who work around them.

Sod: squares of grass that have already matured.

Softwood: wood produced by evergreens or conifers that is generally used for construction.

Solar: energy that comes from the sun. It supplies over 90 percent of all the energy we use on earth.

Span: the total width of the building that is to be covered by the roof.

Specialize: to concentrate your efforts on things you enjoy most or at which you are best.

Specification: a written statement that informs the builder what is to be built, what materials are to be used, and how the job is expected to be done.

Square: the unit of measure for roofing materials.

Stadia rod: an instrument that is placed over a point being established by the transit.

Stain: a substance applied to wood to give it a variety of different colors.

Staking: marking of the boundaries of a project being constructed.

Stocks: ownership rights to a company, bought by people investing in that company.

Storm sewer: a sewer system in which the drains for the sewers drain off the water into large underground pipes. These pipes are connected to the sewage treatment plant where the water is treated and returned to the river.

Straight gravity dam: a straight, concrete dam with a foundation composed of the earth's bedrock. It holds back the water simply by its massive weight.

Strike: a concerted withdrawal from work by a group of workers employed in the same economic enterprise to force the employer to be aware of their demands.

Structural drawing: a sketch that shows details of the structural parts of the project.

Stucco: a masonry product that makes an excellent exterior covering for buildings.

Stud: the vertical framing member to which horizontal pieces are nailed.

Subcontractor: a person who specializes in a particular type of construction enterprise, such as plumbing, electrical, concrete, or framing.

Subfloor: the material that is placed on top of the joists once the floor joists and bridging are installed.

Substructure: the part of construction that ties the construction project to the earth.

Subsystem: a separate but necessary part of a structure.

Subway: a rail transportation system that moves large numbers of people relatively short distances in a very efficient manner.

Super insulated home: an energy efficient structure that prevents the loss of heat through use of an air-to-air heat exchanger.

Superstructure: the part of the constructed project that is above the foundation or substructure.

Supplemental agreements: agreements that are added on previous agreements. They may be classified as either major or minor.

Supply system: the part of the plumbing system that supplies water to the various fixtures such as the sink, washing machine, and water closet.

Support column: the part of a structure that transfers the load from above to the foundation or substructure below.

Survey: a large-scale geometrical measurement that establishes the location and size of tracts of land according to given requirements.

Surveyor: a person who finds exact locations to build structures such as roads, dams, or buildings.

Suspended ceiling: a ceiling affixed to a framework that is hung on wires from the ceiling joists.

Suspension bridge: a bridge that is supported by smaller cables that are attached to large main cables suspended from high towers.

Switch: a mechanism used to open and close an electric circuit.

Systems approach: the process of dividing larger components into smaller parts in order to understand any technology or construction activity.

T

Tape: the process of applying a coat of taping compound along shallow channels formed by the tapered edges where the boards are butted against each other and along both sides of inside corners.

Taping compound: an adhesive that is used to conceal fasteners and level edges when installing drywall.

Technical provision: a document listing the kinds of materials and processes to be used in construction. Generally, the provision includes architectural, civil, structural, plumbing, electrical, heating/cooling, and mechanical specifications.

Technology: the science of using tools and techniques in their most efficient manner.

Tempered glass: a glass heated almost to its melting point and then chilled rapidly. This produces a glass three to five times as strong as ordinary glass.

Tensile strength: a material's ability to resist being pulled apart (tension).

Tensive forces: the forces at the lower edge of the girder that pull the girder apart.

Terra-cotta: a nonloadbearing burned clay building unit. Literally means baked earth.

Thermal envelope: a highly energy efficient building shell.

Tilt up construction: a type of construction in which the walls are either built in a plant and transported to the construction site, or they are poured flat on the construction site and then lifted into place by cranes.

Timber frame construction: a heavy timber frame consisting of columns and girders that are used to support the walls and roof of a structure.

Time: the interval in which construction takes place.

Toe space: a small space between the bottom of a cabinet and the floor. Also called a toe strip.

Toe strip: a small space between the bottom of a cabinet and the floor. Also called a toe space.

Tool: a device or machine used to aid in the construction process.

Total heat loss: the amount of heat that passes through a structure such as a door, wall, or room.

Training sessions: meetings to prepare workers for their jobs.

Transfer: to change a job to another of the same level and pay.

Transformer: a regulatory device used to increase voltage and lower the amperage for transfer purposes.

Transit: an instrument that establishes angles and straight lines.

Transport: the moving of construction items from where they are manufactured to where they are needed.

Transportation system: any method used to move people, equipment, materials, or supplies.

Triple glazing: a window unit containing three panes of glass.

Tripod: an instrument that supports the transit while a line or point is being established.

Trowel: a steel piece of equipment that leaves a smooth finish on the concrete.

Truss: a group of beams formed into a rigid frame.

Truss bridge: a bridge in which a group of beams formed into a rigid frame support the decking.

Truss framing: a roof construction in which the rafters are cut and assembled into large triangles and installed in units.

Tunnel: a structure consisting of a wall with two sides as well as a top and bottom.

U

Uniform Plumbing Code: a set of rules or codes published by the International Association of Plumbing and Mechanical Officials. It ensures the finished plumbing meets the needs of the customer and provides a safe and long-lasting system.

Unit: a metal, wood, or plastic window or door frame that secures glass panels.

Unit rise: a number in inches that determines the slope of the roof. It is used in conjunction with the unit run.

Unit run: the standard unit used to determine the slope of a roof. It is always equal to 12 in.

Universal Systems Model: a four-part system consisting of inputs, processes, outputs, and feedback that works together to meet our needs, wants, and goals.

Uranium: a radioactive material that is used in nuclear power plants.

U value: a unit of measurement of the total heat loss of a structure.

V

Varnish: a transparent liquid that protects the wood's surface while allowing the natural beauty of the wood to be seen. It dries to a clear and tough finish.

Vehicle: the liquid portion of paint.

Vertical vent: a pipe that provides air circulation throughout the drainage system.

Vibrator: a piece of equipment that works out the air that is trapped in the concrete. It also works the smaller aggregate particles in between the larger aggregate.

Voltage: the electromotive force along the conductor of a circuit.

W

Waferboard: a panel produced from flakes of wood that are about 1 1/2 in. square. They are bonded into sheets of varying thicknesses with a waterproof adhesive.

Waffle pattern: the concrete flooring pattern that occurs when reinforcement steel is placed between special metal pans and then concrete is placed in the forms.

Wallpaper: a wall covering made from paper, cloth, plastic, or foil.

Want ads: newspaper advertisements that publicize job openings.

Warranty: a written guarantee of a product's performance and of the maker's responsibilities concerning defective parts.

Waste system: the drain and waste pipes that carry away the wastewater to the sewer system.

Water base paint: paint that uses water as the main vehicle.

Water hammer: a pounding noise coming from the plumbing system. The noise occurs when water is flowing through supply pipes at pressures in excess of 70 pounds per square inch and the valve is suddenly closed.

Water main: pipelines that transport water many miles to bring an adequate supply to a growing population.

Water pressure storage tank: a tank used for storing water in rural homes. Air pressure is used as the switching mechanism for the tank.

Water trap: a U-shaped bend in a waste pipe that stops the flow of sewer gases back into the fixture where they could be released into the home.

Water treatment plant: a plumbing structure that provides water free of bacteria and other organisms.

Watt: the unit of electrical power.

Weatherstripping: the process of covering the joint of a door or window and the sill with a strip of material to provide protection from the elements.

Welded pipe: a pipe produced by forming a flat strip into a U-shape, then closing it into a tube. The seam is welded by passing an electric current between the edges.

Well: a large tank that pumps in and stores well water to supply a city with needed water pressure.

Wholesale distributor: a company that markets and sells a product to other companies.

Winch: a machine with a cable that is attached to a powerful drum and motor and is used to transport construction materials.

Wind farm: a facility built to utilize the energy of the wind.

Wood frame construction: structures made of wood framing members.

Wood siding: a material made from species of wood that have a natural resistance to decay such as western red cedar or redwood.

Wood trim: interior trim made from common soft or hard woods, or from very exotic woods.

Working drawing: a final sketch that includes all the necessary views, dimensions, and details of the construction project.

Workmen's compensation insurance: insurance that reimburses a contractor for damages through an industrial accident paid to injured workmen.

Wrap-around mortgage: a mortgage made up partly by assuming an old mortgage on the property that is at a lower rate of interest. Then, a new mortgage that is at a higher interest rate is added. The interest rate is adjusted so that it lies somewhere between the two rates.

USING THE INDEX/GLOSSARY REFERENCE

The Index/Glossary Reference serves two purposes. It can be used as an index for finding topics in the text. It also provides an easy method of locating definitions in the text.

A **bold typeface** is used to give page numbers of definitions. If you need the definition of a technical term, simply turn to the page number printed in bold type. The defined word is also printed in bold type.

The Index/Glossary Reference has several advantages over a conventional glossary because it allows you to obtain more information about the terms you want to define. You can read the definition in the context of the book and also refer to the illustrations relating to the new terms.

INDEX/GLOSSARY REFERENCE

D

E

I

J

K

L

M

N

O

P

Q

R

S

T

U

V

W